PHYSICS

Foundation for Class IX

S A S Kirmani
Chairman
Erose Educational Infotech Private Limited
New Delhi

Universities Press

PHYSICS FOUNDATION FOR CLASS IX

UNIVERSITIES PRESS (INDIA) PRIVATE LIMITED

Registered Office
3-6-747/1/A & 3-6-754/1, Himayatnagar, Hyderabad 500 029, Telangana, India
info@universitiespress.com; www.universitiespress.com

Distributed by
Orient Blackswan Private Limited

Registered Office
3-6-752 Himayatnagar, Hyderabad 500 029, Telangana, India

Other Offices
Bengaluru / Chennai / Guwahati / Hyderabad / Kolkata
Mumbai / New Delhi / Noida / Patna

First published 2024
Reprinted 2025

Cover and book design

ISBN: 978-93-93330-51-2

Typeset in Times New Roman PS Std 10 *by*
SRS Publishing Services, Puducherry

502214

Printed in India by
Deccan Press, Hyderabad 500 020

Published by
Universities Press (India) Private Limited
3-6-747/1/A & 3-6-754/1, Himayatnagar, Hyderabad 500 029, Telangana, India

Disclaimer
Care has been taken to confirm the accuracy of the information presented in this book. The author and the publisher, however, cannot accept any responsibility for errors or omissions or for consequences from application of the information in this book, and make no warranty, express or implied, with respect to its contents.

Endorsements

As a field grows, it becomes harder for a newcomer, and sometimes even for an expert, to find relevant information in a quick and convenient fashion. This book captures the theories and presents them masterfully with a lucidity and refreshing clarity I have not encountered so far… When reading this book, I was captivated by its concise explanation and illustration of real-life applications.

The book is an ideal Reference Book for Physics and is suitable for Grade 9 students across any board, with many examples and well-designed self-study problems. This well-structured and informative book is a valuable companion for students looking to delve into different fascinating branches of physics. It is a welcome addition to the field.

Chandra Shekhara PS
Director, Marigold International School, Bengaluru
Former Principal, DPS Damanjodi Odisha, and DPS Ghana

I would like to express my sincere appreciation for the outstanding Physics Foundation books authored by SAS Kirmani for aspirants of competitive examinations. I extend my gratitude for his remarkable contribution to the field of education. Having reviewed his Physics Foundation books, I am truly impressed with the depth of knowledge, clarity of explanations, and strategic organization that he has demonstrated throughout its contents.

The author's commitment to excellence in education is truly inspiring and sets a high standard for educational resources in the field of physics. I hope his expertise will continue to enrich the learning journey of countless students. I, along with the HoD of physics and other teachers of the physics department highly recommend this book for board and competitive exam aspirants.

Dr HS Vashistha
Principal, DPS, Navi Mumbai

The book is an awesome and concrete source of fundamental concepts of Physics for aspirants of competitive examinations like IITJEEE/ NEET/NTSE. It covers all aspects of these examinations and provides crystal-clear idea to the aspirants, especially in the grey areas of Physics. The book is a great help for composed and constructive preparations. The users will definitely reap the advantages by coming out with flying colours.

R S Choudhary
Principal, DPS, NFL Township, Vijaipur, Guna (MP)

The hallmark of this book is to be really able to catch the attention and imagination of students and inculcate foundational scientific principles in their minds.

Sections on 'Common Misconceptions', 'Practice Questions' (both Subjective and Objective), and the carefully crafted 'Hints and Solutions' will generate adequate interest and confidence in the students.

Kumar Supravin
B.Tech., Ph.D. (IIT DELHI)
Ex-IRS_97, IFS and IAS_99

Strongly recommend for Class 9 foundation students, The language is very clear and concise, making complex concepts more accessible to students.

Sameer Gupta
Senior Faculty in Physics and Director
EDUNOVA
Ahmedabad | Vadodra | Dahod | Godhra

The book gives a vivid knowledge about the fundamental aspects of Physics based upon which a solid foundation for the advanced concepts can be built. The treatment makes learning the subject an enriching experience.

Ashok Kumar
Ph.D. IIT Delhi
HoD, Department of Physics, AIAS
Amity University, Noida, UP, 201313

The book is an excellent resource for the students and teachers of Class 9 across all education boards in India. It has very good examples and clears common misconceptions the subject. There are many solved and unsolved numerical problems with varying levels of difficulty to guide and help the students. The book also has a vast range of questions to help students prepare for their examinations.

Mrs Suchitra Radhakrishnan,
Former Senior Physics Teacher, DPS, R K Puram, New Delhi
HOD in C S Academy, Coimbatore

This book will help students to understand the basic concepts and expose them to the pattern of JEE, NEET, NTSE and other competitive examinations at their level.
The book is equally useful for various board examinations.

Prashant Sharma
Head Master (Physics)
DPS, Sahibabad, Ghaziabad

Physics Foundation books for Classes IX and X have been well-crafted to cater to a foundation course in Physics for JEE/NEET as well as CBSE boards. The books are packed with comprehensive theory supported by illustrations, tables, well-structured concepts, solved problems and ample number of MCQs. Highly recommended for students of Classes 9 and 10.

Goldy Banerjee
M.Sc. (Physics), M.Tech., B.Ed.
PGT Physics, DPS, Duliajan, Assam

With immense pleasure, I want to say that the Physics Foundation books for Classes VIII–X written by SAS Kirmani are among the best in the subject. The content is well-organized, following a logical sequence that aligns with the typical Physics curriculum. It helps in removing the constraints of misconceptions in the topics of Physics. The book has numerous practice questions and problems for students to test their understanding and concepts. I strongly recommend this book.

Swati Yadav
Subject Enrichment Developer (Physics)
Bharat Ram Global School
Indirapuram, Ghaziabad

Contents

Preface

I am extremely happy to present this book "Physics Foundation" for the students of Class IX.

In the course of our teaching Physics and Mathematics, I came upon the idea of writing a book which aims to build a firm foundation in students on the basic concepts of Physics, and provide them an exposure to the types of questions commonly asked in competitive examinations, be it IIT JEE, NEET, NTSE, KVPY or Olympiad, while also helping them prepare for their boards.

In recent years, competitive examinations have become more challenging and dynamic in nature. Every year it surprises aspirants, and the way question papers are set is becoming increasingly uncertain and unpredictable. This innovative and evolving nature of the examination requires students to quickly adapt to the questions and arrive at the solution in the shortest possible time in order to have an edge over their competitors.

Such quick-thinking skills require first-hand knowledge of the basic facts and principles of the subject, clarity of concepts, and a clear understanding of the various dimensions and aspects of examination. Lastly, repeated practice helps to solve application-based questions.

In order to meet above objectives, this foundation book on Physics is structured in a logical sequence.

The book contains 7 chapters, of which Chapters 1 to 5 are based on CBSE syllabus. The ICSE syllabus topics that are not included in the CBSE syllabus are presented as separate Chapters 6 and 7. Each chapter begins with Learning Objectives and Introduction.

The chapters are informally arranged in the following order:

- The first part contains theory and illustration along with Points to Remember and Misconceptions, if any.
- The second part of each chapter targets CBSE/ICSE exams and consists of Exercises in the form of very-short, short and descriptive type questions, Numerical Problems, Fill in the Blanks and True or False questions.
- The third part is Competition Vista, which covers questions asked in competitive examinations. These are – Match the Columns, Assertion–Reason, Comprehension, Objective Questions (one option and more than one option correct), and Scholar Zone.
- The last part of each book provides Hints and Solutions for the Practice Questions and questions in the Competition Vista section.

I am highly indebted to the authors and publishers of various books that I have consulted during the preparation of the manuscript.

I thank and acknowledge the following scholars for their suggestions, criticism and encouragement:

Chandra Shekhara	Director, Marigold International School, Bengaluru Former Principal, DPS Damanjodi Odisha, and DPS Ghana
H S Vashista	Principal, DPS, Navi Mumbai
R S Choudhary	Principal, DPS, NFL Township, Vijaipur, Guna (MP)
Kumar Supravin	B. Tech., Ph. D. (IIT DELHI) Ex-IRS_97, IFS and IAS_99
Sameer Gupta	Senior Faculty in Physics and Director EDUNOVA Ahmedabad | Vadodra | Dahod | Godhra
Ashok Kumar	PhD, IIT Delhi HoD, Department of Physics, AIAS
Suchitra Radhakrishnan	Former Senior Physics Teacher, DPS, R K Puram, New Delhi HOD in C S Academy, Coimbatore

Prashant Sharma	Head Master (Physics) DPS, Sahibabad Ghaziabad
Goldi Banerjee	M.Sc. (Physics), M.Tech., B.Ed. PGT Physics DPS, Duliajan, Assam
Swati Yadav	Subject Enrichment Developer (Physics) Bharat Ram Global School Indirapuram, Ghaziabad

I sincerely hope that this book will prove to be of great help for aspirants to various competitive examinations and equally for students appearing for their board examination as the book has been written taking into account all the requirements.

I do not know how adequately express my thanks to my wife Fatma Kirmani, for her constant motivation, support and long-term sacrifices. Also, my love and blessing to my granddaughter Aaira for her support and enthusiasm while writing the book. My love and affection to my grandson Izhaan, who has not tired of repeatedly asking "Dadu aap kya likh rahee hain". After all, the book consumed a lot of time which was exclusively dedicated to him.

I am obliged to Anas, Akshay and Aditi Basu for their transcription and the PATS Team for their proof reading and editing.

I am grateful to my colleagues and friends who helped me in bring the book to fruition.

Finally, special thanks and recognition goes to Thomas Mathew Rajesh, Kalol Das and Ramesh MR of Universities Press (India) Private Limited for their attention and diligence to detail.

I would be very blessed to receive constructive suggestion and comments from the reader for further improvement of the book. These may please be sent to the publisher or directly to me at saskirmani@gmail.com.

S A S KIRMANI

About the Author

SAS Kirmani is the Founder of Erose Educational Infotech Private Limited, a PAN India Educational and Research Company headquartered in New Delhi. He founded Erose in 1992 and is the chairman of the organisation. Under his leadership and vision, Erose has successfully guided and trained aspirants for IIT-JEE and other competitive exams such as NEET and Civil Services Examination. Kirmani articulated, designed and implemented the Global Education and Training model to prepare students for various entrance examinations, which has become a huge success. Under the model, he teaches students to focus on answering the maximum number of questions in optimal time by using the concepts for both straightforward application and lateral thinking.

An electrical engineer from HBTI Kanpur, Kirmani is on the advisory board of many schools and educational institutions to provide academic and administrative support. He was a member of the Draft Formulation Committee of Computer Literacy Awareness Program (CLASS) jointly sponsored by NCERT and Ministry of Education in the year 1984. In his initial years, he worked as senior physics faculty at Delhi Public School, R K Puram, New Delhi.

CHAPTER 1 Motion

LEARNING OBJECTIVES

At the end of this chapter, you will be able to learn about:

- State of Motion and Rest
- Speed, Velocity and Acceleration
- Equations of Motion
- Velocity vs. Time and Distance vs. Time graph
- Vectors and addition of vectors

Introduction

The physics of motion focuses on the language, principles and laws, which describe and explain the motion of objects. The efforts should concentrate on the meaning of the information and its applications. Mechanics is the branch of physics that focuses on the motion of objects and the cause of motion to change. There are two branches of Mechanics: Kinematics and Dynamics. The word kinematics comes from the Greek word "Kinema" meaning motion. The word dynamics comes from the Greek word "Dynamis" meaning power. Kinematics deals with the concepts that are needed to describe motion, without any reference to forces or the cause of the motion. The goal of study of kinematics is to develop sophisticated mental models, which serve us is describing the motion of real-world objects.

The language of kinematics deals with the words Vectors, Scalars, Distance, Displacement, Speed, Velocity, Acceleration, which are used frequently to describe the motion of objects.

The underlying concepts and principles of Physics have a mathematical basis. Throughout the course of study of physics, the emphasis will often be upon the conceptual nature of physics. We will give considerable and persistent attention to its mathematical aspect.

The motion of objects can be described by words such as distance, displacement, speed, velocity, and acceleration. These physical quantities can be divided into two categories.

The quantity is either a vector or a scalar. These two categories can be distinguished from one another by their distinct definitions.

Scalar Quantity: Any physical quantity, which can be measured on some scale, is said to be a scalar quantity. Any scalar quantity has a magnitude. Distance is a scalar quantity. The values of physical quantities (such as mass, distance, etc.) that have only magnitude add according to the rules of arithmetic. Mass, length, time, density, volume etc. are examples of scalar quantities.

Vector Quantity: Any physical quantity, which has a magnitude as well as the direction, is called vector quantity. Displacement is a vector quantity. Vectors can be represented as a straight-line segment. Vectors are added by geometric addition. Physical quantities are those, which can be defined and measured. To understand a physical quantity, we should know its

- Definition
- Measuring technique or formula
- Scalar or vector nature
- Units

Review of Basic Concepts

Particle

An object of negligible dimensions, i.e., a point mass is known as a particle. This is only a mathematical idealization. The earth can be treated as a particle if we consider the motion of the earth around the sun. The radius of the earth is very small, when we compare the distance between the earth and sun.

Rest and Motion

The concept of motion is a relative one and a body that may be in motion relative to one reference system, may be at rest relative to another.

Rest: *If the position of an object does not change as time passes, it is said to be at rest with reference to observer.*

Motion: *If the position of the object changes as time passes, it is said to be in motion with respect to observer. Or a continuous change in the position of a body relative to other bodies is called a mechanical motion.*

Motion is Relative

While sitting in a moving train, your distances from the walls, roof and the floor of the compartment do not change. That is, with respect to the compartment, your position does not change. You are at rest with respect to the compartment. But your distance from the platform from which you boarded the train, changes as time passes. So you are moving with respect to the platform. This means that an object can be at rest with respect to one thing and in motion with respect to some other thing at the same time. So *motion is not absolute; it is relative*.

Is the platform at rest, or is it moving with respect to the compartment? The distance of the platform from the compartment is changing as time passes. So the platform is moving with respect to the compartment. Also, the compartment is moving with respect to the platform.

Suppose two trains are moving on parallel tracks in the same direction. Both started together and are moving equally fast. The distance of a person A sitting in the first train from another person B sitting on the other train will not change. So, A is at rest with respect to B. Similarly, B is at rest with respect to A. Both are moving with respect to the platform, but they are at rest with respect to each other. We are going to deal with the rectilinear motion of an object in this chapter. A broad range of physical situations are examples of straight line or rectilinear motion.

Types of Motion

All bodies in motion do not move the same way; for example, a car moves straight along the road; a spinning top spins round and round; the strings of a sitar moves up and down and the pendulum clock moves sideways in either direction. The motion of bodies can be classified under the following types.

Translational Motion

The motion in which all the particles of a body move through the same distance in the same time is called translation motion.

Examples: An automobile moving along a road, a freely falling body, firing of a bullet from a gun, etc., are some of the examples, of translational motion.

Translational motion is of two types.

Rectilinear Motion

When a body moves in a straight line, the motion described by the body is a rectilinear motion (Fig. 1.1).

Fig. 1.1

Examples of rectilinear motion: A train moving along a straight track; a car moving on a straight road.

Curvilinear Motion

When a body moves along a curved path, the motion described by the body is called curvilinear motion (Fig. 1.2).

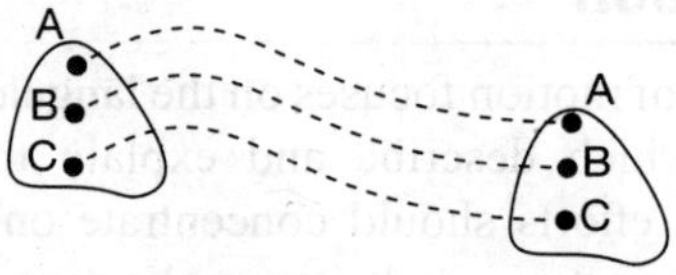

Fig. 1.2

Examples of curvilinear motion: The motion of a baseball hit in the air; a car negotiating a bend; a ball thrown upward at an angle.

Rotational Motion

A motion in which a body moves about a fixed axis without changing its position is called *rotational motion* (Fig. 1.3).

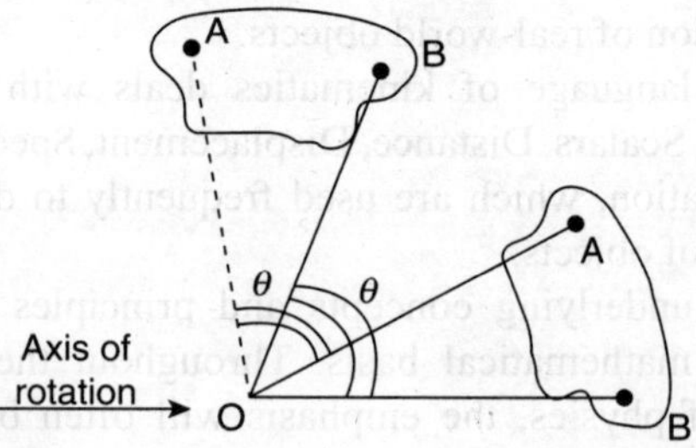

Fig. 1.3

Examples: The turning of a merry-go-around, Ferris wheel, a fan, a potter's wheel and the blades of a food processor are all examples of rotational motion.

Combination of Translation and Rotational Motions

In some cases, the motion of rotation and motion of translation occur at the same time. When a drill cuts through a metal piece, it performs the motion of translation as well as the motion of rotation at the same time. Other examples of this kind of motion are the wheels of a moving bicycle or a car, the ball delivered by a spin bowler.

Circular Motion

When a body moves around a circular path, the way the athlete runs around a circular track, its motion is called circular motion. When you tie a string around an object and whirl around, the motion of the object is circular.

Note: In circular motion, the position of the body itself keeps changing with time, but in rotational motion, the position of the body does not change.

Oscillatory Motion

Oscillatory motion is a repetitive, to-and-fro motion of a body about its position of rest. The oscillations of a pendulum, the needle of a sewing machine moving up and down, the movement of a swing and the swinging of a bell or chandelier are examples of oscillatory motion.

Vibratory Motion

The vibratory motion is a kind of oscillatory motion in which the moving object undergoes change in shape or size.

Examples: When you pluck a string of a guitar or sitar, it vibrates to and fro about its position of rest. When you strike the membrane of a drum, it vibrates up and down about its position of test. These vibrations are like oscillations of a pendulum, except that they are faster. There is another difference. The pendulum swings freely, while the string of a musical instrument is fixed at two ends, and the membrane of a drum is fixed around the circumference. Even during breathing, when our chest expands or contracts, the motion is vibratory in nature, as all the points on the chest do not move out equally.

Periodic Motion

A repetitive motion, which repeats itself after a regular interval of time, is called the periodic motion. The rotation of the earth about its axis is repeated every 24 hours, so it is periodic. The revolution of the earth around the sun is also periodic. The swinging pendulum of a wall clock is also an example of periodic motion.

Non-Periodic Motion

A repetitive motion, which repeats itself, but not at fixed intervals of time, is called non-periodic motion. If you swing your arms to and fro or twirl a pen round and round, the movements are non-periodic. The beating of the heart and the expansion and contraction of lungs are repetitive, but non-periodic. The motion of a cyclist around a circular track with varying speed is non-periodic.

Different motions at the same time: Often a body has different kinds of motion at the time. The earth rotates about its axis while revolving around the sun. A spinning top rotates, wobbles and moves along the floor at the same time, and so on.

Measuring Position

Reference Systems

The motion of a particle is always described with respected to a reference system. A reference system is 'made' by taking an arbitrary point as origin and imagining a co-ordinate system to be attached to it. This co-ordinate system chosen for a given problem constitutes the reference system for it. We generally choose a coordinate system attached to the earth as the reference system for most of the problems.

Coordinate Systems and Frames of Reference

The location of a point on a line can be described by one coordinate; a point on a plane can be described by two coordinates; a point in a three-dimensional volume can be described by three coordinates. In general, *the number of coordinates equals the number of dimensions.* A coordinate system consists of:

1. A fixed reference point (origin)
2. A set of axes with specified directions and scales
3. Instructions that specify how to *label* a point in space *relative to the origin and axes.*

Position: The position of a particle moving along a straight line is the distance of a particle specified in magnitude and direction from the origin of a reference system. Consider a particle moving along a straight line. Choose a convenient point '*O*' as origin to describe its position. There are two directions along the line. The choice of positive and negative directions of the two directions is entirely ours. Any one of the directions may be taken as positive and other negative. In general, left-to-right direction is positive direction.

Since the line shown in the figure resembles the *x*-axis, the position of a particle moving along a straight line is described by *x*. The details of the origin, positive and negative directions are shown in Fig. 1.4.

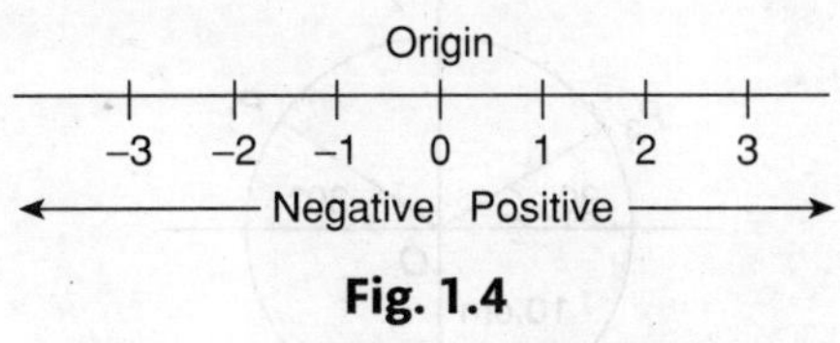

Fig. 1.4

To describe the position of the particle at a given time, we have to specify:

- its distance from the origin, i.e., numerical value with unit
- whether the particle is moving towards right seen from the origin or moving towards left seen from the origin, i.e., the positive direction or the negative direction with the sign.

ILLUSTRATION 1.1

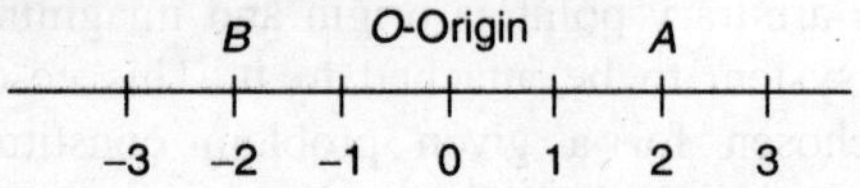

Consider the situation shown in figure. (a) What is the position of a particle when it is at A and when it is at *B*? (b) Are the two positions the same? (c) Are the two distances of the particle from the origin the same? (d) What is the position of particle at '*O*'?

Solution

(a) The position of the particle is $x = 2$ m when it is at *A*, and $x = -2$ m when it is at *B*.

(b) The two positions are not the same.

(c) The distances of the particles from the origin on the two positions are the same, and equal to 2 m.

(d) The position of the particle is $x = 0$ m when it is at *O*.

If the motion of a particle is not along a straight line, its position cannot be described by just a single quantity. In the case of a straight line, only two directions originate from the origin and the symbols '+' and '–' are enough to indicate the direction in which the particle is situated.

But if we consider a plane and choose an origin in it, there are infinite number of directions from the origin. To describe the position of the particle, we have to specify its distance from the origin in which the particles is situated. If the particle is confined to a plane, the concept of north-south east-west may be used to describe it.

ILLUSTRATION 1.2

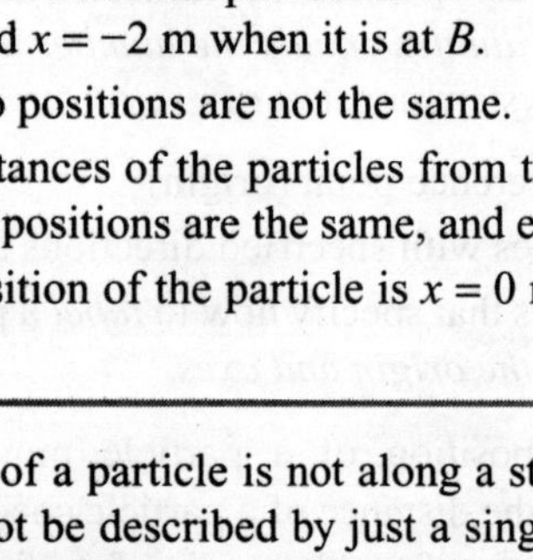

The above figure shows three different positions of a particle moving along a circle of radius 10 cm. Describe each position by giving its distance and direction from the origin.

Solution

The position at P_1 is "10 cm, 30° north of east". The position of P_2 is 10 cm north.

The position of P_3 is 10 cm 30° north of west.

Uniform Motion

A body has a uniform motion if it travels equal intervals of time, no matter how small these time intervals may be.

Non-Uniform Motion

A body has a non-uniform motion if it travels unequal distances in equal intervals of time.

Speed

Speed of a body is the distance travelled by it per unit time.

$$\text{Speed} = \frac{\text{Distance travelled}}{\text{Time taken}}$$

$$\text{Speed, } v = \frac{s \text{ metres}}{t \text{ seconds}}$$

$$\text{Speed, } v = \frac{s}{t} \text{ metres per second}$$

Speed is a scalar quantity.

Average Speed

The average speed of a body is the total distance travelled divided by the total time taken to cover this distance.

Uniform Speed or Constant Speed

A body has a uniform speed if it travels equal distances in equal intervals of time, no matter how small these time intervals may be. A car is said to have a uniform speed of say, 60 km per hour, if it travels 30 km every half hour, 15 km every quarter of an hour, 1 km every minute or $\frac{1}{60}$ of a km every second.

$$v = \frac{s \text{ metres}}{t \text{ seconds}}$$

Uniform Velocity or Constant Velocity

A body has a uniform velocity if it travels in a straight line and moves over equal distances in equal intervals of time, no matter how small these time intervals may be. Velocity is a vector quantity.

Acceleration

Acceleration of a body is defined as the rate of change of its velocity with time.

$$\text{Acceleration} = \frac{\text{Change in velocity}}{\text{Time taken for change in velocity}}$$

Acceleration is a vector quantity.

Uniform Acceleration

A body has a uniform acceleration if it travels in a straight line and its velocity increases by equal amounts in equal intervals of time. In other words, a body has a uniform acceleration if its velocity changes at a uniform rate.

Examples of uniformly accelerated motion

(i) The motion of a freely falling body is an example of uniformly accelerated motion.

(ii) The motion of a bicycle going down the slope of a road when the rider is not pedaling and wind resistance is negligible, is also an example of uniformly accelerated motion.

Equations of Uniformly Accelerated Motion

There are three equations for the motion of bodies which travel with a uniform acceleration. These equations give the relationship between initial velocity, final velocity, time taken, acceleration and distance travelled by the bodies.

First equation of motion: $v = u + at$

Second equation of motion: $s = ut + \frac{1}{2}at^2$

Third equation of motion: $v^2 = u^2 + 2as$

To solve problems on motion we should remember that

(i) If a body starts from rest, its initial velocity, $u = 0$

(ii) If a body comes to rest (it stops), its final velocity, $v = 0$

(iii) If a body moves with uniform velocity, its acceleration, $a = 0$.

ILLUSTRATION 1.3

A car travels a distance A to B at a speed of 60 km/h and returns to A at a speed of 40 km/h.

(a) What is the average speed for the whole journey?

(b) What is the average velocity?

Solution

(a) Let S be the distance between AB

Time taken to go from A to B

$$t_1 = \frac{S}{60}$$

Time taken to come from B to A

$$t_2 = \frac{S}{40}$$

Total time taken

$$t_1 + t_2 = \frac{S}{60} + \frac{S}{40} = \frac{5S}{120}$$

Total distance travelled $= S + S = 2S$

$$\text{Average speed} = \frac{\text{Total distance travelled}}{\text{Total time taken}}$$

$$= \frac{2S \times 120}{5S}$$

$$= 48 \text{ km/hr}$$

(b) $\text{Average velocity} = \frac{\text{Displacement}}{\text{Time}}$

$$= \frac{0}{2t} = 0$$

Problem-Solving Strategy

The following procedure is recommended for solving problems that involve an object undergoing a constant acceleration.

1. Think about what is going on physically in the problem. Establish a mental representation.
2. Simplify the problem as much as possible. Confirm that the problem involves either a particle or an object that can be modeled as a particle and that it is moving with a constant acceleration.
3. Construct an appropriate pictorial representation, such as a motion diagram, or a graphical representation.
4. Make sure all the units in the problem are consistent. That is, if positions are measured in metres, be sure that velocities are in units of m/s and acceleration in m/s^2.
5. Choose a coordinate system which should be used throughout the problem.
6. Set up the mathematical representation. Choose an instant to call the "initial" time $t = 0$ and another the "final" time t. The initial instant need not to be when the particle stops moving. Identify all the quantities given in the problem. Select from the list of kinematic equations the one or ones that enables you to determine the unknowns. Solve the equations.
7. Once you have determined your result, check to see if your answers are consistent with the mental and pictorial representations and that your results are realistic.

ILLUSTRATION 1.4

A body starting from rest has an acceleration of 20 m/s². Calculate the distance travelled by it in 6 seconds.

Solution

Given $u = 0$

$t = 6$ s

$a = 20$ m/s²

$$s = ut + \frac{1}{2}at^2$$

$$= 0 \times t + \frac{1}{2} \times 20 \times (6)^2$$

$$= 0 + \frac{1}{2} \times 20 \times 36$$

$$= 360 \text{ m}$$

ILLUSTRATION 1.5

A train running at a velocity of 36 km/h is brought to rest in a distance of 200 m. Calculate the retardation produced in the train.

Solution

$u = 36$ km/hr

$$= \frac{36 \times 1000}{60 \times 60}$$

$$= 10 \text{ m/s}^2$$

$v = 0$

$s = 200$ m

$$v^2 = u^2 + 2as$$

$$0 = (10)^2 + 2a \times 200$$

$$a = \frac{100}{400} = -0.25 \text{ m/s}^2$$

Retardation is 0.25 m/s².

ILLUSTRATION 1.6

A man walks on a straight road from his residence to his office 2.5 km away at a speed of 5 km/h. Finding the office closed, he immediately turns and walks back to his residence with a speed 0 7.5 km/h. Find the magnitude of average velocity and average speed of the man over the interval of time:

(i) 0 to 30 min (ii) 0 to 50 min (iii) 0 to 40 min.

Solution

(a) Given that distance of office from the residence, $s = 2.5$ km.

Speed of the man for forward journey, $v_1 = 5$ km/h

∴ Time taken to reach the office

$$t_1 = \frac{s}{v_1} = \frac{2.5 \text{ km}}{5 \text{ km/h}} = \frac{1}{3} \text{ h} = 30 \text{ min}$$

Speed of the man for the return journey, $v_2 = 7.5$ km/h

∴ Time taken to return to his residence

$$t_2 = \frac{s}{v_2} = \frac{2.5 \text{ km}}{7.5 \text{ km/h}} = \frac{1}{3} \text{ h} = 20 \text{ min}$$

(i) Over the interval of time from 0 to 30 min

During this time interval, the man covers the distance from his residence up to his office, i.e., distance covered = 2.5 km.

Also displacement = 2.5 km.

Time taken = 30 min = $\frac{1}{2}$ h

$$\text{Magnitude of average velocity} = \frac{\text{Displacement}}{\text{Time}} = \frac{2.5}{0.5} = 5 \text{ km/h}$$

$$\text{Average speed} = \frac{\text{Distance}}{\text{Time}} = \frac{2.5 \text{ km}}{0.5 \text{ h}} = 5 \text{ km/h}$$

(ii) Over the interval of time from 0 to 50 min

During this time interval, the man returns to his residence, i.e., displacement = 0 km, distance covered = 5 km

$$\text{Time taken} = \frac{50}{60} = \frac{5}{6} \text{ h}$$

$$\therefore \text{Magnitude of average velocity} = \frac{0 \text{ km}}{(5/6)\text{h}} = 0 \text{ km/h}$$

$$\text{and Average speed} = \frac{5 \text{ km}}{(5/6)\text{h}} = 6\frac{\text{km}}{\text{h}}$$

(iii) Over the interval of time from 0 to 40 min

During the first 30 min of this interval, the man covers a distance of 2.5 km and in the next 10 min, he covers 2.5/2, i.e., 1.25 km in the opposite direction.

$\therefore$ Displacement = 2.5 – 1.25 km

and Distance covered = 2.5 + 1.25 – 3.75 km

$$\text{Time taken} = 40 \text{ min } \frac{2}{3} \text{ h}$$

$$\text{Hence, magnitude of average velocity} = \frac{1.25 \text{ km}}{(2/3)\text{h}} = 1.875 \text{ km/h}$$

$$\text{and average speed} = \frac{3.75 \text{ km}}{(2/3)\text{h}} = 5.625 \text{ km/h}$$

Free Fall

We are familiar with falling objects – for example, a paperweight that is accidentally knocked off the edge of a desk. Often in describing the motion of the paperweight, we may neglect air resistance. If air resistance has a negligible effect on a falling object, then it is valid to assume that the object's acceleration is entirely due to gravity.

Galileo Galilei made several experiments and studies of free-fall and determined that the acceleration due to gravity is constant and is same for different objects. The magnitude of this acceleration is represented by the symbol g. Although g varies slightly from place to place on the earth's surface, the value that is accurate for our use is

$$g = 9.8 \text{ m/s}^2$$

Numerical problems of free-fall utilize the four speed equations, replacing the value of a by g.

$$v = u + gt$$

$$v^2 = u^2 + 2gh$$

$$h = ut + \frac{1}{2}gt^2$$

$$h_n = u + \frac{g}{2}(2n-1)$$

ILLUSTRATION 1.7

A body is projected upwards with a velocity of 98 m/s. Calculate:

(a) The maximum height reached

(b) Time taken to reach the maximum height

(c) Its velocity at a height of 196 m form the point of projection

(d) The velocity with which it will cross down the point of projection and

(e) The time taken to reach the point of projection. [$g = 9.8$ m/s^2]

Solution

(a) Given that initial velocity $u = 98$ m/s (upwards).

Acceleration $a = -g = -9.8$ m/s^2

Let H be the maximum height, then

$$0 = (98)^2 - 2 \times 9.8 \times H$$

$$H = \frac{(98)^2}{29.8} = 490 \text{ m}$$

(b) The time taken to reach the highest point,

$$v = 0$$

$$t = \frac{u}{g} = \frac{98}{9.8} = 10 \text{ s}$$

(c) Velocity at a height of 196 m

$$v^2 = u^2 + 2as$$

$$v^2 = (98)^2 + 2(9.8) \times 196$$

$$\therefore \quad v = \pm 75.91 \text{ m/s}$$

+75.91 m/s while crossing the height upwards and –75.91 m/s while crossing it downward.

(d) Velocity with which it will cross the point of projection

$$v^2 = u^2 + 2gs \text{ or } v = \pm u \qquad \text{(as } s = 0 \text{ at the point of projection)}$$

While crossing the point of projection downwards, $v = -u = -98$ m/s

(e) The time taken to reach back the point of projection

$$t = \frac{2u}{g} = \frac{2 \times 98}{9.8} = 20 \text{ sec}$$

ILLUSTRATION 1.8

A ball is dropped from a height of 19.6 m above the ground. It rebounds from the ground and raises itself up to the same height. Taking the starting point as the origin and vertically downward direction as the positive x-axis, draw approximate plots of x versus t, v versus t and a versus t. Assume the small interval during which the ball was in contact with the ground as negligible.

Solution

Displacement

For the downward motion:

As $$a = +9.8 \text{ m/s}^2$$

$$x = ut + \frac{1}{2}at^2$$

$$19.6 = 0 + \frac{1}{2} \times 9.8 \times t^2$$

or $$t = 2 \text{ s.}$$

The ball reaches the ground when $x = 19.6$ m. This gives $t = 2$ s.

After that it moves up, x decreases at $t = 4$ s, x becomes zero, the ball reaches at the initial point. i.e., at $t = 0$, $x = 0$.

$$t = 2 \text{ s}, x = 19.6 \text{ m};$$

$$t = 3\text{s}, x = 4.9 \text{ m}$$

$$t = 4 \text{ s}, x = 0$$

Plot of x versus t is shown in figure.

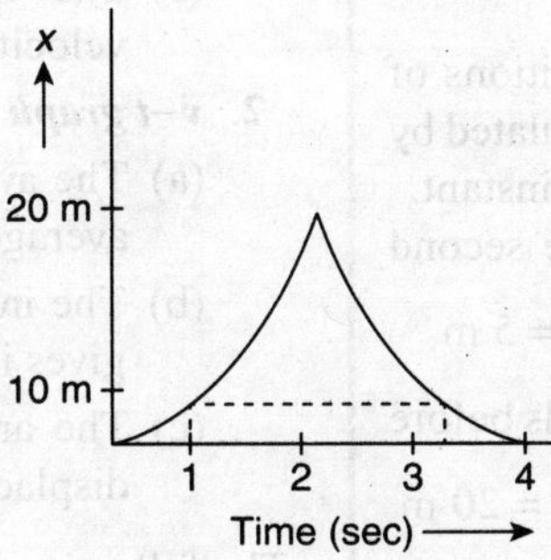

Velocity

During the first two seconds,

$$v = u + at$$

$$= 0 + 9.8\, t$$

At $t = 0, v = 0$

At $t = 1 \text{ s}, v = 9.8 \text{ m/s}$

At $t = 2 \text{ s}, v = 19.6 \text{ m/s}$

During the next two seconds, the ball goes upwards, velocity is –ve, magnitude decreasing and at $t = 4$ s, $v = 0$. Thus

At $t = 2 \text{ s}, v = -19.6 \text{ m/s}$

At $t = 3 \text{ s}, v = -9.8 \text{ m/s}$

At $t = 4 \text{ s}, v = 0$

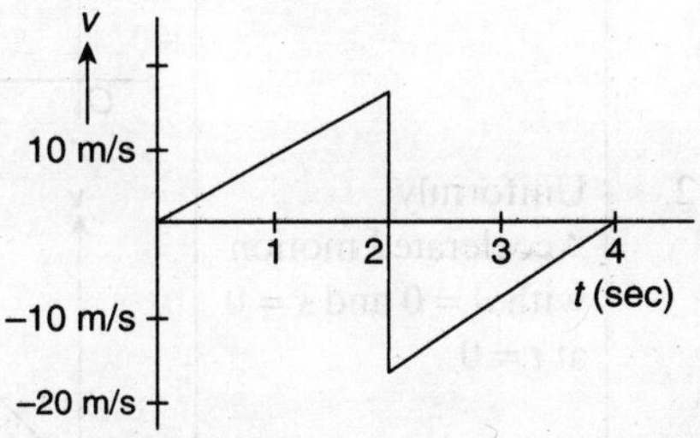

At $t = 2$ s, there is an abrupt change in velocity from 19.6 m/s to –19.6 m/s. In fact, this change in velocity takes place over a very small interval during which the ball remains in contact with the ground.

Acceleration

Acceleration remains constant (= 9.8 m/s²) throughout the motion except at $t = 2$ s. Plot a versus t is shown in figure.

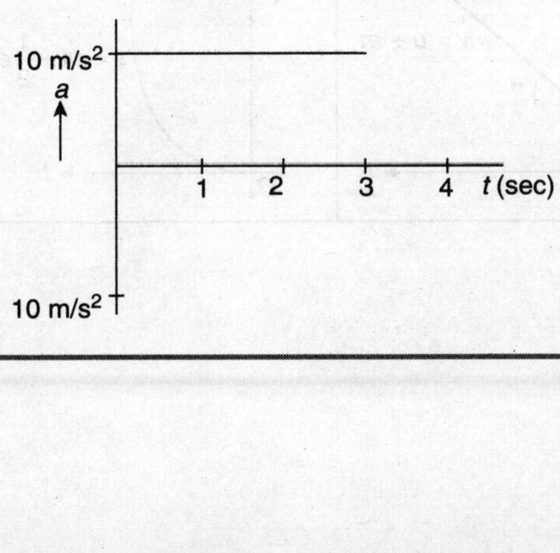

ILLUSTRATION 1.9

A person sitting on the top of a tall building is dropping balls at regular intervals of one second. Find the positions of the 3rd, 4th and 5th ball when the 6th ball is being dropped. Take g =10 m/s^2

Solution

When the 6th ball is being dropped, the positions of the other (previously fallen) balls can be calculated by using the time of falling of each ball till this instant.

For the 5th ball, it was dropped just one second before. Thus it has fallen a distance = $\frac{1}{2}gt^2 = 5$ m

For the 4th ball, it was dropped two seconds before this instant. It has fallen a distance = $\frac{1}{2}(10)^2 = 20$ m

For the 3rd ball, it was dropped two seconds before this instant. Its depth is = $\frac{1}{2}(10)3^2 = 45$ m

Graphical Interpretation of motion

1. ***$\vec{x}$–t graph***
   (a) The point on the graph is called an event.
   (b) The average slope of the graph (or curve) gives average velocity.
   (c) The instantaneous slope gives instantaneous velocity.
2. ***$\vec{v}$–t graph***
   (a) The average slope of a curve in $\vec{v}$–t graph gives average acceleration.
   (b) The instantaneous slope of a curve in $\vec{v}$–t graph gives instantaneousacceleration.
   (c) The area under a curve in $\vec{v}$–t graph gives the displacement.

The following are the graphs for one-dimensional motion with uniform velocity or uniform acceleration.

S.No	Cases	v-t Graph	s-t Graph	Conclusion
1.	Uniform motion v = constant $a = 0$	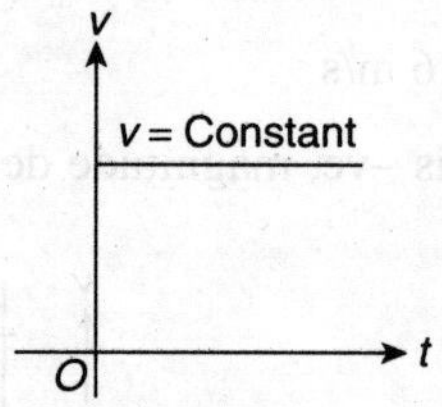	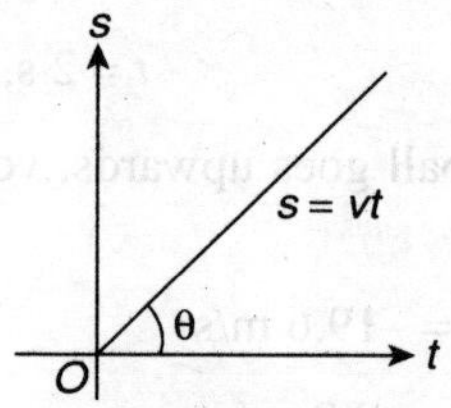	1. Acceleration = slope of v–t graph = 0 2. Slope of s–t graph= v = $\tan\theta$ 3. At $t = 0$, $s = 0$
2.	Uniformly Accelerated motion with $u = 0$ and $s = 0$ at $t = 0$	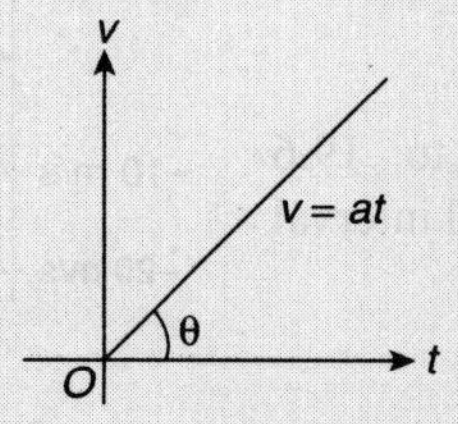	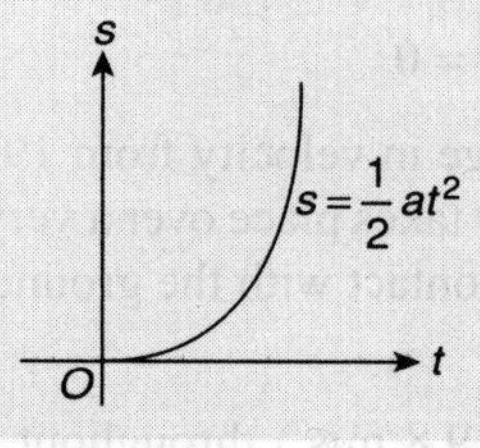	1. $\tan\theta = a$ = constant 2. Slope of s–t graph is zero at $t = 0$
3.	Uniformly accelerated motion with non-zero initial velocity ($u \neq 0$)	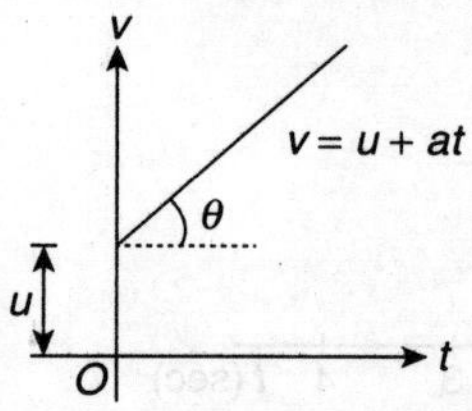	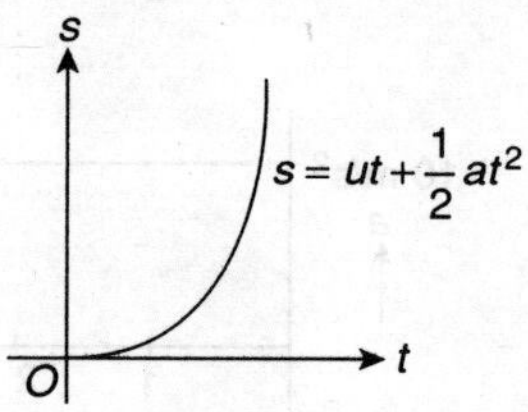	1. $\tan\theta = a$ = constant 2. $u \neq 0$, so slope of s–t graph at $t = 0$ is non-zero and equals u. 3. Slope of s–t graph is non-uniform.

(Continued)

S.No	Cases	v-t Graph	s-t Graph	Conclusion
4.	Uniformly accelerated motion with $u \neq 0$, and $s = s_0$ at $t = 0$	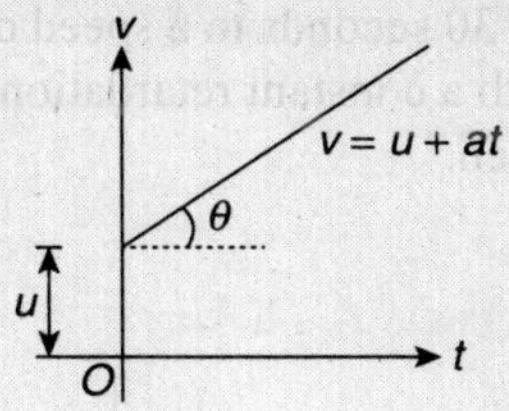	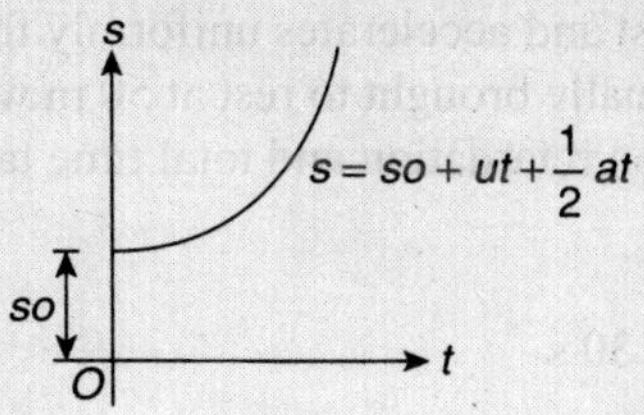	1. $\tan\theta = a =$ constant 2. Still slope is gradually increasing.
5.	Uniformly retarded motion with initial velocity u and final velocity zero.	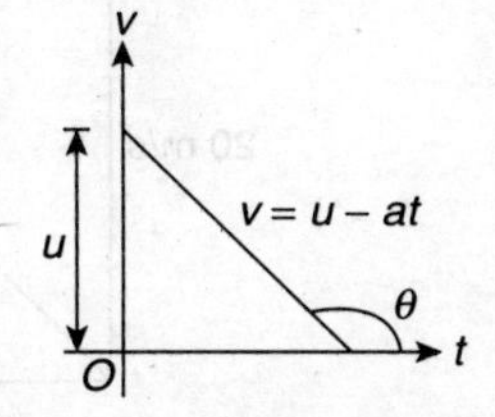	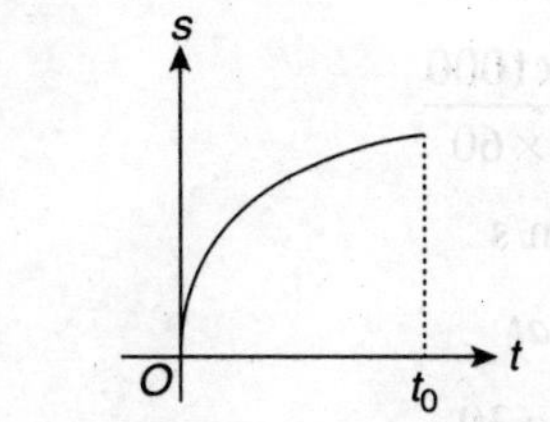	1. $\tan\theta = -a =$ constant 2. Slope of s–t graph at $t = 0$ is u. 3. Slope of s–t graph at $t = t_0$ is zero.
6.	Uniformly retarded, then accelerated motion	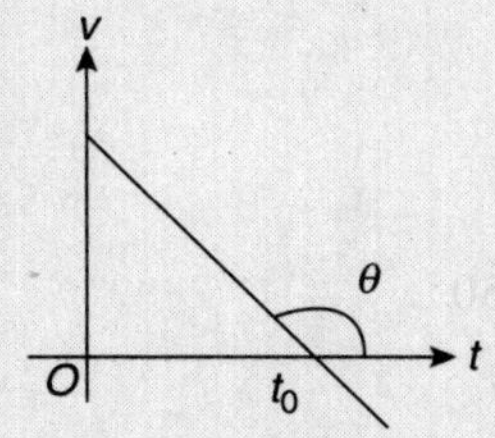	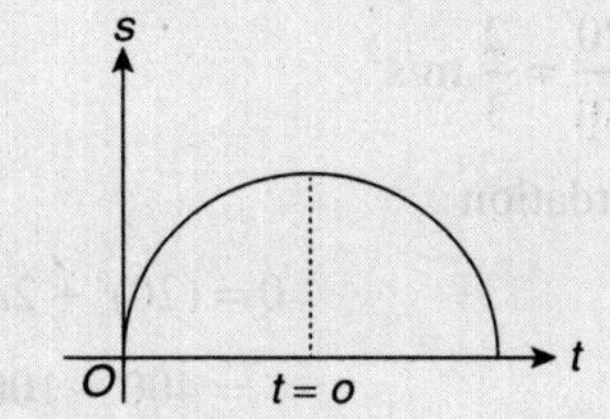	1. $\tan\theta = -a =$ constant 2. In s–t graph, slope or velocity first decreases, then increases with a reverse sign.

ILLUSTRATION 1.10

A body starting from rest accelerates uniformly along a straight line, at the rate of 10 m/s² for 5 s. It moves for 2 s with uniform velocity of 50 m/s, then it retards uniformly and comes to rest in 3 s. Draw the velocity–time graph of the body and find the total distance travelled by the body.

Solution

$v = u + at$

$u = 0, t = 5$ s

$a = 10$ m/s²

$v = 0 + 10 \times 5$

$= 5.0$ m/s

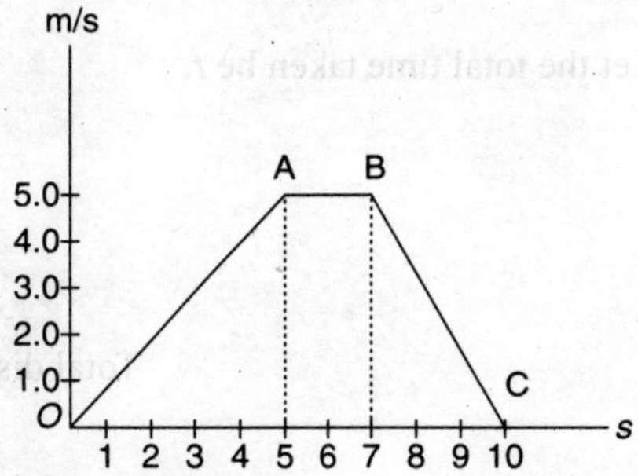

Total distance travelled by the body during motion = Area within the velocity – time graph

$$= \frac{1}{2}(\text{AB} + \text{OC}) \times h$$

$$= \frac{1}{2}(2 + 10) \cdot 50$$

$$= 300 \text{ m}$$

ILLUSTRATION 1.11

An automobile starts from rest and accelerates uniformly for 30 seconds to a speed of 72 km/h. It then moves with a uniform velocity and it is finally brought to rest at 50 m with a constant retardation. If the total distance travelled is 950 m. Find the acceleration retardation and total time taken.

Solution

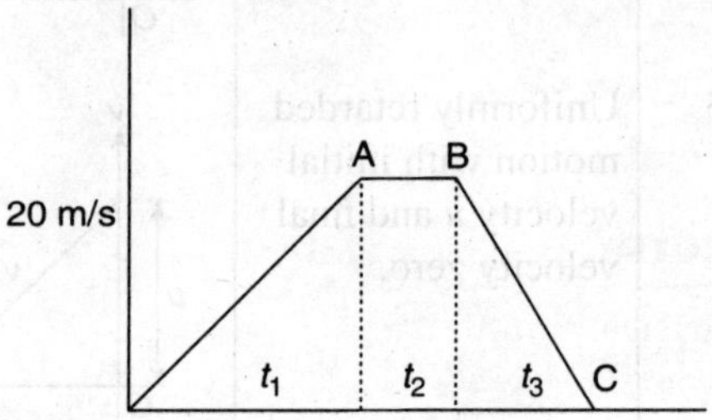

$$t = 30\text{ s}$$

$$v = 72\text{ km/hr}$$

$$= \frac{72 \times 1000}{60 \times 60}$$

$$= 20\text{ m/s}$$

$$v = u + at$$

$$20 = 0 + a.30$$

$$\Rightarrow \quad a = \frac{20}{30} = \frac{2}{3}\text{ m/s}^2$$

Using $v^2 = u^2 + 2as$ during retardation

$$0 = (20)^2 + 2a.50$$

$$= 400 + 100a$$

$$a = -\frac{400}{100} = -4\text{ m/s}^2$$

Now,

$$v = u + at$$

$$0 = 20 - 4t$$

$$\Rightarrow \quad t = \frac{20}{4} = 5\text{ sec}$$

Let the total time taken be t.

$$t_1 = 30\text{ s}$$

$$t_2 = \text{time for uniform motion}$$

$$t_3 = \text{time for retardation}$$

$$\text{Total distance} = \text{Area of trapezium OABC}$$

$$950 = \frac{1}{2} \times \text{AD(AB+OC)}$$

$$= \frac{1}{2} \times 20(t_2 + t_1 + t_2 + t_3)$$

$$= \frac{1}{2} \times 20(t_1 + 2t_2 + t_3)$$

$$= 10\,(30 + 2t_2 + 5)$$

$$\Rightarrow \quad 950 = 10\,(35 + 2t_2)$$

$$35 + 2t_2 = 95$$

$$2t_2 = 95 - 35$$

$$= 60$$

$$t_2 = \frac{60}{2} = 30 \text{ s}$$

Hence, total time $t = t_1 + t_2 + t_3$

$$= 35 + 30 + 5 = 65 \text{ s}$$

Vectors

All physical quantities can be divided into two categories — **scalars and vectors.**

Scalars are quantities which have only magnitude but no direction and can be completely specified if its magnitude is given.

Mass, volume, density, distance, speed, energy, current, temperature, gravitational potential, electric and magnetic flux are some of the examples of scalar quantities.

Vectors are quantities which have both magnitude and direction and can be completely specified only if both its magnitude and direction are given. In addition, a vector must also obey the rules of vector algebra.

Displacement, velocity, force, angular momentum, intensity of gravitational field, intensity of electric and magnetic fields, and moment of a force are some of the examples of vector quantities.

A vector quantity is represented by putting an arrow over the symbol representing it. Hence, for example, $\vec{A}$ stands for a vector quantity.

Graphical Representation of a Vector

A vector is denoted by $\overrightarrow{OA}$, i.e., with an arrow over OA. The point O is known as origin while A as the terminus as shown in the figure below. The magnitude of the vector is given by the length OA.

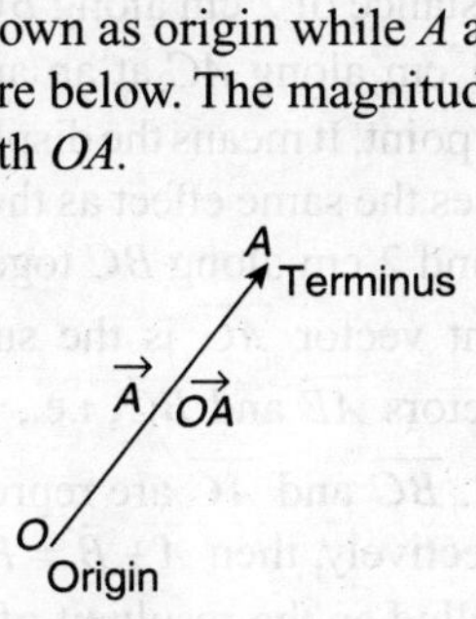

Some Specific Vectors

Like Vectors

Two vectors are said to be like vectors if they have the same direction but different magnitude. The figure below shows two vectors $\vec{A}$ and $\vec{B}$ which have different magnitudes but are parallel to each other.

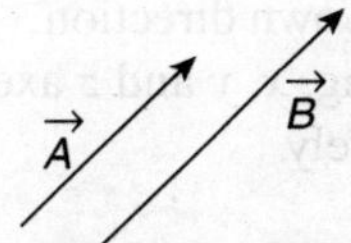

Unlike Vectors

The vectors having opposite direction and different magnitude are called unlike vectors. In the below figure, two such vectors $\vec{A}$ and $\vec{B}$ have different magnitudes and are antiparallel to each other.

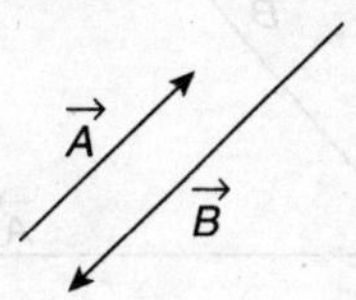

Equal Vectors

Two vectors are said to be equal, if they have the same magnitude and direction.

The below figure shows two vectors $\vec{A}$ and $\vec{B}$ having the same magnitude and same direction and therefore, $\vec{A} = \vec{B}$.

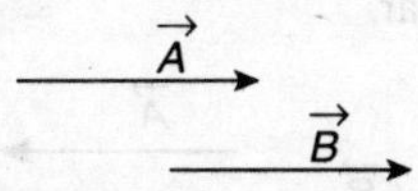

Opposite Vectors

The vectors having same magnitude but opposite direction, are known as opposite vectors. The figure below shows two such vectors $\vec{A}$ and $\vec{B}$ having the same magnitude and opposite direction and therefore,

$$\vec{A} = -\vec{B}$$

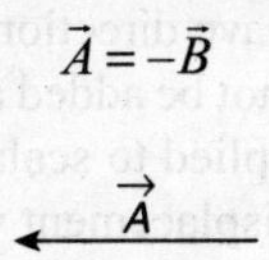

Null Vector

It is defined as a vector having zero magnitude. It has a direction which is indeterminate as its magnitude is zero.

Unit Vector

A vector divided by its magnitude is called a unit vector along the direction of the vector. Obviously, the unit vector has unit magnitude and direction is the same as that of the given vector.

A unit vector in the direction of some vector $\vec{A}$ is written as $\hat{A}$ and is read as 'A cap' or 'A hat'. Therefore, by definition,

$$\hat{A} = \frac{\vec{A}}{A} \text{ or } \vec{A} = \hat{A}A$$

Thus, any vector can be expressed as magnitude times the unit vector along its own direction.

Unit vectors along, x, y and z axes are represented by $\hat{i}$, $\hat{j}$ and $\hat{k}$ respectively.

Co-initial Vectors

Vectors are said to be co-initial, if they have a common initial point. In figure, $\vec{A}$ and $\vec{B}$ originating from the same point O as their origin are called co-initial vectors.

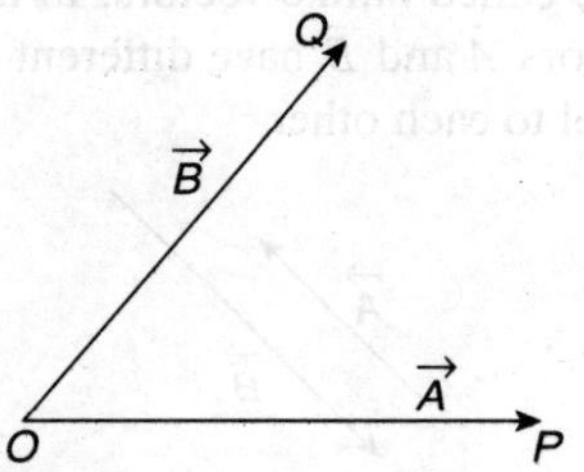

Co-linear Vectors

Two vectors having equal or unequal magnitudes, which either act along the same line [figure (i)] or along the parallel lines in the same direction [figure (ii)] or along the parallel lines in opposite direction [figure (iii)], are called collinear vectors. Like, unlike, equal and opposite vectors are collinear.

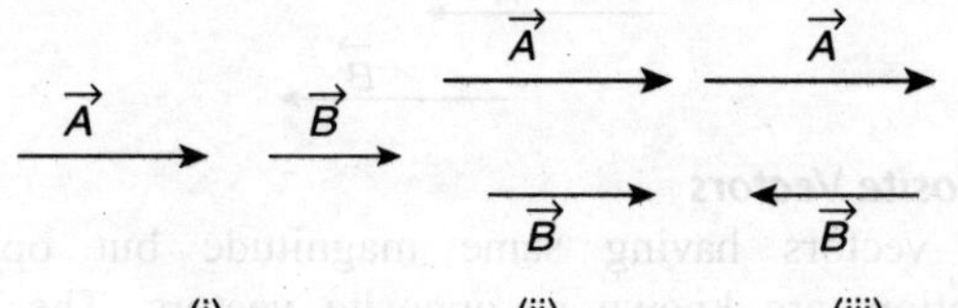

Addition of Vectors

Because the vectors have directions also along with their magnitudes, they cannot be added according to the simple rules of algebra as applied to scalars. In order to make it clear, let us add two displacement vectors $\vec{A}$ and $\vec{B}$ having different possible directions with respect to each other.

Case I: When Two Vectors $\vec{A}$ and $\vec{B}$ are in the Same Direction

If a body moves a distance of 1 cm first and then 2 cm, due east in both the cases and the two displacement vectors are represented by $\vec{A}$ and $\vec{B}$ (figure), then the sum of two displacement vectors i.e. $\vec{A}+\vec{B}$, will be equal to a displacement of 3 cm along east.

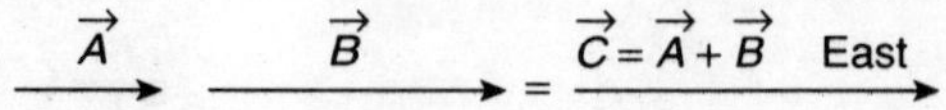

Case II: When Two Vectors $\vec{A}$ and $\vec{B}$ are in Opposite Directions

If the body moves a distance of 1 cm due west and then through 2 cm due east, (figure), then the sum $(\vec{A}+\vec{B})$ a of two displacement vectors will be equal to a displacement of 1 cm due east.

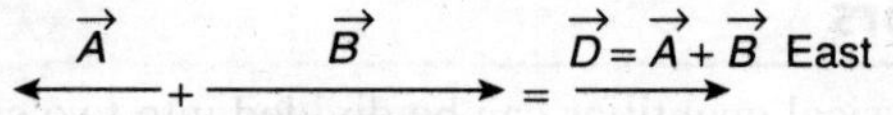

Case III: When Two Vectors are Inclined to Each Other

We consider here the simple case of two mutually perpendicular vectors. Suppose the body moves a distance of 1 cm due east and then a distance of 2 cm due north. In the figure given below the two displacement vectors are shown by AB and BC, according to some scale. It is clear that the magnitude of the displacement vector AC is $\sqrt{5}$ cm and the direction of displacement is due north of east measured in terms of θ.

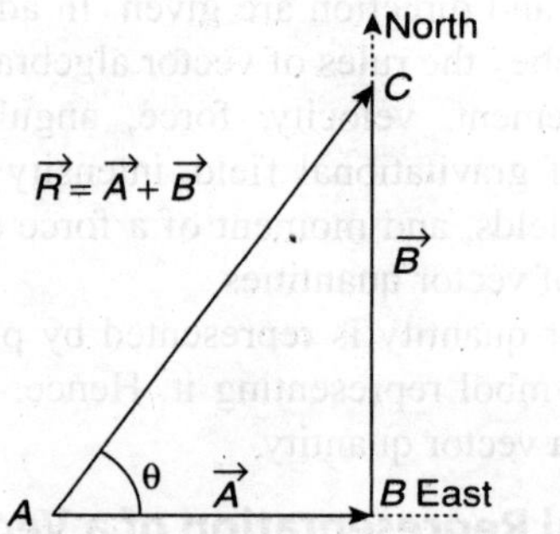

Therefore, if instead of moving a distance of 1 cm along AB and then a distance of 2 cm along BC, the body moves a distance of $\sqrt{5}$ cm along AC at an angle θ with AB, it reaches the same point. It means the displacement of $\sqrt{5}$ cm along AC produces the same effect as the displacements of 1 cm along AB and 2 cm along BC together. This implies that displacement vector $\overrightarrow{AC}$ is the sum or resultant of displacement vectors $\overrightarrow{AB}$ and $\overrightarrow{BC}$, i.e., $\overrightarrow{AB} + \overrightarrow{BC} = \overrightarrow{AC}$

If vectors $\overrightarrow{AB}$, $\overrightarrow{BC}$ and $\overrightarrow{AC}$ are represented by vectors $\vec{A}$, $\vec{B}$ and $\vec{R}$ respectively, then $\vec{A}+\vec{B}=\vec{R}$

where $\vec{R}$ is called as the resultant of vectors $\vec{A}$ and $\vec{B}$.

Thus we conclude that:

(i) The resultant of two or more vectors is a single vector which produces the same effect as the individual vectors do together.

(ii) Vectors are not added algebraically but are added geometrically.

POINTS TO REMEMBER

1. Rest and motion are relative; there is nothing like absolute rest or absolute motion.
2. The change in the position of a body in a particular specified direction is called its displacement.
3. Distance is the total length of the path travelled by a body in a certain time while displacement is the shortest distance, in a specified direction between the final and initial positions of the body.
4. If a body covers equal distances in equal intervals of time, its motion is said to be uniform.
5. If a body covers unequal distances in equal intervals of time, its motion is said to be non-uniform.
6. The speed of a body is defined as the distance covered by it in a unit interval of time, i.e. speed is the rate of change of distance.
7. If the motion of a body is not uniform, we define its average speed as the total distance travelled divided by the total time taken.
8. The distance–time graph for uniform motion is a straight line whose slope gives the speed of the moving body.
9. The velocity of a body is defined as the distance travelled by the body in a particular direction (i.e., its displacement) in a unit interval of time. The direction of velocity is the direction of the displacement, i.e., the direction in which the body is moving.
10. If a body moving in a particular direction travels equal distances in equal intervals of time, its velocity is said to be uniform. In other words, an object has a uniform velocity if its magnitude, i.e., its speed, and its direction do not change with time. If either its speed or direction or both change with time, the velocity of the object is said to be variable.
11. The rate of the change of the velocity of a body is called its acceleration.

$$\text{Acceleration} = \frac{\text{Change in velocity}}{\text{Time}}$$

If the velocity of a body increases with time without change in direction, the acceleration will have the same direction as that of velocity and is taken as positive. If the velocity of a body decreases with time without change in direction, the direction of acceleration is opposite to that of velocity and is regarded negative. Negative acceleration is called deceleration or retardation.
12. If the velocity of a moving object increases by equal amounts in equal intervals of time, its acceleration is said to be uniform.
13. The velocity–time graph for uniformly accelerated motion is a straight line whose slope gives the acceleration.
14. The distance travelled = area under the velocity–time graph.

COMMON MISCONCEPTIONS

1. Average speed and average velocity

The magnitude of the average velocity is not the average sped. Consider a particle moving from the origin to $x = 5$ m and then back to the origin in a time interval of 2.0. The magnitude of the average velocity is zero, because the particle ends the time interval at the same position at which it started – the displacement is zero. The average speed, however, is the total distance divided by the time interval: 10 m/2.0 = 5.0 m/s.

2. Instantaneous speed and instantaneous velocity

In misconception 1, we argued that the magnitude of the average velocity is not the average speed. Notice the difference when discussing instantaneous values. The magnitude of the instantaneous velocity is the instantaneous speed. In an infinitesimal time interval, the magnitude of the displacement is equal to the distance travelled by the particle.

3. Negative acceleration

Keep in mind that negative acceleration does not necessarily mean that an object is slowing down. If the acceleration is negative, and the velocity is negative, the object is speeding up.

Deceleration

The word deceleration has a common popular connotation as slowing down. When combined with the misconception 3 that negative acceleration means slowing down, the situation can be further confused by the use of the word deceleration.

4. *g* and altitude

The value of g decreases with increasing altitude, furthermore, slight variations occur in g with latitude at the surface of the Earth. As long as your problem involves something moving close to the surface of the Earth, a constant value of 9.80 m/s² is a good approximation for g.

5. **The sign of g**

Keep in mind that g is a positive number – it is tempting to substitute – 9.80 m/s² for g, but resist this temptation. The fact that the gravitational acceleration is downward is indicated explicitly by stating the acceleration as $a_y = -g$.

6. **Acceleration at the top of the motion**

Imagine throwing a baseball straight up into the air. It is a common misconception that the acceleration of a projectile at the top of its trajectory is zero. This generally arises due to confusion between velocity and acceleration. While the velocity at the top of the motion of an object thrown upward momentarily goes to zero, the acceleration is still that due to gravity at this point. Remember that acceleration is proportional to force, and the gravitational force still acts at the moment that the object has stopped. If the velocity and acceleration were both zero, the projectile would be stay at the top.

PRACTICE QUESTIONS

Very-short Answer Type Questions

1. Which speed is measured by the speedometer of your scooter?
2. What does the tangent at a point on the position–time graph for an object in non-uniform along a straight line represent?
3. Mention the condition when an object in motion
   a. Can be considered point object.
   b. Cannot be considered point object.
4. Under what condition will the distance and displacement of a moving object have the same magnitude?
5. What will be nature of velocity–time graph for a uniform motion?
6. What will be the nature of position–time graph for a uniform motion?
7. What does slope of position–time graph represent for a uniform motion?
8. What is the numerical ratio of velocity to speed of an object?
9. If the displacement–time graph of a particle is parallel to
   a. displacement axis
   b. the time axis
   what will be the velocity of the particle?

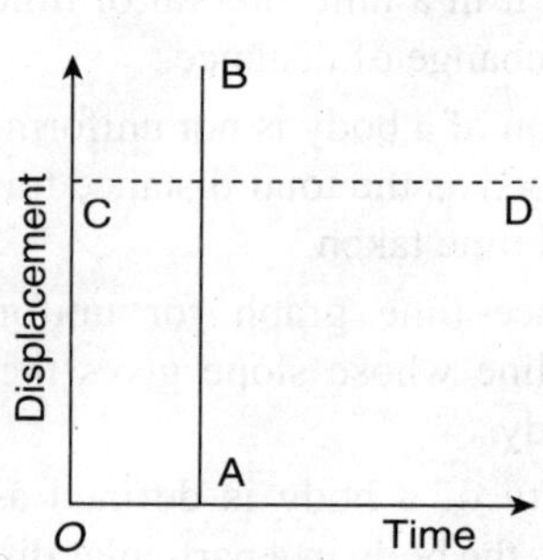

10. Can position–time graph have negative slope?
11. Draw the position–time graph for a stationary object.
12. What do you understand by positive and negative time?
13. Under what condition is the average velocity equal to instantaneous velocity?

Short-answer Type Questions

1. Is it possible that displacement is zero but not the distance?
2. Can a body have a constant velocity but a varying speed?
3. Are the magnitude of average speed and velocity equal?
4. Under what condition is the average velocity equal to instantaneous velocity?
5. Can a body be at rest at any moment of its motion?

6. Can the velocity of an object ever be in the direction other than the direction of the acceleration of the object?

7. In a given time interval, is the total displacement of a particle equal to the product of the average velocity and the time interval, even when the velocity is not constant?

8. A car manufacturer advertises that the brakes are so perfect that the car stop instantaneously. Comment.

9. Is the acceleration of a car is greater when the accelerator is pushed to the floor or when brake pedal is pushed hard?

10. Is the time variation of position shown in figure observed in nature?

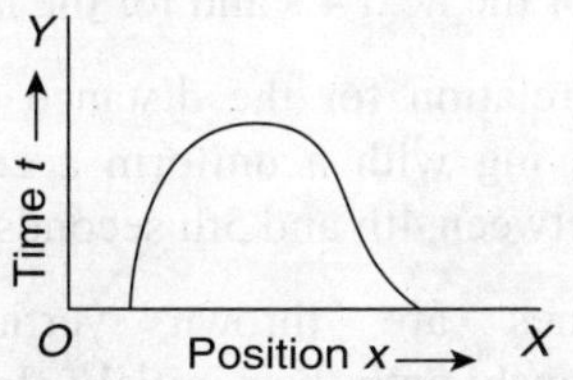

11. Is it true that in one-dimensional space, a particle with zero speed may have non-zero velocity?

12. If the displacement of a body is zero, is the distance covered by it necessarily zero? Comment with illustrations.

13. Is magnitude of the displacement of an object and total distance covered by it in a certain time interval, same? Explain.

14. Explain that a body can have zero average velocity but not zero average speed.

15. Show that average velocity of the object over an interval of time is either smaller than or equal to the average speed of the object over the same interval.

16. When can an object be considered as a point object? Explain in brief.

17. Explain the basic rule for determination of the relative velocity of one object with respect to another.

18. Two particles are moving with constant speed v such that they are always at a constant distance d apart and their velocities are always equal and opposite. After what time will they return to their initial positions?

19. A cyclist moving on a circular track of radius 100 m completes one revolution in 4 minutes. What is his
    i. average speed
    ii. average velocity in one full revolution?

20. An athlete completes one round of a circular track of radius R in 40 seconds. What will be the displacement at the end of 2 min. 20 second?

21. When two bodies move uniformly towards each other, the distance between them decreases by 6 metres/second. If both the bodies move in the same direction with the same speed, the distance between them increases by 4 metres/second. What are the speeds of the two bodies?

Descriptive Type Questions

1. Explain distance and displacement with illustrations.

2. Define uniform velocity and variable velocity of an object moving along a straight line. What will be the velocity–time graphs of such a motion?

3. Derive the second equation of motion $s = ut + \frac{1}{2}at^2$ by graphical method.

4. The graph given below is the distance–time graph of an object.
   (i) Find the speed of the object during the first four seconds of its journey.
   (ii) How long was it stationary?
   (iii) Does it represent a real situation? Justify your answer.

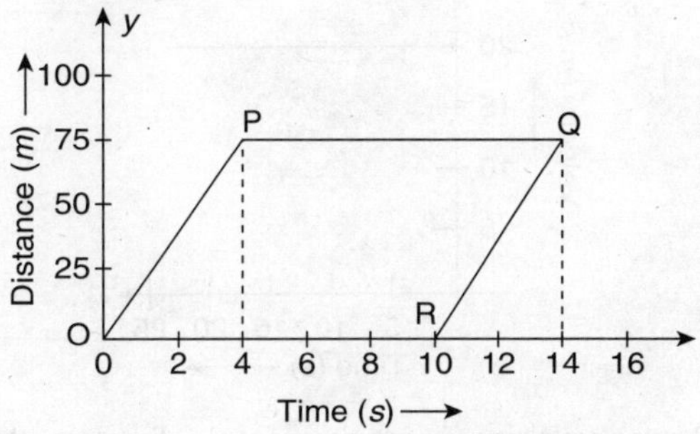

NCERT Corner

1. The displacement of a moving object in a given interval of time is zero. Would the distance travelled by the object also be zero? Justify your answer.

2. How will the equations of motion for an object moving with a uniform velocity change?

3. A girl walks along straight path to drop a letter in a letterbox and comes back to her initial position. Her displacement–time graph is shown in figure. Plot a velocity–time graph for the same.

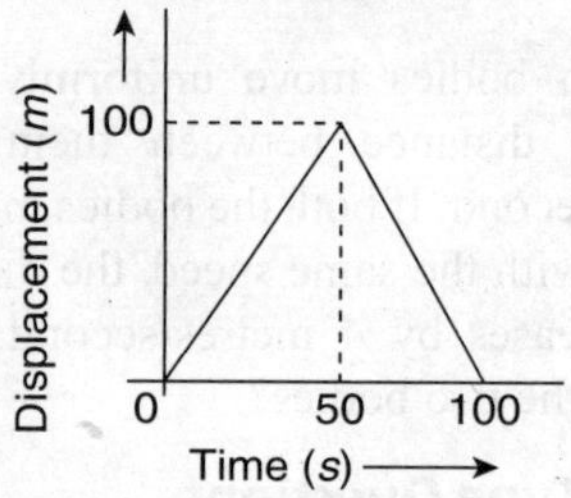

4. A car starts from rest and moves along the x-axis with constant acceleration 5 ms^{-2} for 8 seconds. If it then continues with constant velocity, what distance will the car cover in 12 seconds since it started form the rest?

5. A motorcyclist drives from A to B with a uniform speed of 30 km h^{-1} and returns with a speed of 20 km h^{-1}. Find the average speed.

6. The velocity–time graph (figure) shows the motion of a cyclist. Find (i) the acceleration (ii) the velocity and (iii) the distance covered by the cyclist in 15 seconds.

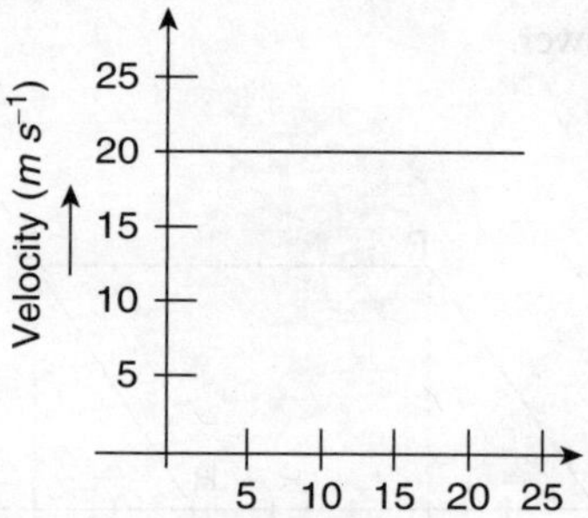

7. Draw a velocity versus time graph of a stone thrown vertically upwards that comes down after attaining the maximum height.

8. An object is dropped from rest at a height of 150 m and simultaneously another object is dropped from rest at a height 100 m. What is the difference in their heights after 2 s if both the objects drop with same accelerations? How does the difference in heights vary with time?

9. An object starting from rest travels 20 m in first 2 s and 160 m in next 4 s. What will be the velocity after 7 s from the start?

10. Using the following data, draw time–displacement graph for a moving object.

Time (s)	0	2	4	6	8	10	12	14	16
Displacement (m)	0	2	4	4	4	6	4	2	0

Use this graph to find the average velocity for the first 4 s, for the next 4 s and for the last 6 s.

11. Obtain a relation for the distance travelled by an object moving with a uniform acceleration in the interval between 4th and 5th seconds.

12. Two stones are thrown vertically upwards simultaneously with their initial velocities u_1 and u_2 respectively. Prove that the heights reached by them would be in the ratio of $u_1^2 : u_2^2$ (Assume upward acceleration is $-g$ and downward acceleration to be $+g$).

Numericals

Level – I

1. A bike starts from rest and accelerates at 4 m/s² for 5.0 s. It then moves at constant velocity for 25.0 s, and then decelerates at 2.0 m/s² until it stops. Find the total distance that the motorcycle has moved.

2. A car can accelerate from 0 to 48 kph in 3.6 s and from 0 to 96 kph in 10.2 s. Also, under constant acceleration from rest, it crosses the 0.4 km marker at a speed of 140 kph.
   a. Calculate the average acceleration needed to get the speed of 48 kph.
   b. Calculate the average acceleration during the time it requires to go from 48 to 96 kph.
   c. What constant acceleration would be required to get a speed of 140 kph over 0.4 km run starting from rest?

3. Two friends start bikes from one corner of a square field of edge L towards the diagonally opposite corner at the same time t. They both start from the same place and take different routes. One travels

along the diagonal with constant acceleration a, and the other accelerates momentarily and then travels along the edge of the field with constant speed v. What is the relationship between a and v?

4. A truck travelling along a straight road at a constant speed of 72 kph passes a car at time $t = 0$ moving much slower. At the instant the truck passes the car, the car starts accelerating at constant 1 m/s^2 and overtakes the truck 0.6 km further down the road, from where the car moves uniformly. Find the distance between them at time $t = 50$ s.

5. What is the relative velocity of a car moving with 40 km/h due east with respect to itself?

6. A particle moving at constant acceleration covers distance of 30 m in the 3rd second. It covers a distance of 50 m in the 5th second. What is the acceleration of the particle?

7. If a body starts from rest and travels 120 m in the 8th second, then what is the acceleration?

8. If a train travelling at 72 km/h is to be brought to rest in a distance of 200 m, then what should be its retardation?

9. A car accelerates from rest at constant rate of 2 m/s^2 for some time. Then, it retards at constant rate of 4 m/s^2 and comes to rest. If the total time for which it remains in motion is 3 seconds, what is the total distance travelled?

10. A bullet fired into a wooden block loses $\frac{3}{4}$ of its velocity after penetrating 15 cm into it. How much further distance it has to penetrate to be stopped completely?

11. A bird is tossing (flying to and fro) between two cars moving towards each other on a straight road. One car has a speed of 18 km/h while the other has the speed of 27 km/h. The bird starts moving from the first car towards the other and is moving with the speed of 36 km/h our when the two cars were separated by 36 km. What is the total distance covered by the bird till the cars collide?

12. The engineer of a train moving at a speed v_1 sights a freight train a distance d ahead of him on the same track moving in the same direction with a slower speed v_2. He pulls on the brakes and gives his train a constant deceleration α. Find the minimum value of d at which brakes are applied so as to avoid collision.

13. A drunkard walking in a narrow lane takes 5 steps forward and 3 steps backward, followed again by 5 steps forward and 3 steps backward, and so on. Each step is 1 m long and requires 1 s. After what time he will fall into the pit 13 m away from him?

Level – II

1. The below figure shows displacement (x) versus time (t) graph of a particle. Find the time t such that the average velocity of the particle during the period 0 to t is zero.

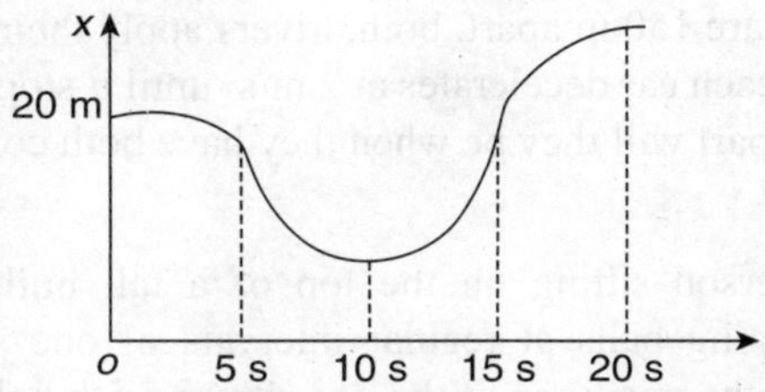

2. A police jeep (speed 90 km/h) is trying to overtake a culprit going on a bike (speed 72 km/h). The bike crosses a turning and the jeep crosses the same turning 10 s later. Assuming that they move at constant speed, find how far from the turning will the police catch up with the bike.

3. A body travels 200 cm in the first two seconds and 220 cm in the next 4 sec with deceleration. Find the velocity of the body at the end of the 7th second.

4. The engine of a motor cycle can produce a maximum acceleration 5 m/s^2. It brakes can produce a maximum retardation 10 m/s^2. What is the minimum time in which it can cover a distance of 1.5 km?

5. A motorcycle and a car start from rest at the same place at the same time and they travel in the same direction. The cycle accelerates uniformly at 1 m/s^2 up to a speed of 36 kph and the car at 0.5 m/s^2 up to a speed of 54 kph. Calculate the time and the distance at which the car overtakes the motorcycle.

6. A driver, having a defined reaction time, is capable of stopping his car over a distance of 30 m on seeing a red traffic signal, when the speed is 72 kph and over a distance of 10 m when the speed is 36 kph. Find the distance over which he can stop the car if it was running at a speed of 54 kph. Assume that his reaction time and the deceleration of the car remain same in all the three cases.

7. A point moving with constant acceleration from A to B in the straight line AB has velocities u and v at A and B respectively. Find its velocity at C, the midpoint of AB. Also show that if the time from A to C is twice that from C to B, then $v = 7u$.

8. A train is targeted to run from Delhi to Pune at an average speed of 80 kph but due to repairs of track loses 2 hours in the first part of the journey. It then accelerates at a rate of 20 kph^2 till the speed reaches 100 kph. Its speed is now maintained till the end of the journey. If the train now reaches station in time, find the distance from when it started accelerating?

9. Two cars travelling towards each other on a straight road at velocity 10 m/s and 12 m/s respectively. When they are 150 m apart, both drivers apply their brakes and each car decelerates at 2 m/s^2 until it stops. How far apart will they be when they have both come to a stop?

10. A person sitting on the top of a tall building is dropping balls at regular intervals of one second. Find the positions of the 3rd, 4th and 5th ball when the 6th ball is being dropped. Take $g = 10$ m/s^2

Fill in the Blanks

1. The change in velocity may be due to change in magnitude or direction or due to ______.
2. If the acceleration of a body is plotted on y-axis and time on x-axis and the graph is parallel to x-axis the acceleration is ______.
3. A cyclist moving on a circular track of radius 100 m completes one revolution in 2 minutes. Then its average speed is ______ and average velocity in full revolution is ______.
4. The change in velocity of a body in a particular direction is called ______.
5. The velocity of a body is ______ quantity.
6. The negative acceleration is known as ______.
7. The weight of a body in projectile motion is ______.
8. The speedometer in an automobile measures its ______.
9. A food packet is released from a helicopter which is rising steadily at 2 m/s. After 2 s the velocity of the packet is ______. ($g = 9.8$ m/s^2)

True or False

1. The speed acquired by a body when falling in vacuum for a given time is dependent on the mass of the falling body.
2. A rocket moves forward by pushing surrounding air backwards.
3. A vector quality has both magnitude and direction.
4. An inflated balloon will rise to a definite height in the atmosphere and then it will come to rest.
5. Two balls of different masses are thrown vertically upwards with the same speed. They pass through the point of projection in their downward motion with the same speed (neglect air resistance).
6. Two balls of different masses are thrown vertically upward with same initial speed. They rise to the same maximum height above the ground.
7. In the absence of air resistance it is found that all bodies, regardless of their size, weight, or composition, fall with the same acceleration at the same point of the earth's surface. Now if the distance covered is not great, the acceleration varies throughout the fall.
8. Two identical trains are moving on rails along the equator on the earth in opposite directions with the same speed. They will exert the same pressure in the rails.
9. From the top of a tower, two stones are thrown – one straight up with initial speed u and the second straight down with the same speed u. The two stones will then have the same speed when they hit the ground. (Air resistance is assumed negligible).
10. A stone thrown vertically upward with speed u, reaches a height 50 metres. If the stone is thrown with a speed 2 u, it reaches a height 100 metres.
11. The distance moved by a freely falling body (starting from rest) during the 1st, 2nd, 3rd, … nth second of its motion are proportional to $1 : 4 : 9 : \ldots : n^2$
12. Two vectors having same magnitude but opposite in direction are known as opposite vectors.
13. An object has constant speed but variable velocity.
14. Earth can be regarded as a point object when it is describing its yearly journey around the sun.
15. The speed of a body be can negative.
16. Displacement can be greater than the distance travelled by an object.
17. A body have a constant velocity but a varying speed.
18. The magnitude of average velocity is equal to average speed.
19. The speed acquired by a body when falling in vacuum for a given time is dependent on the mass of the falling body.

COMPETITION VISTA

Match the Columns

1. Match Column I with Column II.

Column I	Column II
(A) Displacement	(E) m/s^2
(B) Acceleration	(F) m/s
(C) Distance	(G) Vector
(D) Velocity	(H) Scalar

2. Match Column I with Column II.

Column I	Column II
(A) Uniform acceleration with initial velocity	(E) 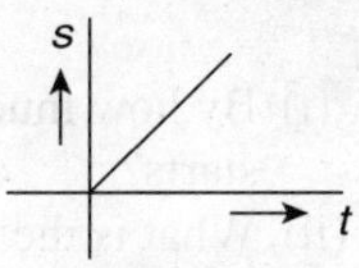
(B) Uniform Velocity	(F) 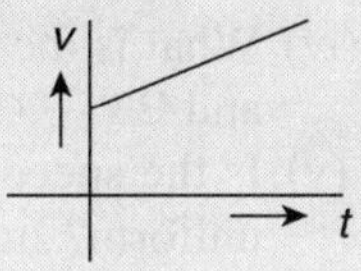
(C) Body at rest	(G) 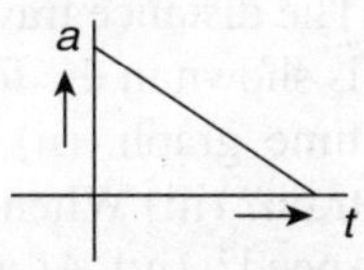
(D) Decreasing acceleration at uniform rate	(H) 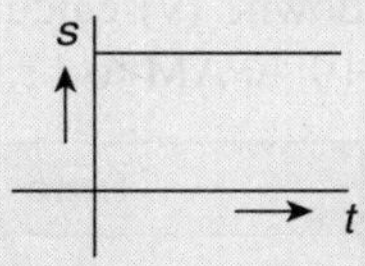

3. Match Column I with Column II.

Column I	Column II
(A) Velocity–time graphs	(E) Acceleration are used
(B) Acceleration vs. time graphs are used	(F) Downward
(C) km/s.h	(G) Velocity
(D) Highest point acceleration	(H) Displacement

4. Match Column I with Column II.

Column I	Column II
(A) Object is dropped from rest	(E) 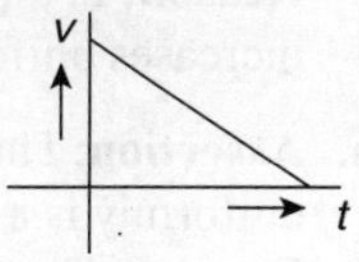
(B) Velocity vs. time graph for a body projected upward	(F) 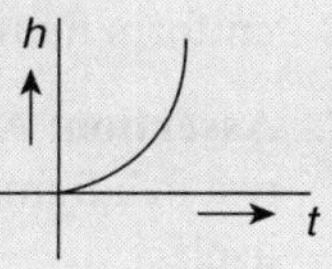
(C) Freely falling body	(G) 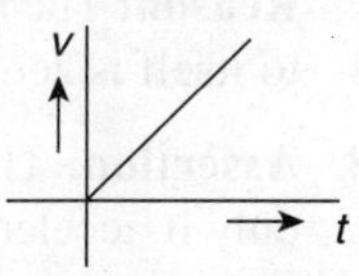
(D) Velocity vs. time graph for a freely falling body	(H) 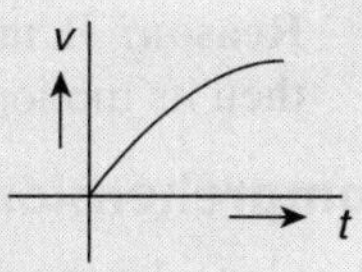

Assertion–Reason Questions

Read the assertions and reasons carefully to mark the correct option out of the options given below:

(a) If both assertion and reason are true and the reason is the correct explanation of the assertion.
(b) If both assertion and reason are true but reason is not the correct explanation of the assertion.
(c) If assertion is true but reason is false.
(d) If the assertion and reason both are false.
(e) If assertion is false but reason is true.

1. Assertion: A body falling freely may do so with constant velocity.
Reason: The body falls freely when acceleration of a body is equal to acceleration due to gravity.

2. Assertion: The average velocity of the object over an interval of time is either smaller than or equal to the average speed of the object over the same interval.
Reason: Velocity is a vector quantity and speed is a scalar quantity.

3. Assertion: The position–time graph of a uniform motion in one dimension of a body can have negative slope.
Reason: When the speed of body decreases with time, the position–time graph of the moving body has negative slope.

4. **Assertion:** A positive acceleration of a body can be associated with a 'slowing down' of the body.
Reason: Acceleration is a vector quantity.

5. **Assertion:** The average and instantaneous velocities have same value in a uniform motion.
Reason: In uniform motion, the velocity of an object increases uniformly.

6. **Assertion:** The position-time graph of a body moving uniformly is a straight line parallel to position axis.
Reason: The slope of position–time graph in a uniform motion gives the velocity of an object.

7. **Assertion:** A body, whatever its motion, is always at rest in a frame of reference which is fixed to the body itself.
Reason: The relative velocity of a body with respect to itself is zero.

8. **Assertion:** The equation of motion can be applied only if acceleration is along the direction of velocity and is constant.
Reason: If the acceleration of a body is constant then its motion is known as uniform motion.

Comprehension Type Questions

1. Consider the distance-time graph as shown for three friends A, B and C. All of them move along a straight road in a fixed direction. Observe the graph, state which of the following statements are correct.

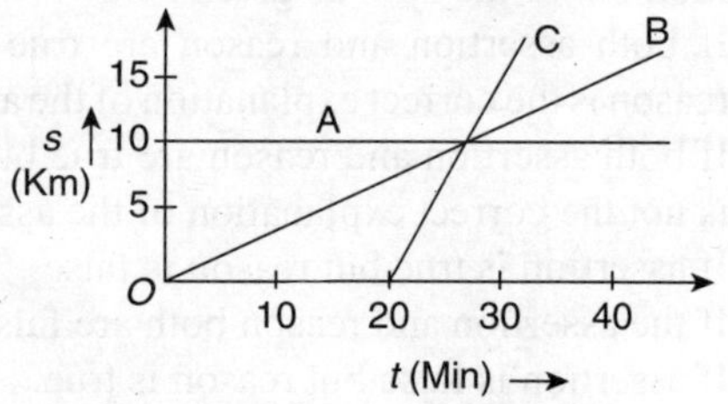

(I) Is any of them, one or more, at rest?
(a) A (b) B
(c) C (d) Both B and C

(II) When did C starts moving?
(a) At 10 min
(b) At 20 min
(c) When A started moving
(d) At 25 min

(III) Did the three ever meet at the same time? If yes, when?
(a) Yes, at 25 min
(b) Yes, at 20 min
(c) Yes at 30 min
(d) No, they never meet

(IV) Who was moving the fastest at $t = 25$ min?
(a) A (b) B
(c) C (d) Both B and C

2. The distance–time graph of two trains are shown in figure. The trains start simultaneously in the same direction.

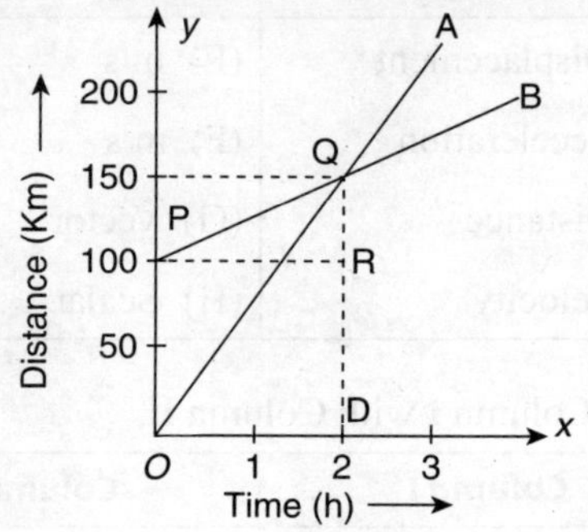

(i) By how much is B ahead of A when the motion starts?
(ii) What is the speed of B?
(iii) When and where will A catch B?
(iv) What is the difference between the speeds of A and B?
(v) Is the speed of both the trains uniform or non-uniform? Justify your answer.

3. The distance travelled by a train and time taken by it is shown in the following table. (i) Plot the distance–time graph. (ii) What is the average speed of the train? (iii) When is the train travelling at the highest speed? (iv) At what distance does the train slow down? (v) calculate the speed of the train between 10:40 AM to 11:00 AM.

Time	Distance (in km)
10:00 AM	0
10:30 AM	25
10:40 AM	28
11:00 AM	40
11:15 AM	42
11:30 AM	50

4. Anirban and his sister Mansi go to school together in their car. Anirban drives much faster than Mansi. Mansi tells Anirban not to take the risk of over-speeding. She tells him that the time taken to

reach the school would depend upon average speed. By over-speeding for a little while, the risk involved is much greater compared to the little time saved.

Read the above passage and answer the following questions:

(i) Which values are displayed by Mansi?

(ii) Is Mansi right in her statement?

(iii) How do you define average speed?

5. Aman and Lakshya went to Agra through Yamuna-Expressway. Aman started the car and accelerated so highly that the car was running at 108 km/h within 10 s. Lakshya stopped him from doing so and told him that overspeeding on road was a straight invitation to life-taking risk situation. Though Aman wanted the adventure of speeding, but he was convinced by Lakshya.

Read the above passage and answer the following question:

(i) Why do you think it is dangerous to drive fast on road?

(ii) Which values of Lakshya are worth appreciating?

(iii) What is the acceleration of the car?

Objective Questions

One Option Correct

1. A locomotive begins to move with uniform acceleration at the moment when a boy runs past it with a uniform velocity of 2 ms^{-1}. The velocity of the locomotive when it overtakes the boy is ________.

(a) 2 ms^{-1} (b) 4 ms^{-1}
(c) 6 ms^{-1} (d) 8 ms^{-1}

2. A body starts from rest and moves with constant acceleration. The ratio of distance covered by the body in *n*th second to that covered in *n* seconds is ________.

(a) $1 : n$ (b) $\frac{2n-1}{n^2}$
(c) $\frac{n^2}{2n-1}$ (d) $\frac{2n-1}{2n^2}$

3. A train runs past a telegraph pole in 15 seconds, and through a tunnel 450 metres long in 45 seconds. When it meets another train 300 metres long, both trains cross each other in 21 seconds. Then the velocity of the second train is ________.

(a) 10 ms^{-1} (b) 15 ms^{-1}
(c) 20 ms^{-1} (d) 30 ms^{-1}

4. Three particles *A*, *B* and *C* are thrown from the top of a tower with the same speed. *A* is thrown straight up, *B* is thrown straight down and *C* is thrown horizontally. They hit the ground with speeds v_A, v_B and v_C respectively.

(a) $v_A = v_B = v_C$ (b) $v_B > v_C > v_A$
(c) $v_A = v_B > v_C$ (d) $v_A > v_B = v_C$

5. A point traversed half the distance with a velocity v_0. The remaining part of the distance was covered with velocity v_1 for half the time and with velocity v_2 for the other half of the time. The mean velocity of the point averaged over the whole time of motion is ________.

(a) $\frac{v_0 + v_1 + v_2}{2}$ (b) $\frac{2v_0 + v_1 + v_2}{3}$
(c) $\frac{v_0 + 2v_1 + 2v_2}{3}$ (d) $\frac{2v_0(v_1 + v_2)}{2v_0 + v_1 + v_2}$

6. Which of the following figures represents uniform motion of a moving object correctly?

(a)
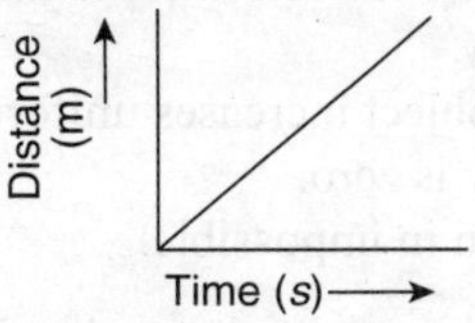

(b)
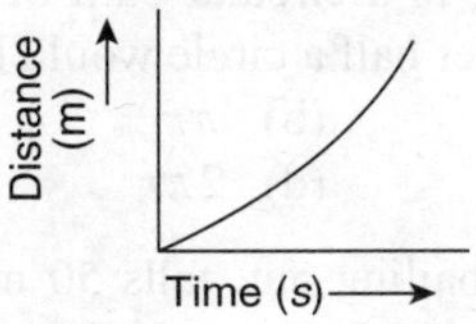

(c)
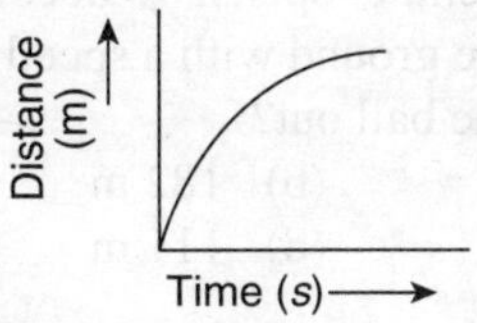

(d)
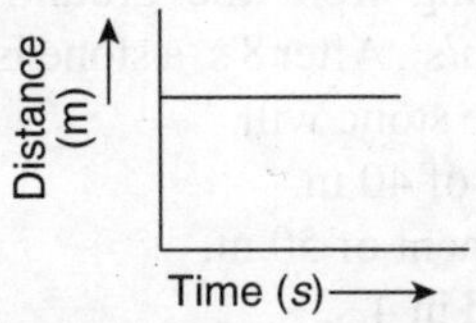

7. Two cars A and B, each having a speed of 30 km hr^{-1}, are heading towards each other along a straight path. A bird that can fly at 60 km hr^{-1} flies off car A when the distance between the cars is 60 km, heads directly towards car B; on reaching B, the bird directly flies back to A and so forth, then the total distance the bird travels till the cars meet is ________.

(a) infinite (b) 30 km
(c) 60 km (d) 120 km

8. A ball is thrown vertically upwards from the ground with a speed of 25 m/s. The ball will be at a height of 30 m above the ground after time ______.
(a) 2.0 s only (b) 2.0 and 3.0 s
(c) 3.0 s only (d) 1.0 sec and 2.0 s

9. Which of the following options is correct for the object having a straight line motion represented by the following graph?

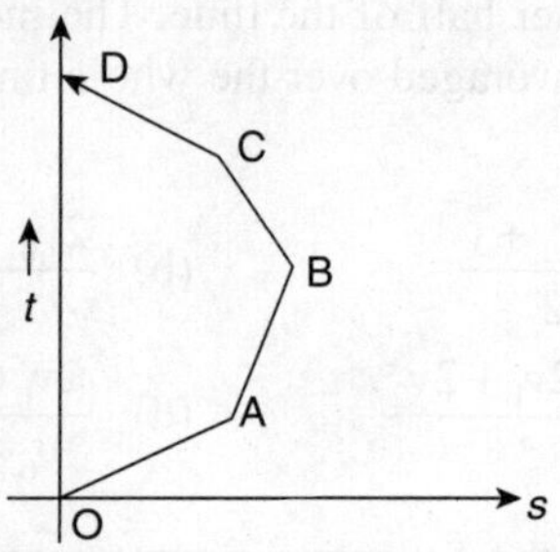

(a) The object moves with constantly increasing velocity from O to A and then it moves with constant velocity.
(b) Velocity of the object increases uniformly.
(c) Average velocity is zero.
(d) The graph shown in impossible.

10. A particle is moving in a circular path of radius r. The displacement after half a circle would be:
(a) zero (b) πr
(c) $2r$ (d) $2\pi r$

11. A parachutist, after bailing out, falls 50 m without friction. When parachute opens, it decelerates at 2 m/s^2. He reaches the ground with a speed of 3 m/s. At what height, did he bail out?
(a) 91 m (b) 182 m
(c) 293 m (d) 111 m

12. A balloon starts rising from the ground with an acceleration of 1.25 m/s^2. After 8 s, a stone is released from the balloon. The stone will
(a) cover a distance of 40 m.
(b) have a displacement of 50 m.
(c) reach the ground in 4 s.
(d) begin to move down after being released.

13. A body covers 200 cm in the first 2 seconds and 220 cm in the next 4 seconds. The velocity at the end of 7th second is ______.
(a) 20 cm s^{-1} (b) 10 cm s^{-1}
(c) 115 cm s^{-1} (d) –10 cm s^{-1}

14. Slope of a velocity–time graph gives ______.
(a) the distance (b) the displacement
(c) the acceleration (d) the speed

15. A stone falls freely from rest and the total distance covered by it in the last second of its motion equals $\frac{9}{25}$ of its motion. The stone remains in air for
(a) 4s (b) 5s
(c) 6s (d) 7s

16. The distance travelled by a body falling under gravity in the first, second and third second is in the ratio ______.
(a) 1 : 8 : 27
(b) 1 : 2 : 3
(c) 1 : 4 : 9
(d) 1 : 3 : 5

17. The V–t graph for a particle is as shown. The distance travelled in the first four seconds is ______.

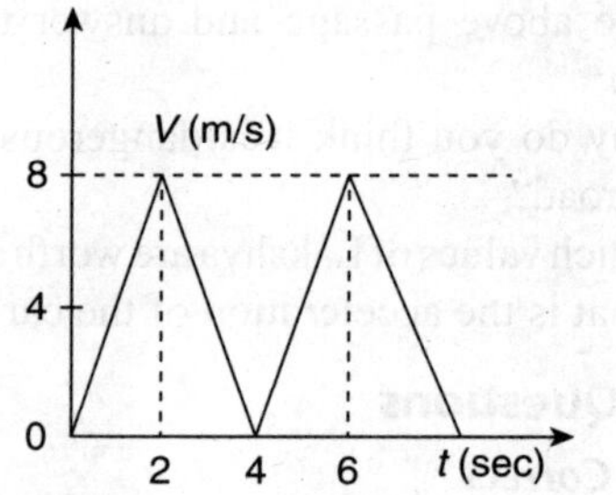

(a) 12 m (b) 16 m
(c) 20 m (d) 24 m

18. The figure shows a particle moving along the x-axis subjected to three periods of acceleration. Rank the periods according to the increases they produce.

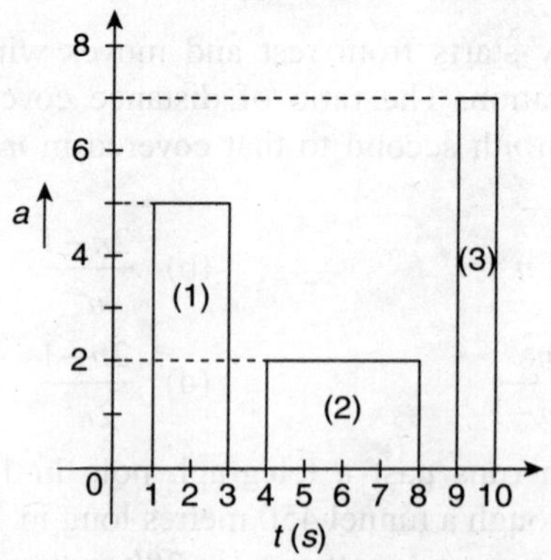

(a) (2) > (1) > (3) (b) (2) > (3) > (1)
(c) (1) > (3) > (2) (d) (1) > (2) > (3)

19. A body is thrown vertically upward with velocity u, the greatest height h to which it will rise is ______.
(a) u/g (b) $u^2/2g$
(c) u^2/g (d) $u/2g$

20. If the figure below represents a parabola, identify the physical quantities representing Y and X for constant acceleration.

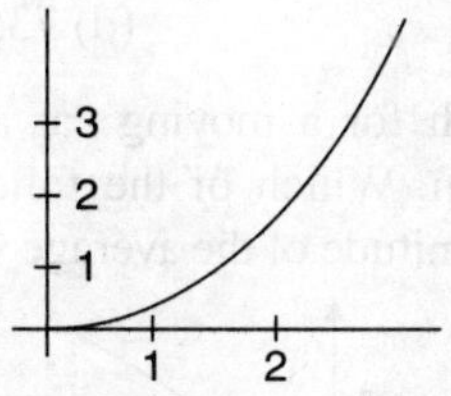

(a) X = time, Y = velocity
(b) X = velocity, Y = time
(c) X = time, Y = displacement
(d) X = time, Y = acceleration

21. A body travels in a straight line from A to B. It travels half the time with velocity v_1 and the other half of the time with velocity v_2. The average velocity is given by

(a) $\frac{2v_1v_2}{v_1+v_2}$ (b) $\frac{v_1+v_2}{v_1v_2}$
(c) $\frac{v_1+v_2}{2}$ (d) $\frac{v_1v_2}{v_1+v_2}$

22. A particle travels half the distance with speed v_1 and the other half the distance with speed v_2. The average speed is ________.

(a) $\frac{2v_1v_2}{v_1+v_2}$ (b) $\frac{v_1+v_2}{v_1v_2}$
(c) $\frac{v_1+v_2}{2}$ (d) $\frac{v_1v_2}{v_1+v_2}$

23. A particle experiences a constant acceleration for 20 s after starting from rest. If it travels a distance S_1 in the first 10 s and S_2 in next 10 s, then

(a) $S_1 = \frac{S_2}{4}$ (b) $S_1 = \frac{S_2}{3}$
(c) $S_1 = \frac{S_2}{2}$ (d) $S_1 = S_2$

24. A particle shows distance–time curve as given in this figure. The maximum instantaneous velocity of the particle is around the point ________.

(a) D (b) A
(c) B (d) C

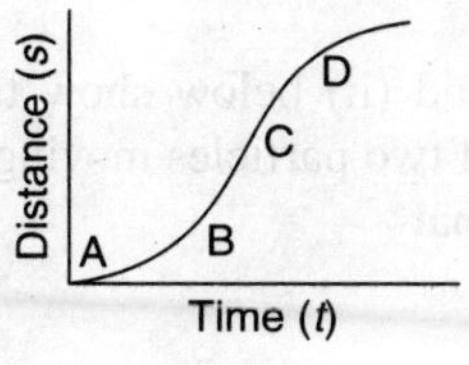

25. The numerical ratio of displacement to distance for a moving object is

(a) always less than 1 (b) always equal to 1
(c) always more than 1 (d) equal or less than 1

26. A bus is moving with a speed of 10 ms^{-1} on a straight road. A scooterist wishes to overtake the bus in 100 s. If the bus is at a distance of 1 km from the scooterist, with what speed should the scooterist chase the bus?

(a) 40 ms^{-1} (b) 25 ms^{-1}
(c) 10 ms^{-1} (d) 20 ms^{-1}

27. The distance travelled by a particle starting from rest and moving with an acceleration $\frac{4}{3}$ ms^{-2}, in the third second is ________.

(a) $\frac{10}{3}$ m (b) $\frac{19}{3}$ m
(c) 6 m (d) 4 m

28. A car runs at a constant speed on a circular track of radius 100 m, taking 62.8 seconds for every circular lap. The average velocity and average speed for each circular lap respectively is ________.

(a) 10 m/s, 0
(b) 0, 0
(c) 0, 10 m/s
(d) 10 m/s, 10 m/s

29. A man throws balls with the same speed vertically upwards, one after the other, at an interval of 2 seconds. What should be the speed of the throw so that more than two balls are in the sky at any time? (Given $g = 9.8$ m/s^2)

(a) more than 19.6 m/s
(b) at least 9.8 m/s
(c) any speed less than 19.6 m/s
(d) only with speed 19.6 m/s

30. If a ball is thrown vertically upwards with speed u, the distance covered during the last t seconds of its ascent is ________.

(a) ut (b) $\frac{1}{2}gt^2$
(c) $ut - \frac{1}{2}gt^2$ (d) $(u+gt)t$

31. If the displacement of an object is proportional to square of time, then the object moves with ________.

(a) uniform velocity
(b) uniform acceleration
(c) increasing acceleration
(d) decreasing acceleration

32. A person travels along a straight road for the first half-length with a velocity v_1 and the second half-length with a velocity v_2. The mean velocity v is given by ________.

(a) $v = \dfrac{v_1 + v_2}{2}$ (b) $v^2 = v_1 v_2$

(c) $\dfrac{2}{v} = \dfrac{1}{v_1} + \dfrac{1}{v_2}$ (d) $v = \dfrac{v_1}{v_2}$

33. A ball is thrown vertically upwards. Which of the following graph/graphs represent the velocity–time graph of the ball during its flight (air resistance is neglected).

(a)

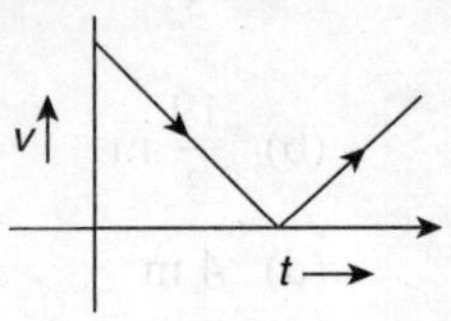

(b)

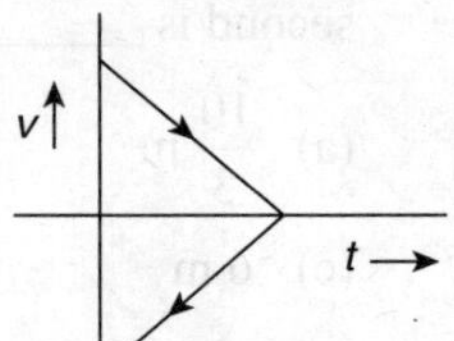

(c)

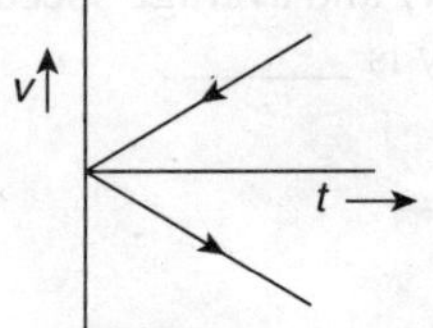

(d)

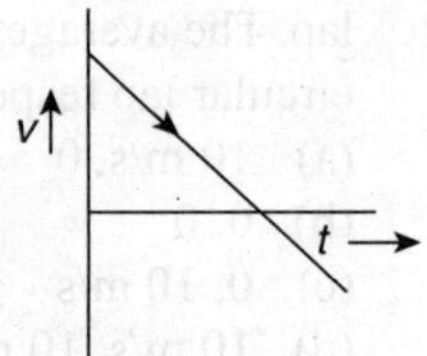

34. In which of the following cases of motions, the distance moved and the magnitude of displacement are equal?

(a) If the car is moving on straight road.
(b) If the car is moving in circular path
(c) A pendulum is moving to and fro
(d) The earth is revolving around the sun.

35. A car accelerates from rest at a constant rate α for some time, after which it decelerates at a constant rate β and comes to rest. If the total time elapsed is t then the maximum velocity acquired by the car is ________.

(a) $\dfrac{\alpha^2 + \beta^2}{\alpha\beta} \cdot t$ (b) $\dfrac{\alpha^2 - \beta^2}{\alpha\beta} \cdot t$

(c) $\dfrac{\alpha - \beta}{\alpha\beta} \cdot t$ (d) $\dfrac{\alpha\beta t}{\alpha + \beta}$

36. A rocket is fired upward from the earth's surface such that it creates an acceleration of 19.6 ms^{-2}. After 5 sec, its engine is switched off. The maximum height of the rocket from the earth's surface would be ________.

(a) 245 m (b) 490 m
(c) 980 m (d) 735 m

37. The v–t graph for a moving car along a horizontal road is shown. Which of the following expressions gives the magnitude of the average velocity of the car?

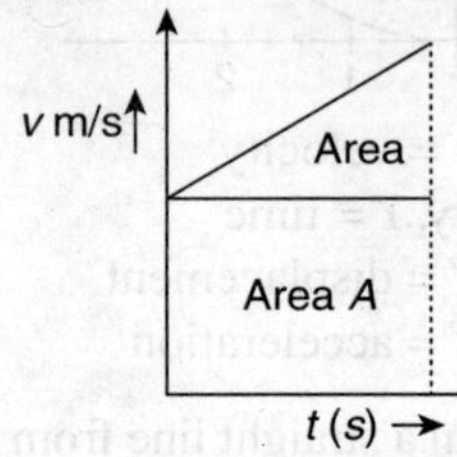

(a) $\dfrac{\text{Area } A}{t}$ (b) $\dfrac{(\text{Area } A + \text{Area } B)}{t}$

(c) $\dfrac{\text{Area } B}{t}$ (d) $\dfrac{(\text{Area } A - \text{Area } B)}{t}$

38. From the given v–t graph, it can be inferred that the object is

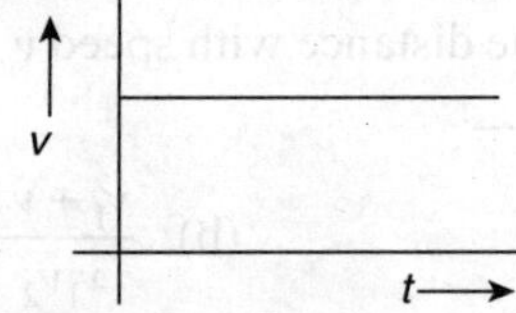

(a) in uniform motion.
(b) at rest.
(c) in non-uniform motion.
(d) moving with uniform acceleration.

39. A particle is moving on a straight-line path with constant acceleration along the direction of instantaneous velocity. Which of the following statements is true about the motion of the particle?

(a) The average velocity is less than the average speed.
(b) The average velocity is equal to the instantaneous velocity.
(c) The distance covered is equal to the magnitude of displacement.
(d) The particle may reverse the direction of motion.

40. Figures (i) and (ii) below show the displacement–time graph of two particles moving along the x-axis. We can say that

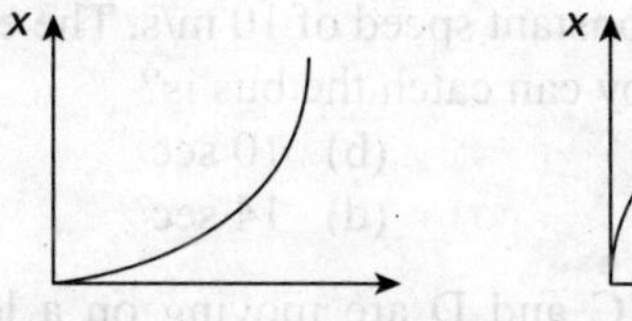

(i) (ii)

(a) both the particles have a uniformly accelerated motion.
(b) both the particles have a uniformly retarded motion.
(c) particle (i) has uniformly accelerated motion while particle (ii) has uniformly retarded motion.
(d) particle (i) has uniformly retarded motion while particle (ii) has a uniformly accelerated motion

41. In 1.0 second, a particle goes from point A to point B moving in a semicircle of radius 1 m. The magnitude of average velocity is

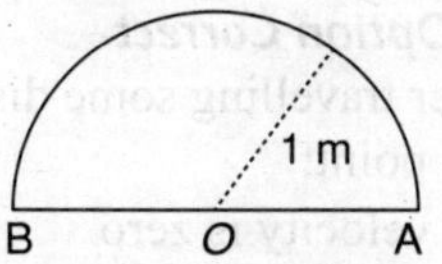

(a) 3.14 m/s (b) 2.0 m/s
(c) 1.0 m/s (d) 2.5 m/s

42. Which of the following curves show that acceleration of the body is zero?

(a)

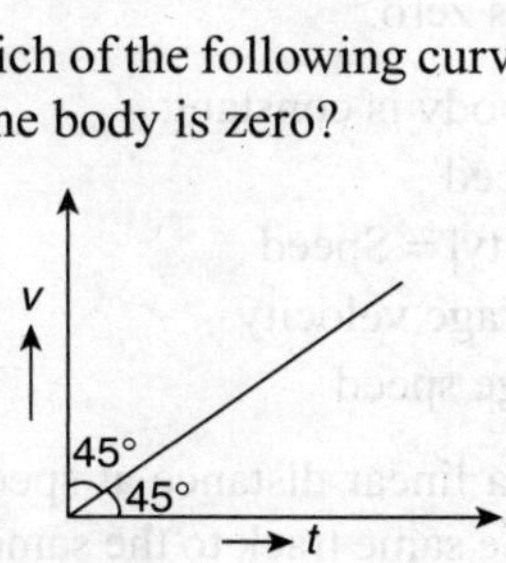

(b)

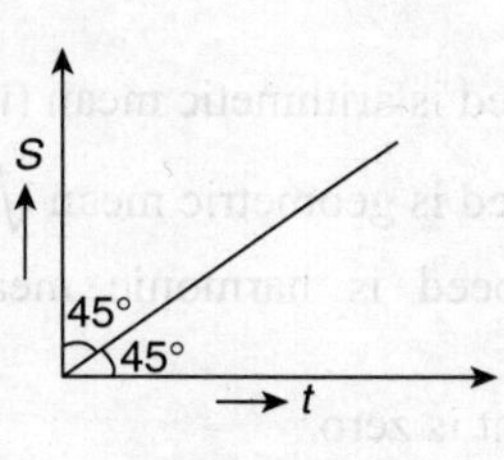

(c)

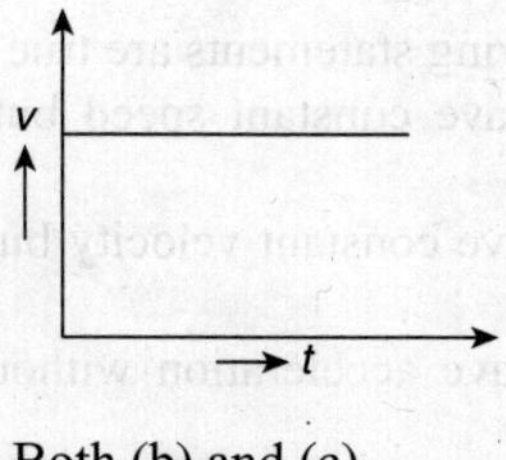

(d) Both (b) and (c)

43. An object is moving with a uniform acceleration which is parallel to its instantaneous direction of motion. The displacement (s) – velocity (v) graph of this object is

(a)

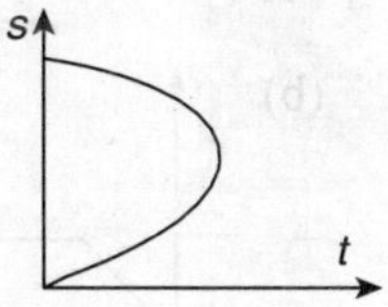

(b)

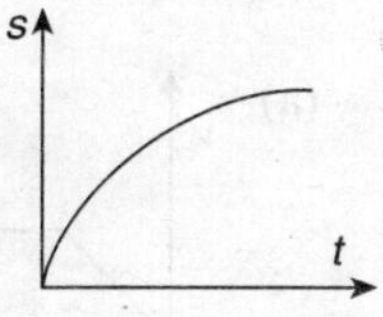

(c)

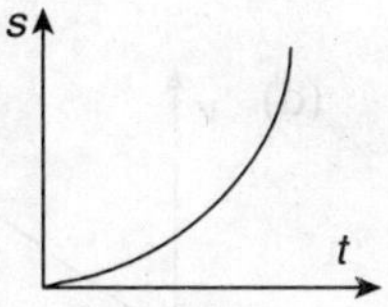

(d)

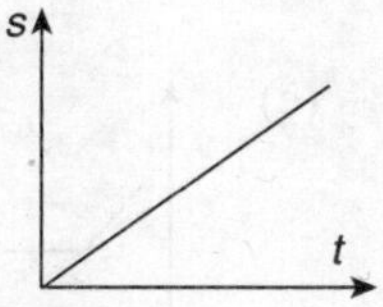

44. Suppose a boy is enjoying a ride on a merry-go-round which is moving with a constant speed of 10 ms^{-1}. It implies that the boy is
(a) at rest.
(b) moving with no acceleration.
(c) in acceleration.
(d) moving with uniform velocity.

45. Which of the following velocity–time graph shows a realistic situation for a body in motion?

(a)

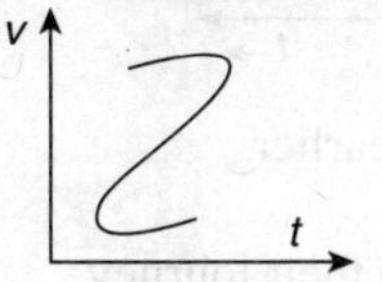

(b)

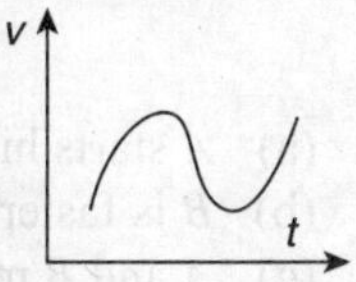

(c)

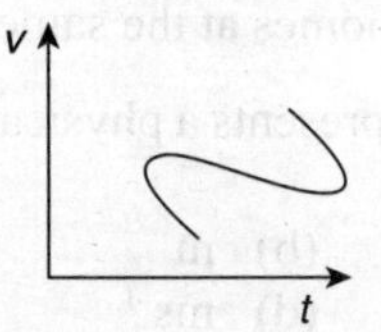

(d)

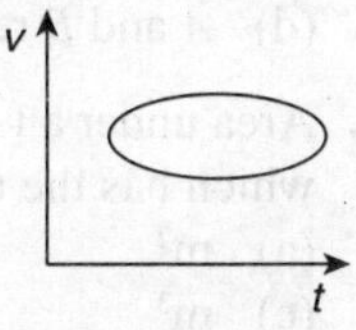

46. Which of the following represents s–t graph for a retarded motion?

(a)

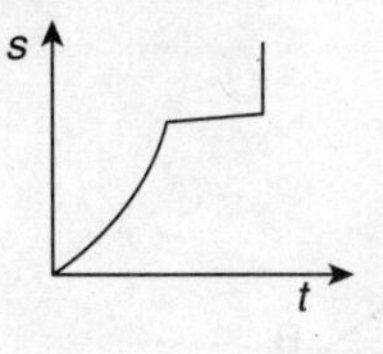

(b)

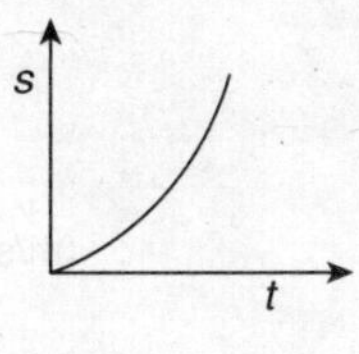

(c)

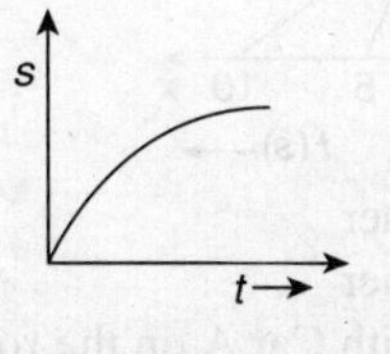

(d)

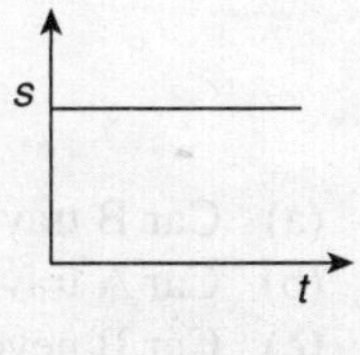

47. Which of the following would show the v–t graph for a body? The acceleration time graph is shown in figure.

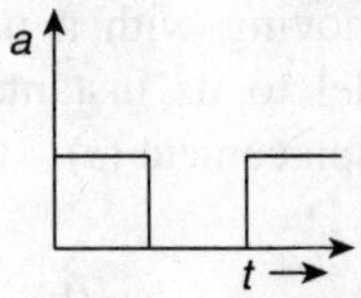

(a)
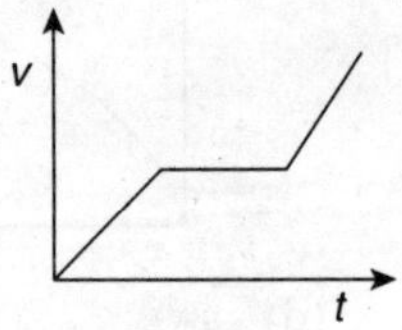

(b)
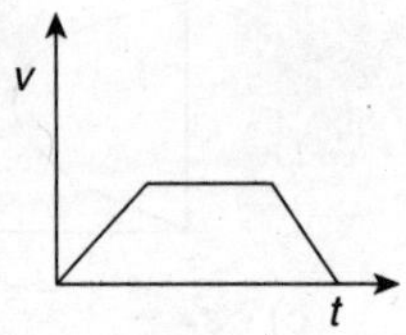

(c)
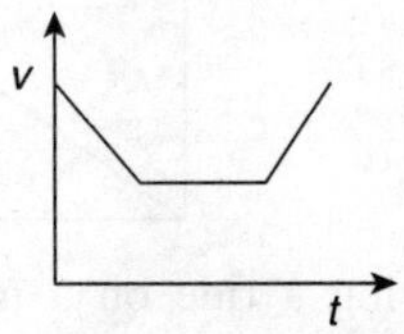

(d)
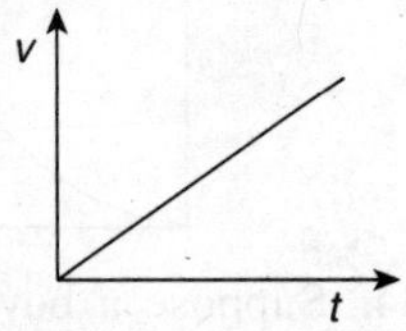

48. A and B start their journey towards their homes P and Q from different points. Which of the following statements is not represented by the graph given here?

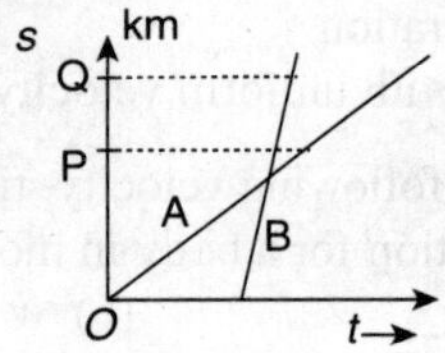

(a) *A* starts his journey earlier.
(b) *B* is faster than *A*.
(c) *A* and *B* meet during their journey.
(d) *A* and *B* reach their homes at the same time.

49. Area under a *v-t* graph represents a physical quantity which has the unit

(a) m^2 (b) m
(c) m^3 (d) ms^{-1}

50. The speed versus time graphs for the two cars *A* and *B* are shown in the figure. The figure shows that

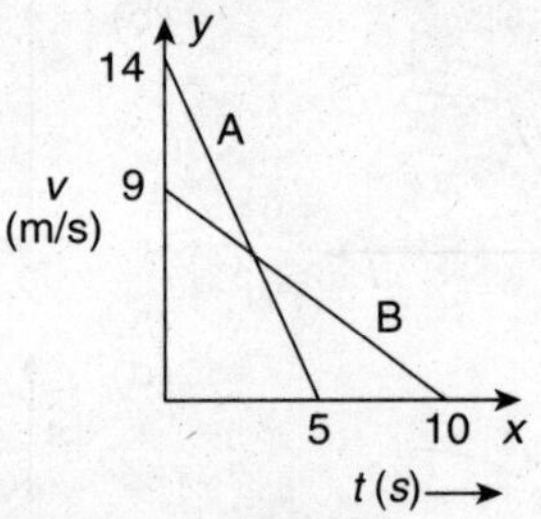

(a) Car B travelled farther
(b) Car A travelled farther
(c) Car B never meet with Car A on the road
(d) Both cars have the same retardation.

51. A bus is beginning to move with an acceleration of 1 m/s^2. A boy who is 48 m behind the bus starts running with a constant speed of 10 m/s. The earliest time when the boy can catch the bus is?

(a) 8 sec (b) 10 sec
(c) 12 sec (d) 14 sec

52. Four cars A, B, C and D are moving on a leveled road. Their distance versus time graphs is shown. Choose the correct statement.

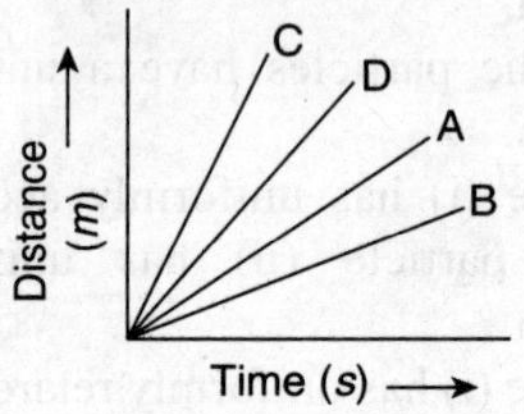

(a) Car A is faster than car D
(b) Car B is the slowest
(c) Car D is faster than car C
(d) Car C is the slowest

More Than One Option Correct

1. If a body, after travelling some distance, comes back to its starting point:

(a) Average velocity is zero
(b) Average speed is zero
(c) Distance travelled is zero
(d) Displacement is zero.

2. If the velocity of a body is constant:

(a) |Velocity| = Speed
(b) |Average velocity| = Speed
(c) Velocity = Average velocity
(d) Speed = Average speed

3. If a particle travels a linear distance at speed v_1 and comes back along the same track to the same point at speed v_2:

(a) Its average speed is arithmetic mean $(v_1 + v_2)/2$
(b) Its average speed is geometric mean $\sqrt{v_1 v_2}$
(c) Its average speed is harmonic mean $2v_1v_2/(v_1 + v_2)$
(d) Its displacement is zero.

4. Which of the following statements are true?

(a) A body can have constant speed but varying velocity.
(b) A body can have constant velocity but varying speed.
(c) A body can have acceleration without having velocity.
(d) A body can have velocity without having acceleration

5. Find the correct statements out of the following:
   (a) If acceleration = 0, the motion is uniform.
   (b) If acceleration = constant, the acceleration is uniform but the motion is not uniform.
   (c) If acceleration ≠ constant, both acceleration and motion are not uniform.
   (d) If acceleration = constant, both acceleration and motion are uniform.

6. If a body is accelerating:
   (a) It may speed up.
   (b) It may speed down.
   (c) It may move with the same velocity.
   (d) It may move with the same speed.

7. If two bodies are in motion with velocity $\vec{v_1}$ and $\vec{v_2}$:
   (a) $|v_{rel}| = \sqrt{v_1^2 + v_2^2}$
   (b) $|v_{rel}| = |v_1 \pm v_2|$
   (c) $v_{rel} = 0$
   (d) $v_{rel} > c$ (speed of light)

8. A particle moves 10 m in the first 2 seconds, 20 m in the next 3 seconds and 30 m in the next 10 seconds. Which of the following statements are true for this motion?
   (a) The particle was uniformly accelerated.
   (b) Decelerating forces were acting on the particle during the motion.
   (c) The average speed of the particle was 4 m/s
   (d) The average accelerating of the particle was 3 m/s²

9. Consider the following velocity–time plot, Which of the following is not true?

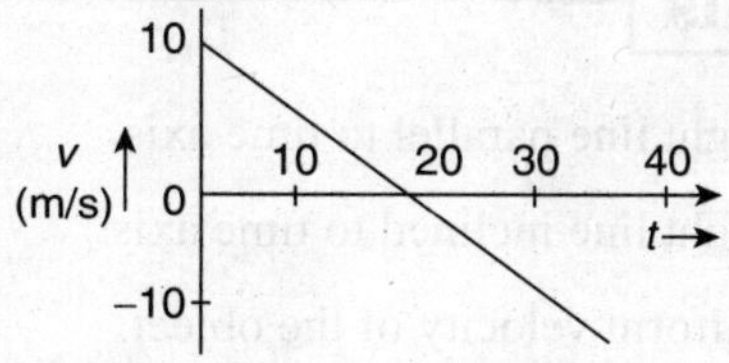

   (a) The particle has a constant acceleration.
   (b) The particle has a never changed direction.
   (c) The particle may have zero displacement.
   (d) The average speed in the interval 0 – 20 s is same as average speed in the interval 20 – 40 s.

Scholar Zone

1. Consider the following motions of an object in one dimension: (a) A ball is thrown directly upward; it rises to its highest point, and falls back into the thrower's hand. (b) A race car starts from rest and speeds up to 120 m/s along a straight line. (c) A Voyager spacecraft drifts through empty space at constant velocity. Are there any instants of time in the motion of these objects at which the instantaneous velocity at the instant and the average velocity over the entire interval are the same? If so, identify the points (s).

2. A sky diver steps out of a stationary helicopter. A few seconds later, another sky diver steps out, so that they both fall along the same vertical line. Ignore air resistance, so that both sky divers fall with the same acceleration. Let us model the sky divers as particles under constant acceleration. Does the vertical separation distance between them stay the same? Does the difference in their speeds stay the same?

3. A car starts moving rectilinearly, first with acceleration $a = 5.0$ m/s² (the initial velocity is equal to zero), then uniformly, and finally decelerates at the same rate a, to come to a stop. The total time of motion equals $t = 25$ s. The average velocity during that time is equal to $v_{average}$ = 72 km/hr. How long does the car move uniformly?

4. During a shuttle launch, the bottom of the launch system clears launch tower at a time of 6.0 s after liftoff. The height of the launch tower from the initial position at the bottom of the launch system is 75 m. Determine (a) the acceleration (assumed constant) of the launch system during these 6.0 s; and (b), the velocity of the launch system as the bottom clears the tower.

5. Study the speed-time graph of a body given here and answer the following questions:

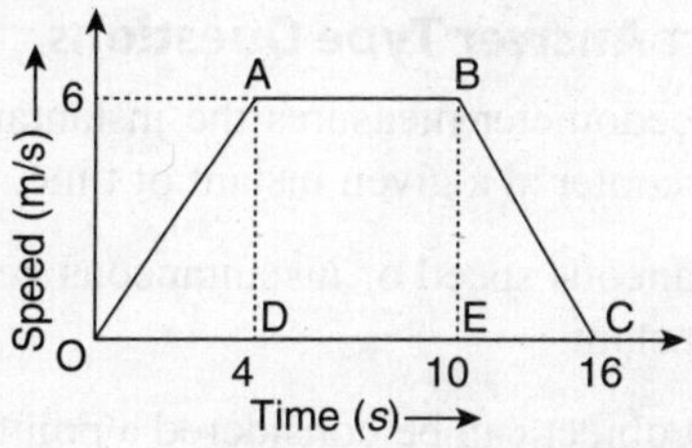

   (i) What type of motion is represented by OA?
   (ii) What type of motion is represented by AB?
   (iii) What type of motion is represented by BC?
   (iv) Find out the acceleration of the body.
   (v) Calculate the retardation of the body,
   (vi) Find out the distance travelled by the body from A to B.

6. The position of a particle moving along x-axis is given by $x = 3t - 4t^2 + t^3$, where x is in metre and t in s.
   (i) Find the position of the particle at $t = 2$ s.
   (ii) Find the displacement of the particle in the time interval from $t = 0$ to $t = 4$ s.
   (iii) Find the average velocity of the particle in the time interval from $t = 2$ s to $t = 4$ s.
7. The driver of train A travelling at a speed of 54 kmh^{-1} applies brakes and retards the train uniformly. The train stops in 5 s. Another train B is travelling on a parallel track with a speed of 36 kmh^{-1}. Its driver applies the brakes and the train retards uniformly; train B stops in 10 s. Plot the speed–time graph for both the trains on the same axis. Which of the trains travelled farther after the brakes were applied?
8. If a point moves in a straight line with uniform acceleration and covers successive equal distances in times t_1, t_2, t_3, then show that
$$\frac{1}{t_1} - \frac{1}{t_2} + \frac{1}{t_3} = \frac{3}{t_1 + t_2 + t_3}$$
9. Starting from rest, a particle moves in a straight line with acceleration $a = \{2 + |t - 2|\}$ m/s^2. The velocity of the particle at the end of 4 s will be
   (a) 16 m/s (b) 20 m/s
   (c) 8 m/s (d) 12 m/s

Answer Keys

Objective Questions

One Option Correct

1. (c)	2. (b)	3. (b)	4. (a)	5. (d)	6. (a)	7. (c)	8. (b)	9. (c)	10. (c)
11. (c)	12. (c)	13. (c)	14. (c)	15. (b)	16. (d)	17. (b)	18. (d)	19. (b)	20. (c)
21. (c)	22. (a)	23. (b)	24. (d)	25. (d)	26. (d)	27. (a)	28. (c)	29. (a)	30. (b)
31. (b)	32. (c)	33. (d)	34. (a)	35. (d)	36. (d)	37. (b)	38. (a)	39. (c)	40. (c)
41. (b)	42. (d)	43. (c)	44. (c)	45. (b)	46. (c)	47. (a)	48. (d)	49. (b)	50. (a)
51. (b)	52. (b)								

More Than One Option Correct

1. (a) and (d) 2. (a), (b), (c) and (d) 3. (c) and (d) 4. (a) and (c) 5. (a), (b) and (c)
6. (a), (b) and (d) 7. (a), (b) and (c) 8. (b) and (c) 9. (a), (c) and (d)

Hints and Solutions

Very-short Answer Type Questions

1. The speedometer measures the instantaneous speed of the scooter at a given instant of time.
2. Instantaneous speed or instantaneous velocity at the given instant.
3. (a) An object can be considered a point object if its size is negligible in comparison to the distance travelled by object
   (b) An object is not considered a point object when its size is not negligible in comparison to the distance travelled by object.
4. When an object is moving with uniform velocity along a straight line.
5. A straight line parallel to time axis.
6. A straight line inclined to time axis.
7. The uniform velocity of the object.
8. Either less than or equal to one.
9. a. infinity
   b. zero
   In fact, it is not a practical situation.
10. Yes, when the velocity of the object is negative.

11. 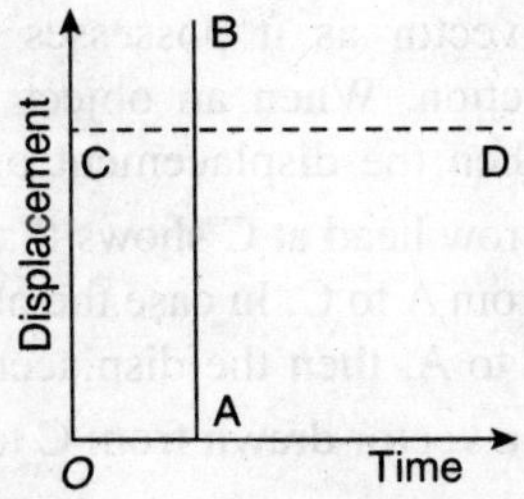

12. The time which is taken after the origin of time (i.e., zero time) is called positive time and the time which is taken before the origin of time is called negative time.

13. It is so when a body is moving with a uniform velocity, i.e., there is a uniform motion of the body in one dimension.

Short-answer Type Questions

1. Yes, for example a man goes out of his residence and after covering distance of 5 km returns back. Here, distance is 5 km while displacement is zero.

2. No, a body can not possess a constant velocity but a varying speed since velocity = speed + direction.

3. No, they are often unequal. However, in straight-line motion they are equal.

4. When the displacement–time curve is a straight line, i.e., when the body is moving with a uniform velocity.

5. Yes, the velocity at a given instant may be zero but due to acceleration, the body can continue to be in motion.

6. Yes, instantaneous acceleration is independent of instantaneous velocity. So the direction of the velocity has no relation to the direction of its acceleration.

7. Consider the case of uniform acceleration where

   Displacement = (Average velocity) × (Time)

   But the above equation does not hold good if the acceleration is not constant.

8. No, it is not possible. To stop motion in a time interval equal to zero, an infinite deceleration is needed, which is impossible.

9. In the latter case, because car suddenly comes to halt, i.e., rate of change of velocity is large.

10. No, it is obvious from the figure that as the position x increases, the time first increases then decreases. This is not possible.

11. It is not possible because, Velocity = Speed + Direction. If speed is zero, the velocity is also zero.

12. No, because the distance travelled by an object is the actual path traversed by the object during motion in a given interval of time. The displacement of an object in a given interval of time is given by the vector drawn from the initial position to final position. Consider the following illustrations:
    i. when an object completes a circular motion on a circle of radius r, then its displacement is zero but distance travelled is $2\pi r$.
    ii. when an object moves in a triangle from A to B, B to C and C to A in a given time, its displacement in time t is zero but the distance travelled is AB + BC + CA.

13. Not necessarily. If an object covers a complete circular track of radius r, then its displacement is zero but distance travelled is $2\pi r$.

14. Average velocity = Total displacement/Total time taken and Average speed = Total distance travelled/Total time taken. If an object completes a circular path of radius r in time t, then its displacement is zero but distance travelled by body is $2\pi r$. Therefore, the average velocity of body = zero but average speed of body = $2\pi r/1$

15. Since, average velocity = $\dfrac{\text{Total displacement}}{\text{Total time taken}}$ and

    $$\text{Average speed} = \frac{\text{Total distance travelled}}{\text{Total time taken}}$$

    Since displacement of an object can be equal or less than total distance travelled in a given time, average velocity of an object over an interval of time is either equal or smaller than the average speed of the object over the same interval of time.

16. An object can be considered a point object if during motion in a given time, it covers a distance much greater than its own size. For example, if a car of size 3 meters is moving a few hundred kilometer distance, then to study this motion, the car can be taken as a point object, since 3 m is less than a few hundred kilometer distance.

17. The basic rule for the relative velocity of a body A with respect to another body B is that it is the vector difference between the velocities of A and B, i.e., $\vec{A} - \vec{B}$. Similarly, the relative velocity of B with respect to A is $\vec{B} - \vec{A}$.

18. Since the particles while moving with constant speed v, are always at a constant distance d, they must be at the two ends of the diameter of a circular path. The diameter of this circular path = d. Each particle will return to its initial position after describing a circular path $2\pi r = \pi d$

$$(\because 2r = d)$$

Time interval after which each particle will return to its initial position = Distance travelled/speed = $\pi d/v$

19. Distance travelled in one revolution = $2\pi r = 2\pi \times 100$ m; Time taken = 4 minute.

Displacement in 4 minute = zero

i. Average speed = $\dfrac{\text{Distance travelled}}{\text{Time taken}}$

$$= \frac{2\pi \times 100}{4} = 50\pi \text{ metre/minute}$$

ii. Average velocity = $\dfrac{\text{Displacement}}{\text{Time taken}} = \dfrac{0}{4} = 0$

20. In 2 minutes 20 seconds, the athlete will complete three complete rotations and one half-rotation, i.e., he will be at the opposite end of the diameter from the starting point. So displacement = $2R$.

21. $u + v = 6$ and $u - v = 4$ or $2u = 10$ or $u = 5$ m/s

$2v = 6 - 4 = 2$ or $v = 1$ m/s

Descriptive Type Questions

1. i. **Distance:** The length of the actual path traversed by an object during motion in a given interval of time is called distance travelled by that object. Suppose an object goes from A to C following the path ABC, in a certain time t, then total distance travelled by object in the time interval t = AB + BC.

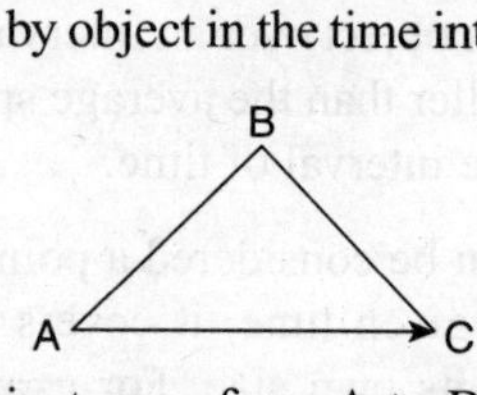

If the object goes from A to B, B to C and C to A in time t, then the total distance travelled by object in time t = AB + BC + CA

Distance is a scalar quantity. Its value can never be zero or negative, during the motion of an object.

ii. **Displacement:** The displacement of an object in a given interval of time is defined as the shortest distance between the two positions of the object in a particular direction during that time and is given by the vector drawn from the initial position to its final position.

Displacement is a vector as it possesses both magnitude and direction. When an object goes on the path ABC, then the displacement of the object is $\overrightarrow{AC}$. The arrow head at C shows that the object is displaced from A to C. In case the object is displaced from C to A, then the displacement of the object is $\overrightarrow{CA}$, (a vector drawn from C to A). Then $\overrightarrow{CA} = -\overrightarrow{AC}$

It means the displacement $\overrightarrow{CA}$ is having the same magnitude as that of $\overrightarrow{AC}$ but in opposite direction. In case the object goes from A to B, B to C and C to A, then the total displacement will be $\overrightarrow{AA} = 0$

2. **Velocity:** The velocity of an object defined as the time rate of change of displacement of the object. It is also defined as the speed of an object in a given direction. Quantitatively,

$$\text{Velocity} = \frac{\text{Displacement}}{\text{Time interval}}$$

Velocity is a vector quantity, as it has both magnitude (speed) and direction. The velocity of an object can be positive, zero and negative according to whether its displacement is positive, zero or negative.

The unit of velocity is cm s^{-1} in cgs system and m s^{-1} in mks system or S.I. The dimensional formula of velocity is $[M^0L^1T^{-1}]$.

(a) **Uniform Velocity:** An object is said to be moving with a uniform velocity if it undergoes equal displacements in equal intervals of time, howsoever small these intervals may be.

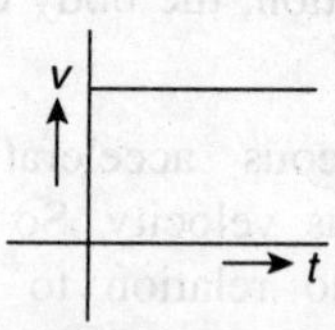

Uniform velocity

(b) **Variable velocity:** An object is said to be moving with a variable velocity if it undergoes equal displacements in unequal intervals of time or changes its direction of motion while moving with a constant speed.

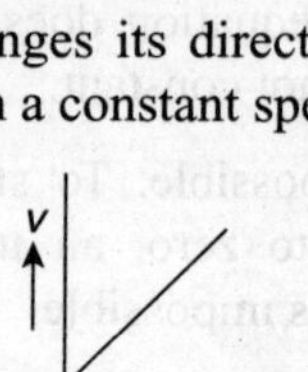

Variable velocity

3. Let us suppose that an object has travelled a distance s in time t under uniform acceleration a. From the given figure, the distance travelled by an object is obtained by the area enclosed within OABC under the velocity–time graph AB.

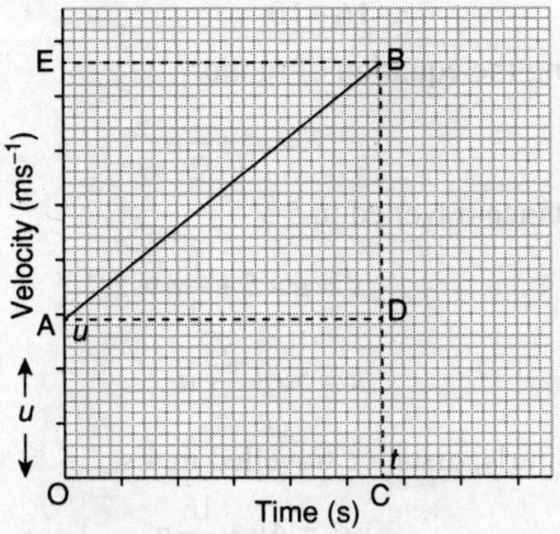

Thus, the distance s travelled by the object is given by

s = area OABC (which is a trapezium)

= area of the rectangle OADC + area of the triangle ABD

$= OA \times OC + \frac{1}{2}(AD \times BD)$ ...(1)

But, OA = u, OC = AD = t, BD = at

Substituting the above values in equation (1), we get

$$S = u \times t + \frac{1}{2}(t \times at) \text{ or } s = ut + \frac{1}{2}at^2$$

Hence, this is the second equation of motion.

4. (i) Speed during first 4 s = Slope of OP part of the (distance)

$$= \frac{75}{4} = 18.75 \text{ m/s}$$

(ii) From 4 s to 14 s, distance is not changing with time, so it is stationery for 10 s.

(iii) No, it is not a real situation because distance travelled cannot decrease with time. Moreover, time cannot flow backward as shown by QR part of the graph.

NCERT Corner

1. No. Though the moving object comes back to its initial position the distance travelled is not zero.

2. Acceleration $a = 0$, $v = u$

$$s = ut$$

$$v^2 - u^2 = 0$$

3.

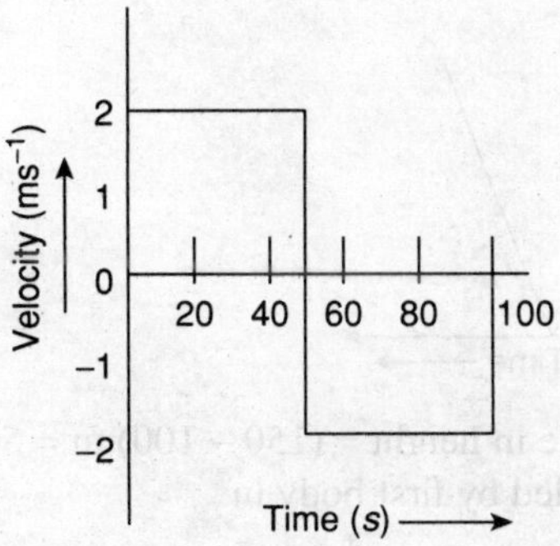

4. The distance travelled in first 8 s,

$$x_1 = 0 + \frac{1}{2}(5)(8)^2 = 160 \text{ m.}$$

At this point, the velocity $v = u + at = 0 + (5 \times 8) = 40 \text{ ms}^{-1}$

Therefore, the distance covered in last four seconds, $x_2 = (40 \times 4)$ m = 160 m.

Thus, the total distance $x = x_1 + x_2 = (160 + 160)$ m = 320 m

5. Let $AB = x$, So $t_1 = \frac{x}{30}$ and $t_2 = \frac{x}{20}$

$$\text{Total time} = t_1 + t_2 = \frac{5x}{60} \text{ h}$$

Average speed for entire journey

$$= \frac{\text{Total distance}}{\text{Total time}} = \frac{2x}{\frac{5x}{60}} = 24 \text{ kmh}^{-1}$$

6. i. Since velocity is not changing, acceleration is equal to zero.

ii. Reading the graph, velocity = 20 ms⁻¹

iii. Distance covered in 15 seconds, $s = u \times t = 20 \times 15 = 300$ m

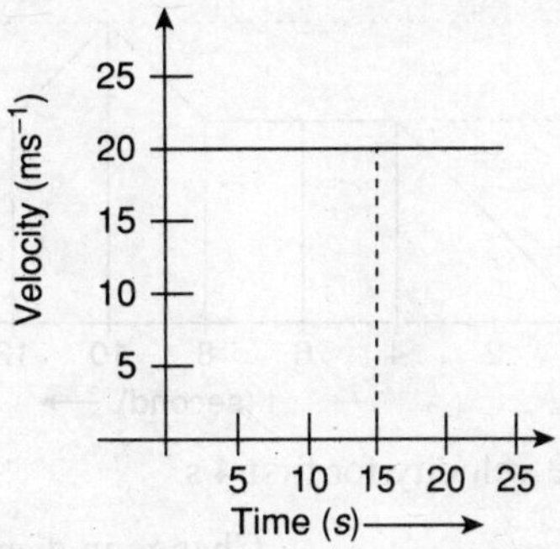

7.

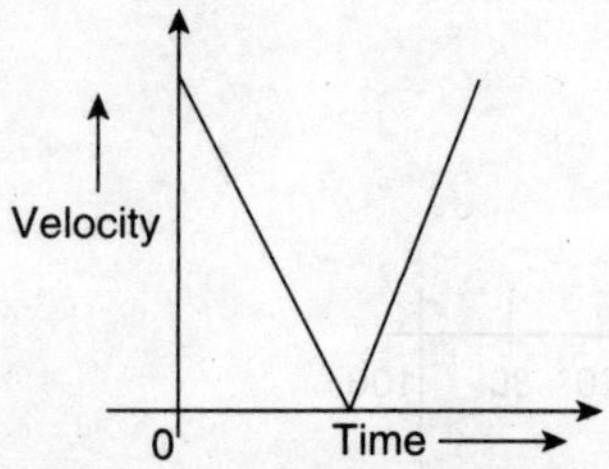

8. Initial difference in height = (150 – 100) m = 50 m
Distance travelled by first body in

$$2\text{ s} = h_1 = 0 + \frac{1}{2}g(2)^2 = 2g$$

Distance travelled by another body in

$$2\text{ s} = h_2 = 0 + \frac{1}{2}g(2)^2 = 2g$$

After 2 s, height at which the first body will be = $h_1' = 150 - 2g$

After 2 s, height at which the second body will be = $h_2' = 100 - 2g$

Thus, after 2 s, difference in height

$$= 150 - 2g - (100 - 2g)$$

= 50 m = Initial difference in height
Thus, difference in height does not vary with time.

9. $s_1 = ut + \frac{1}{2}at^2$ or $20 = 0 + \frac{1}{2}a(2)^2$ or $a = 10\text{ ms}^{-2}$

$v = u + at = 0 + (10 \times 2) = 20\text{ ms}^{-1}$

$$s = 160 = vt' + \frac{1}{2}a'(t)^2 = (20 \times 4) + \left(\frac{1}{2}a' \times 16\right)$$

$= a' = 10\text{ ms}^{-2}$

Since acceleration is the same, we have $v' = 0 + (10 \times 7) = 70\text{ ms}^{-1}$

10.

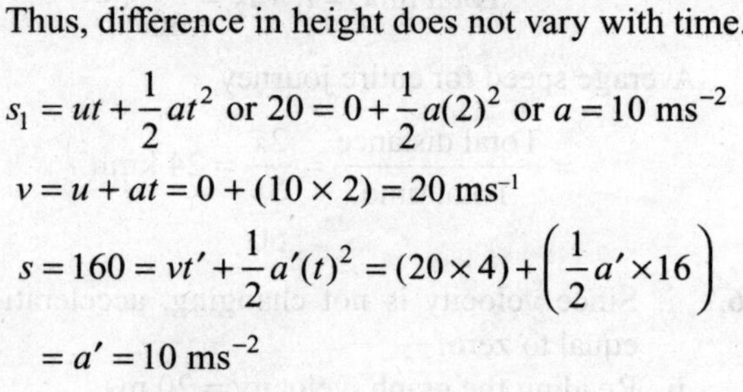

Average velocity for first 4 s.

$$\text{Average velocity} = \frac{\text{Change in displacement}}{\text{Total time taken}}$$

$$v = \frac{4-0}{4-0} = \frac{4}{4} = 1\text{ ms}^{-1}$$

For next 4 s, $v = \frac{4-4}{8-4} = \frac{0}{4} = 0\text{ ms}^{-1}$

(or as x remains the same from 4 to 8 seconds, velocity is zero)

For last 6 s, $v = \frac{0-6}{16-10} = -1\text{ ms}^{-1}$

11. Using the equation of motion

$$s = ut + \frac{1}{2}at^2$$

Distance travelled in 5 s

$$s = u \times 5 + \frac{1}{2}a \times 5^2$$

or $$s = 5u + \frac{25}{2}a \quad \text{(i)}$$

Similarly, distance travelled in 4 s,

$$s' = 4u + \frac{16}{2}a \quad \text{(ii)}$$

Distance travelled in the interval between 4th and 5th second

$$= (s - s') = \left(u + \frac{9}{2}a\right)\text{m}$$

12. We know for upward motion, $v^2 = u^2 - 2gh$

or $$h = \frac{u^2 - v^2}{2g}$$

But at highest point $v = 0$

Therefore, $h = \frac{u^2}{2g}$

For first ball, $h_1 = \frac{u_1^2}{2g}$

And for second ball, $h_2 = \frac{u_2^2}{2g}$

Thus $$\frac{h_1}{h_2} = \frac{\frac{u_1^2}{2g}}{\frac{u_2^2}{2g}} = \frac{u_1^2}{u_2^2}$$

or $h_1 : h_2 = u_1^2 : u_2^2$.

$$\left(\frac{u}{4}\right)^2$$

Numericals

Level – I

1. During acceleration $v = at = 4 \times 5 = 20$ m/s

$$s_1 = \frac{1}{2}at^2 = \frac{1}{2} \times 4 \times (5)^2 = 50\text{ m}$$

During uniform motion

$$s_2 = vt = 20 \times 25 = 500\text{ m}$$

During retardation $t = \frac{v}{a} = \frac{20}{2} = 10$ sec

$$s_3 = vt - \frac{1}{2}at^2$$

$$= 20 \times 10 - \frac{1}{2} \times 2 \times (10)^2 = 100 \text{ m}$$

Total distance = 50 + 500 + 100 = 650 m

2. a. $a = \frac{48 \times \frac{5}{18}}{36} = 3.7 \text{ m/s}^2$

b. $a = \frac{48 \times \frac{5}{18}}{10.2 - 3.6} = 2.01 \text{ m/s}^2$

c. $a = \frac{v^2}{2s} = \frac{\left(140 \times \frac{5}{18}\right)^2}{2 \times 400} = 1.89 \text{ m/s}^2$

3. $t = \sqrt{\frac{2\sqrt{2}L}{a}} = \frac{2L}{v}$

$$\frac{2\sqrt{2}L}{a} = \frac{4L^2}{v^2}$$

$$\Rightarrow \quad a = \frac{v^2}{\sqrt{2}L}$$

4. Overtaking time $= \frac{600}{20} = 30$ s

If the car's initial speed is u,

$$600 = 4(30) + \frac{1}{2}(1)(30)^2$$

$$\Rightarrow \quad u = 5 \text{ m/s}$$

After 30 seconds, the car's speed v = 5 + (1)(30) = 35 m/s

In the next 20 seconds, separation between car and truck = (35 − 20) × 20 = 300 m

5. Relative velocity of car w.r.t itself = 40 + (−40) = 0 km/h

6. We know that

$$s_n = u_n + \frac{a}{2}(2n - 1)$$

Therefore, $30 = u_n + \frac{a}{2}(6 - 1) = u_n + \frac{5a}{2}$ …(i)

and $50 = u_n + \frac{a}{2}(10 - 1) = u_n + \frac{9a}{2}$ …(ii)

Solving the above two equations, we get

$$a = 10 \text{ m/s}^2$$

4. The distance covered by a body in nth second

$$s_n = u_n + \frac{a}{2}(2n - 1)$$

Therefore, $120 = 0 + \frac{a}{2}(2 \times 8 - 1) = \frac{15a}{2}$

or, $a = \frac{120 \times 2}{15} = 16 \text{ m/s}^2$

5. Given that u = 72 km/h = 20 m/s, v = 0 and s = 200 m

Now $v^2 = u^2 + 2as$

or $0 = (20)^2 + 2a \times 200$

Therefore, $a = -1 \text{ m/s}^2$ or retardation = 1 m/s²

9. In the given question,

$$v = at_1 = 2t_1$$
$$v = 4at_2$$
$$ut_2 = 4t_1 \text{ or } t_1 = 2t_2$$
$$t_1 + t_2 = 3$$
$$t_1 = 2 \text{ s}$$
$$t_2 = 1 \text{ s}$$

Now, $s = ut + \frac{1}{2}at^2$

Therefore, $s_1 = 0 + \frac{1}{2} \times 2 \times 4 = 4$ m

and $s_2 = \frac{1}{2} \times 4 \times 1 = 2$ m

Therefore, $s_1 + s_2 = 6$ m

10. Let u be the initial velocity of the bullet and a be the retardation of the wooden board assumed to be constant. Using the formula $v^2 = u^2 + 2as$, we have

$$\left(\frac{u}{4}\right)^2 = u^2 - 2a \times 15$$

or, $\frac{u^2 - 16u^2}{16} = -30a$

Therefore, $u^2 = 32a$ …(1)

Let it travel a further distance s before it is stopped. Then

$$0 = \left(\frac{u}{4}\right)^2 - 2as$$

or $u^2 = 32as$ …(2)

From equations (1) and (2),

$$32as = 32a \quad \text{or} \quad s = 1 \text{ cm}$$

11. Velocity of approach of cars = 18 + 27 = 45 km/h

Distance between the cars = 36 km

Therefore, Time of meeting the cars (t)

$$= \frac{\text{Distance between the cars}}{\text{Relative speed of car}} = \frac{36}{45}$$

$$= \frac{4}{5}\text{h} = 0.8 \text{ h}$$

Speed of the bird (v_b) = 36 km/h

Therefore, distance covered by the bird = $v_b \times t$

$$= 36 \times 0.8$$

$$= 28.8 \text{ km.}$$

12. Collision will be avoided if speed of the train v_1 becomes equal to v_2 in travelling a relative distance d. Therefore, final relative speed of the trains becomes zero. The initial relative speed $v_{12} = v_1 - v_2$. By third equation of motion, we have

$$v_{12}^2 = u_{12}^2 - 2a_{12}s$$

$$0 = (v_1 - v_2)^2 - 2(\alpha - 0)d$$

or, $$d = \frac{(v_1 - v_2)^2}{2\alpha}$$

The collision can be avoided if $d \geq \frac{(v_1 - v_2)^2}{2\alpha}$

13. The time taken to move net 2 steps (5 steps forward and 3 steps backward) is 8s, and so for 8 steps he takes 32 s. In the last 5 steps he will take 5s and fall into the pit.

Therefore, Total time = 32 + 5 = 37 s.

Level - II

1. From the graph it follows that the particle starts at $t = 0$ from $x = 20$ m and again at $t = 15$ s, it passes through the point $x = 20$ m. Therefore, the displacement of the particle during this interval is zero, and so, the average velocity of the particle is zero during the period $t = 0$ to $t = 15$ s.

2. 72 kmph = 20 m/s and 90 kmph = 25 m/s.

The police jeep crosses the turning 10 s after the culprit has crossed it. In these 10 s, the culprit has covered a distance of 20 × 10 = 200 m.

Therefore, the distance between the police jeep and culprit when the jeep reaches the turning is 200 m. Now, both the police jeep and the culprit are moving in the same direction with speed of 25 m/s and 20 m/s, respectively. Hence, the relative velocity of the police jeep with respect to the culprit is 5 m/s. With this relative velocity, the police jeep can catch up with the culprit in time t, if it covers a distance of 200 m in t sec.

Hence, using equation, we get

$$200 = 5 \times t + \frac{1}{2} \times 0 \times t^2$$

$$= t = 40 \text{ s}$$

In 40 s, the culprit will move further a distance of 20 × 40 = 800 m.

Hence, the distance of the culprit from the turning to where he is caught is 200 + 800 = 1000 m

$$= 1 \text{ km.}$$

3. $$200 = u \times 2 - \frac{1}{2}a(2)^2$$

or, $$u - a = 100 \quad \ldots(1)$$

$$(200 + 200) = u(2+4) - \frac{1}{2}a(2+4)^2$$

or $$u - 3a = 70 \quad \ldots(2)$$

Solving equations (1) and (2), we get a = 15 cm/s²

and u = 115 cm/sec.

Further $$v = u - at = 115 - 15 \times 7$$

$$= 10 \text{ cm/sec.}$$

4. Let v be the minimum velocity attained by the motor cycle. Suppose, the motor cycle accelerates for distance s_1 and retards for distance s_2. Then

$$v^2 = 0 + 2a_1s_1 = 2 \times 5 \times s_1 = 10s_1$$

Also $$v^2 = 0 + 2a_2s_2 = 2 \times 10 \times s_2 = 20s_2$$

Therefore, $$s_1 + s_2 = \frac{v^2}{10} + \frac{v^2}{20} = 1500 \text{ metre}$$

Therefore, v = 100 m/s ⇒ v^2 = 10000

Let the motorcycle accelerate for time t_1 and retard for time t_2. Then $v = a_1t_1$ and $v = a_2t_2$

So, $100 = 5t_1$ or $t_1 = 20$ sec.

$100 = 10t_2$ or $t_2 = 10$ sec.

Therefore, $t_1 + t_2 = 20 + 10 = 30$ sec.

5. Time to which motorcycle accelerates

$$t_1 = \frac{v_1}{a_1} = \frac{10}{1} = 105$$

and $$s_1 = \frac{1}{2}(1)(10)^2 = 50 \text{ m}$$

Time to which car accelerates

$$t_2 = \frac{v_2}{a_2} = \frac{15}{0.5} = 30 \text{ s}$$

and $$s_2 = \frac{1}{2}(0.5)(30)^2 = 225 \text{ m}$$

If overtaking occurs at time t, we use

$$50 + 10(t - 10) = 225 + 15(t - 30)$$

$$10t - 50 = 15t - 225$$

$$5t = 175$$

$$t = \frac{175}{5} = 35 \text{ s}$$

At $t = 35$ s

Distance is

$$s = 50 + 10\,(35 - 10)$$

$$= 300 \text{ m}$$

6. $30 = 20t_R + \dfrac{(20)^2}{2a}$

$$10 = 10t_R + \frac{(10)^2}{2a}$$

Solving (1) and (2) we get

$$a = 10 \text{ m/s}^2$$

$$t_R = 0.5 \text{ sec}$$

Thus at 15 m/s

$$s = 15t_R + \frac{(15)^2}{2a}$$

$$15 \times 0.5 + \frac{225}{20}$$

$$7.5 + 1125 = 18.75 \text{ m}$$

7. 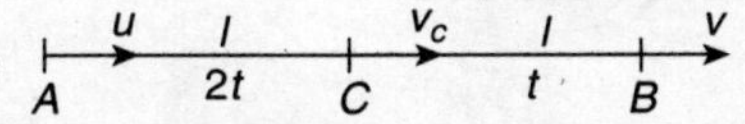

For uniform acceleration $v^2 = v^2 + 2al$

$$v^2 = u^2 + 2al$$

Solving $\quad v_c = \sqrt{\dfrac{v^2 + u^2}{2}}$

Also $\quad v_c = u + 2at$

and $\quad v = v_c + at$

Substituting v_c and eliminating at we get

$$v = 7u$$

8. If t is normal running time, then

$$t = \frac{l}{80}$$

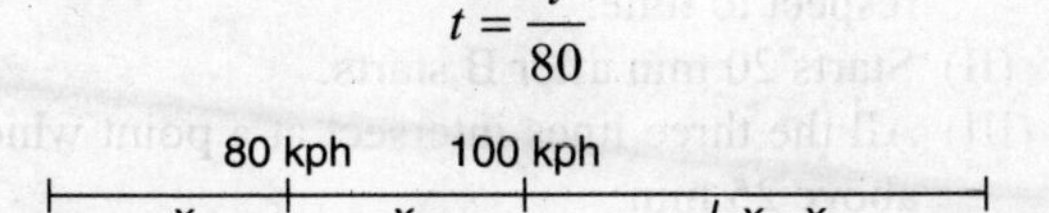

Here,

$$t_1 = \frac{x_1}{80} + 2$$

$$t_2 = \frac{100 - 80}{20} = 1 \text{ hr}$$

$$t_3 = \frac{l - x_1 - x_2}{100}$$

Here,

$$x_2 = 80(1) + \frac{1}{2}(20)(1)^2 = 90 \text{ km}$$

Thus total time

$$t = t_1 + t_2 + t_3 = \frac{l}{80}$$

$$\left(\frac{x_1}{80} + 2\right) + 1 + \left(\frac{l - x_1 - 90}{100}\right) = \frac{l}{80}$$

Solving it gives

$$l - x_1 = 840 \text{ km}$$

9. Stopping distance of car – 1

$$\frac{(10)^2}{2 \times 2} = 25 \text{ m}$$

Stopping distance of car – 2

$$\frac{(12)^2}{2 \times 2} = 36 \text{ m}$$

Separation between cars

$$= 150 - 25 - 36 = 89 \text{ m}$$

10. When 6th ball is being dropped, the positions of the other (previously fallen) balls can be calculated by using the time of falling of each ball till this instant. For 5th ball, it was dropped just one second before. Thus it has fallen a distance

$$= \frac{1}{2}gt^2 = 5 \text{ m}$$

For 4th ball, it was dropped two second before this instant, It has fallen a distance

$$= \frac{1}{2}(10)2^2 = 20 \text{ m}$$

For 3rd ball, it was dropped two second before this instant. Its depth is

$$= \frac{1}{2}(10)3^2 = 45 \text{ m}$$

Fill in the Blanks

1. Both magnitude and direction
2. Constant
3. 100 m/minute, Zero
4. Velocity of boat in still water $v_b = \frac{8+8}{2} = 8$ km/h

   Velocity of boat w.r.t water in upstream flow = v_{bw} = $8 - 4 = 4$ km/h

   Velocity of boat, w.r.t water in downstream flow = $v_{bw} = 8 + 4 = 12$ km/h

   Time in up $= \frac{8}{4} = 2$ h

   Time in down $= \frac{8}{12} = \frac{2}{3}$ h

   Total time $= 2 + \frac{2}{3}$ hr $= 160$ minutes.
5. vector
6. retardation or deceleration
7. zero.
8. instantaneous speed
9. $v = -2 + 9.8 \times 2$

   $= -2 + 19.6 = 17.6$ m/s

True or False

1. False
2. False, the rocket moves by ejecting gases from inside it and not by pushing surrounding air. It can move even in vacuum where there is no surrounding air.
3. True, Projectile is a freely falling body and weight of a freely falling body is zero.
4. True
5. True
6. True.
7. False, the acceleration remains constant.
8. False, the relative speed (with respect to earth) of the trains (and hence the centripetal forces) is different for each train.
9. True
10. False
11. False, the ratio is 1 : 3 : 5 : … $(2n - 1)$
12. True.
13. True, it can be so in case of uniform circular motion of an object.
14. True, because the diameter of the earth is very small as compared to the radius of the orbital path of earth around the sun.
15. False, because the speed of an object is the distance travelled by object in unit time and distance can never be negative.
16. False, because the displacement of an object can be either equal to or less than the distance travelled by the object.
17. False, since Velocity = Speed + Direction; therefore, a body cannot possess a constant velocity but a varying speed.
18. False, they are often unequal. Average velocity and average speed will be equal only in case of a uniform motion of a body along a straight line. Then average velocity = average speed = uniform velocity of the body.
19. False

Match the Columns

1. A → F; B → E; C → H ; D → G
2. A → G; B → E; C → H; D → F
3. A → H; B → G; C → E; D → F
4. A → G; B → E; C → F and D → G

Assertion-Reason Questions

1. (e)
2. (a)
3. (c)
4. (b)
5. (e)
6. (a)
7. (a)
8. (d)

Comprehension Type Questions

1. (I) A is at rest. Position of A does not change with respect to time.

   (II) Starts 20 min after B starts.

   (III) All the three lines intersect at a point which is above 25 min.

   (IV) Slope of s–t graph. C moves the fastest.

2. (i) 100 m as OP = 100 m

(ii) Speed of B = slope of PQ = $\frac{150-100}{2-0}$ = 25 m/s

(iii) A and B meet at Q, i.e., 150 m from origin and 2 h after the start.

(iv) Speed of A = Slope of OQ = $\frac{150}{2}$ = 75 m/s

Difference between the speeds
= 75 − 25 = 50 m/s

(v) Speed of both the trains is uniform as the *s*–*t* graph is a straight line.

3. (i) The distance–time graph is shown below:

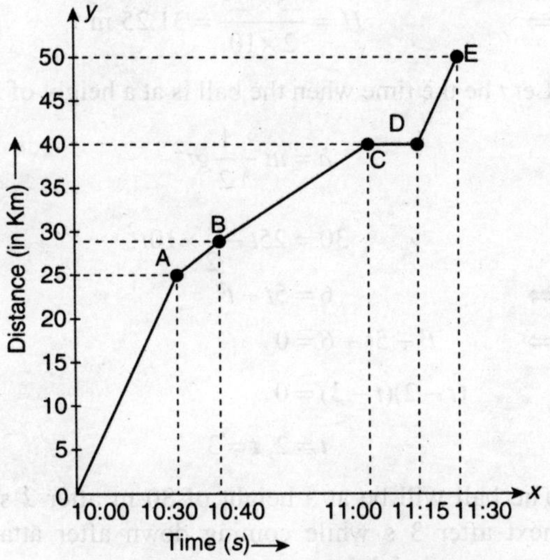

(ii) Average speed = $\frac{\text{Total distance travelled}}{\text{Total time taken}}$

In this problem, total distance travelled = 50 km
Total time taken 10:00 a.m. to 11:30 a.m.

$$= 1\text{ h } 30\text{ min} = 1\frac{1}{2}\text{ h} = \frac{3}{2}\text{ h}$$

Therefore,

Now average speed = $\frac{50\text{ km}}{\frac{3}{2}\text{ h}} = \frac{100}{3} = 33.33\text{ kmh}^{-1}$

(iii) We know, Speed = Slope of distance–time graph. The greater the slope, the greater is the speed.
From the graph, it is clear that slope of the distance–time graph is maximum between 10:00 a.m. to 10:30 a.m., so the train was travelling at the highest speed during this interval of time.

(iii) The part CD of the graph has minimum slope, so the train had minimum speed between 11:00 a.m. and 11:15 a.m. Thus, the train had slowed between 40 km and 42 km.

(iv) Speed between 10:40 a.m. to 11:00 a.m.

$$= \frac{\text{Distance}}{\text{Time}}$$

$$= \frac{(40-28)\text{km}}{20\text{ min}} = \frac{12\text{ km}}{20/60\text{ h}} = 36\text{ kmh}^{-1}$$

4. (i) Mansi is displaying safety concerns for her brother Anirban in particular, and all drivers in general.

(ii) Yes, Mansi's statement is perfectly right.

(iii) Average speed is defined as the ratio of the total distance travelled by an object to the total time taken.

5. (i) Accidents occurring with fast moving vehicles are dangerous and might be fatal.

(ii) Lakshya is caring, concerned, responsible and convincing.

(iii) $a = \frac{v-u}{t} = \frac{30-0}{10} = 3\text{ m/s}^2$

$$[\text{hence, } v = 108\text{ km/h} = 108\times\frac{5}{18} = 30\text{ m/s}]$$

Objective Questions

One Option Correct

1. (c) Let the velocity of the locomotive be *v* when it overtakes after time *t* at a distance *s*; then, the distance covered by the boy in time *t*,

$$s = ut = 2t$$

The distance covered by the locomotive in time *t*

$$s = \frac{1}{2}at^2$$

$$\frac{1}{2}at^2 = 2t \Rightarrow at = 4$$

Now, $v = u + at = 0 + 4 = 4$ m/s

2. (b) The distance covered by body in *n* seconds

$$= \frac{1}{2}an^2 \qquad (u = 0)$$

The distance covered by the body in *n*th second

$$= \frac{1}{2}a(2n-1)$$

$$\text{Ratio} = \frac{\frac{1}{2}a(2n+1)}{\frac{1}{2}an^2}$$

$$= \frac{2n-1}{n^2}$$

3. (b) Let the length of the first train be l_1 and velocity v_1, then

$$\frac{l_1}{v_1} = 15 \quad \text{(i)}$$

$$\frac{l_1 + 450}{v_1} = 45 \quad \text{(ii)}$$

$$\Rightarrow \quad v_1 = \frac{l_1}{15} = \frac{l_1 + 450}{45}$$

$$l_1 = 225 \text{ m and } v_1 = 15 \text{ m/s}$$

Let the velocity of the second train be v_2

Relative velocity $= v_1 + v_2 = 15 + v_2$

$$\frac{225+300}{15+v_2} = 21 \Rightarrow v_2 = 20 \text{ m/s}$$

4. (a) Let the velocity if the particles be u

$$v_A^2 = u^2 + 2gh \quad \text{(i)}$$

$$v_B^2 = u^2 + 2gh \quad \text{(ii)}$$

(u for B also remains same because when it comes back to the tower, it has velocity u in downward direction)

$$v_C^2 = 2gh$$

(u for C is in horizontal direction)

Therefore, $v_A = v_B > v_C$

5. (d) Let total distance be = 4 s

For the second half, the average velocity

$$v = \frac{v_1 + v_2}{2}$$

$$v_{av} = \frac{\text{Total distance}}{\text{Total Time}}$$

Time t_1 for first half of the journey $= \frac{2s}{v_0}$

Time t_2 for second half of the journey $\frac{2s}{v}$

$$v_{av} = \frac{4s}{\frac{2s}{v_0} + \frac{2s}{v}}$$

$$= \frac{2v_0 v}{v_0 + v}$$

$$= \frac{2v_0(v_1 - v_2)}{2\times\left(v_0 + \frac{v_1 - t_2}{2}\right)}$$

$$= \frac{2v_0(v_1 + v_2)}{2v_0 + v_1 + v_2}$$

6. (a) Distance–time graph is a straight line. In option (d) distance is parallel to time axis, here velocity is zero.

7. (c) Their relative velocity = 30 + 30 = 60 km/hr^{-1}

The bird will keep on flying till the two cars meet.

Time = 1 hr

The distance covered by the bird is also 60 km.

8. (b) Let the maximum height to which the ball rises be H, then,

$$v^2 = u^2 - 2gh$$

$$0 = (25)2 - 2 \times 10H$$

$$\Rightarrow \quad H = \frac{25\times 25}{2\times 10} = 31.25 \text{ m}$$

Let t be the time when the ball is at a height of 30 m.

$$h = ut - \frac{1}{2}gt^2$$

$$30 = 25t - \frac{1}{2}\times 10t^2$$

$$\Rightarrow \quad 6 = 5t - t^2$$

$$\Rightarrow \quad t^2 - 5t + 6 = 0$$

$$(t-2)(t-3) = 0$$

$$t = 2, t = 3$$

The ball will be at a height of 30 m after 2 s, and next after 3 s while coming down after attaining maximum height.

9. (c) Velocity is a vector quantity. Average velocity is defined as the displacement divided by time.

In the given graph, displacement is zero.

Hence, Average velocity

$$= \frac{\text{Total displacement}}{\text{Total time}} = \frac{0}{t} = 0$$

10. (c) Half-circle means displacement = $2r$

11. (c) Velocity acquired by parachute after falling 50 m.

$$v^2 = \sqrt{2gh} = \sqrt{2\times 10\times 50}$$

$$= 10\sqrt{10} \text{ m/s}$$

Let h be the height at which the parachutist will bail out, then,

$$v^2 = u^2 - 2ah$$

$$(3)^2 = \left(10\sqrt{10}\right)^2 - 2\times 2\times h$$

$$9 = 1000 - 4h$$

$$h = \frac{1000-9}{4} = 247.75 \text{ m}$$

Total height = 50 + 247.5

= 297.5 m

12. (c) Velocity of balloon after 8 seconds

$$v = u + at$$

$$= 0 + 1.25 \times 8 = 10 \text{ m}$$

The height reached by the balloon,

$$h = \frac{1}{2} \times 1.25 \times 8^2 = 40 \text{ m}$$

According to inertia, the initial velocity of stone will be –10 m/s

$$h' = -10t + \frac{1}{2} \times 10t^2 = 40$$

$$\Rightarrow \quad t = 4, -2$$

13. (c) $200 = u \times 2 + \frac{1}{2} a \times (2)^2$

$$200 = 2u + 2a$$

$$100 = u + a \quad \text{...(i)}$$

The distance covered in 6 seconds

$$420 = 6u + \frac{1}{2} a \times 6^2$$

$$70 = u + 3a \quad \text{...(ii)}$$

From (i) and (ii),

$$2a = -30$$

$$\Rightarrow \quad a = -15 \text{ m/s}^2$$

From (i)

$$\Rightarrow \quad 100 = u - 15$$

$$u = 115$$

Velocity at the end of 7 seconds

$$v = 115 + (-15 \times 7)$$

$$= 10 \text{ m/s}$$

15. (b) The distance travelled in tth second

$$s_t = ut + \frac{1}{2} a(2t-1)$$

$$\frac{9}{25} h = 0 + \frac{1}{2} g(2t-1) \quad \text{...(i)}$$

$$h = \frac{1}{2} gt^2 \quad \text{...(ii)}$$

From (i) and (ii),

$$\frac{25}{9} = \frac{t^2}{2t-1}$$

$$\Rightarrow \quad 9t^2 - 50t + 25 = 0$$

$$\Rightarrow \quad t = 5 \text{ or } \frac{5}{9}$$

16. (d) $s = ut + \frac{1}{2} g(2n-1)$

where u is initial velocity, t is time, g is acceleration due to gravity and n is the particular second.

Given $u = 0$

For $n = 1, s_1 = 0 + \frac{1}{2} g(2-1) = \frac{g}{2}$ ...(i)

$n = 2, s_2 = 0 + \frac{1}{2} g(2 \times 2 - 1) = \frac{3g}{2}$ ...(ii)

$n = 3, s_3 = 0 + \frac{1}{2} g(2 \times 3 - 1) = \frac{5g}{2}$ ...(iii)

From Eqs (i), (ii) and (iii),

$$s_1 : s_2 : s_3 = \frac{g}{2} : \frac{3g}{2} : \frac{5g}{2}$$

17. (b) Distance covered = Area enclosed by v–t graph

= Area of triangle = $\frac{1}{2} \times 4 \times 8 = 16$ m

= 1 : 3 : 5

18. (d) The area of a–t represents change in velocity. Thus,

$$A_1 = 5 \times 3 = 15;$$

$$A_2 = 2 \times 4 = 8;$$

$$A_3 = 7 \times 1 = 7;$$

$\therefore$ (1) > (2) > (3)

19. (b) $v^2 = u^2 + 2gh$

$= u^2 - 2gh$

$$\Rightarrow \quad h = \frac{u^2}{2g}$$

20. (c) Since the body has uniform acceleration, the velocity of the particle increases with time. Hence, this is displacement time graph with X as time and Y as displacement axis.

21. (c) Average speed $= \dfrac{S_1 + S_2}{2t}$

$$S_1 = v_1 t,\ S_2 = v_2 t$$

$$\Rightarrow \quad v_{av} = \frac{v_1 t + v_2 t}{2t} = \frac{v_1 + v_2}{2}$$

22. (a) Average speed $= \dfrac{2S}{t} = \dfrac{2S}{t_1 + t_2}$

$$t_1 = \frac{S}{v_1},\ t_2 = \frac{S}{v_2}$$

$$t_1 + t_2 = \frac{S}{v_1} + \frac{S}{v_2} = \frac{S(v_1 + v_2)}{v_1 + v_2}$$

$$\frac{2S}{t_1 + t_2} = \frac{2v_1 v_2}{v_1 + v_2}$$

23. (b) $S_1 = \dfrac{1}{2} \times a \times 10^2 = 50a$ m

$$S_1 + S_2 = \frac{1}{2} a \times 20^2 = 200a \text{ m}$$

$$\Rightarrow \quad S_2 = 150a \text{ m}$$

$$\therefore \quad S_2 = 3S_1,\ S_1 = \frac{S_2}{3}.$$

24. (d) Because the slope is highest at C,

$v = \dfrac{ds}{dt}$ is maximum

26. (d) Let v_s be the velocity of the scooter, $s = 1000$ m, $t = 100$

The velocity of the bus = 10 ms^{-1}

Time taken to overtake = 100 s

Relative velocity of the scooter with respect to the bus = $((v_s - 10))$

$$\therefore \quad \left(\frac{1000}{v_s - 10}\right) = 100 \text{ s} \Rightarrow v_s = 20 \text{ ms}^{-1}$$

27. (a) The distance travelled in nth second

As, $u = 0$

$$S_n = \frac{1}{2} a(2n - 1)$$

$$= \frac{1}{2} \times \frac{4}{3}(2 \times 3 - 1)$$

$$= \frac{1}{2} \times \frac{4}{5} \times 5$$

$$= \frac{10}{3} \text{ m}$$

28. (c) Distance travelled in one rotation (lap) = $2\pi r$

$$\therefore \quad \text{Average speed} = \frac{\text{Distance}}{\text{Time}} = \frac{2\pi r}{t}$$

$$= \frac{2 \times 3.14 \times 100}{62.8} = 10 \text{ ms}^{-1}$$

Net displacement in one lap = 0

$$\text{Average velocity} = \frac{\text{Net displacement}}{\text{Time}} = \frac{0}{t} = 0$$

29. (a) If we want minimum three (more than two) balls to remain in air, then time of flight of first ball must be greater than 4 sec.

$$t > 4 \text{ sec or } \frac{2u}{g} > 4 \text{ sec} \Rightarrow u > 19.6 \text{ m/s}$$

30. (b) Let total height = H, Time of ascent = T

So, $$H = uT - \frac{1}{2} gT^2 \qquad \text{(i)}$$

Distance covered by ball in time $(T - t)$ sec.

$$h = u(T - t) - \frac{1}{2} g(T - t)^2 \qquad \text{(ii)}$$

So distance covered by ball in last t sec.

$$h' = H - h$$

$$= \left[uT - \frac{1}{2} gT^2\right] - \left[u(T - t) - \frac{1}{2} g(T - t)^2\right] \qquad \text{(iii)}$$

We have $\quad 0 = u - gt \quad$ or $\quad T = \dfrac{u}{g}$

Putting this value in Eq. (iii)

$$h' = \frac{1}{2} gt^2$$

31. (b) $S = \dfrac{1}{2} at^2$

$S \propto t^2$

32. (c) Let total distance = $2x$

$$x \rightarrow v_1,\ x \rightarrow v_2$$

$$\therefore \quad \text{Mean velocity} = \frac{x + x}{\dfrac{x}{v_1} + \dfrac{x}{v_2}}$$

$$\Rightarrow \quad v = \frac{2}{\frac{1}{v_1} + \frac{1}{v_2}}$$

$$\frac{2}{v} = \frac{1}{v_1} + \frac{1}{v_2}$$

33. (d) In the positive region, the velocity decreases linearly (during rise) and in the negative region, the velocity increases linearly (during fall). The direction is opposite to each other during rise and fall. Hence, the fall is shown in negative direction.

35. (d) $V_{max} = 0 + \alpha t_1 = \alpha t_1$

$$0 = V_{max} - \beta t_2 \Rightarrow V_{max} = \beta t_2$$

and $\quad t = t_1 + t_2 = \frac{V_{max}}{\alpha} + \frac{V_{max}}{\beta}$

$$= V_{max}\left(\frac{1}{\alpha} + \frac{1}{\beta}\right)$$

or $\quad = V_{max} = \frac{\alpha\beta t}{\alpha + \beta}$

36. (d) $h_1 = ut + \frac{1}{2}ut^2$

$$= 0 + \frac{1}{2} \times 19.6 \times (5)2$$

$$= 245 \text{ m}$$

The velocity of the rocket when engine is switch of

$$v = u + gt$$

$$= 0 + 19.6 \times 5$$

$$= 19.6 \times 5$$

Now the motion is under gravity

So $\quad v^2 = u^2 + 2gh^2$

$$0 = (19.6 \times 5)^2 - 2 \times 9.8 \times h_2$$

$$= 490 \text{ m}$$

$$\Rightarrow \quad h_2 = \frac{19.6 \times 19.6 \times 5 \times 5}{2 \times 9.8}$$

$$h = h_1 + h_2 = 245 + 490 = 735 \text{ m}$$

37. (b) The magnitude of average velocity

$$= \frac{\text{Total displacement}}{\text{Total time}}$$

$$= \frac{\text{Area } A + \text{Area } B}{t}$$

40. (c) In graph (i), slope increases as velocity of the particle increases.
In graph (ii) for retardation motion, velocity of particle decreases, so slope of x–t graph decreases.

41. (b) Average velocity $= \frac{\text{Displacement}}{\text{Time}}$

$$= \frac{2.0}{1.0} = 2 \text{ m/s.}$$

42. (d) In option (b) S–t graph is straight line. Therefore, velocity is constant. Hence, acceleration is zero.
In option (c) V–t graph is parallel to time axis. Hence, velocity is constant. Therefore, acceleration is zero.

43. (c) $v^2 = u^2 + 2as$. If $u = 0$, $v^2 = 2as$
i.e., the graph should be a parabola symmetric to the displacement axis.

44. (c) Speed is constant but direction is changing. Hence, velocity is changing. Therefore, it is accelerated motion.

45. (b) The other graphs show more than one velocity of the particle at a single instant of time which is not practically possible.

46. (c) Distance decreasing with equal interval of time shows retarded motion. While (b) represents accelerated motion of then body.

47. (a) When acceleration is constant, $v \propto t$
When acceleration is zero, $\Delta v = 0$

48. (d) Graph shows insufficient data to reach any other conclusion.

49. (b) Distance, therefore the unit is m.

50. (a) The area within velocity–time graph for car

$$A = \frac{1}{2} \times 14 \times 5$$

$$= 35 \text{ m}$$

The area within velocity–time graph for car

$$B = \frac{1}{2} \times 9 \times 10$$

$$= 45 \text{ m}$$

Hence car B travelled further

51. (b) The distance moved by the bus $= \frac{1}{2}at^2$

The boy moves the distance $x + 48 = 10t \frac{1}{2} \times 1 \times t^2 + 48 = 10t$

$$t^2 - 20t + 96 = 0$$

$$(t - 12)(t - 8) = 0$$

$$t = 8, t = 12$$

So the earliest time is $t = 8$ seconds,
The correct option is (a)

52. (b) The slope of B is least, means velocity is the least.

Scholar Zone

1. (a) The average velocity over the entire interval for the thrown ball is zero – the ball returns to the thrower's hand at the end of the time interval. There is one point at which the instantaneous velocity is zero – at the top of the motion. (b) The average velocity for the motion of the race car cannot be evaluated unambiguously with the information given, but its magnitude must be some value between 0 and 120 m/s. Because the magnitude of the instantaneous velocity of the car will have every value between 0 and 120 m/s at some time during the interval, there must be some instant at which the instantaneous velocity is equal to the average velocity over the entire interval. (c) Because the instantaneous velocity of the spacecraft is constant, its instantaneous velocity at any point of time and its average velocity over any time interval are the same.

2. At any given point of time, the speeds of the jumpers are different because one had a head start over the other. In any time interval, however, each jumper increases his or her speed by the same amount, because they have the same acceleration. Thus, the difference in speeds remains the same. The first jumper will always be moving with a higher speed than the second. In a given time interval, then, the first jumper will have a larger displacement than the second. Thus, the distance of separation between them increases.

3. **First Solution:**

Average velocity of the car over total time of motion

$$= \frac{72 \text{ km/hr}}{60 \times 60 \text{ s}} = 72 \times 1000 \text{ m} = 20 \text{ m/s}.$$

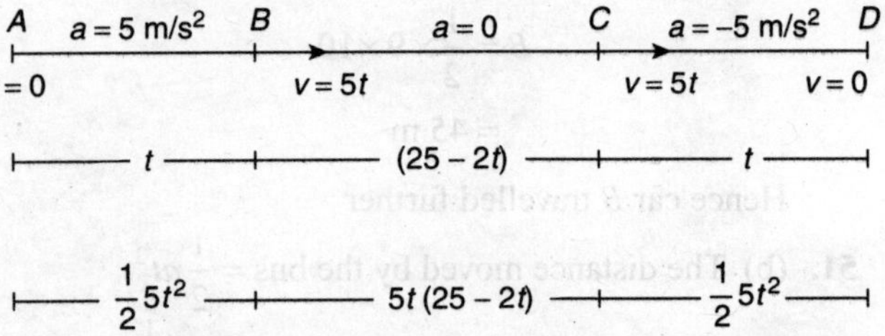

Let the car move with constant acceleration of $a = 5$ m/s² for time t.

Velocity of the car at point B,

$$v = 5t \text{ m/s} \qquad (\text{Using } v = u + at)$$

$$\text{Displacement } AB = 0.\, t + \frac{1}{2} 5t^2 \qquad \left(\text{using } s = ut + \frac{1}{2} at^2\right)$$

Velocity of the car at point $C = 5t$ m/s. (How?)

The time car takes in moving from point C to D:

$$0 = 5t - 5t_1 \text{ or } t_1 = t. \qquad (\text{Using } v = u + at)$$

$$\text{Displacement } CD = 5t \cdot t - \frac{1}{2} \cdot 5 \cdot t^2 = \frac{1}{2} \cdot 5 \cdot t^2$$

The time car takes to travel from B to C is $(25 - 2t)$ s.

Displacement $BC = (25 - 2t)(5t)$.

Now, total displacement $AB + C + CD$ = Average velocity × Total time

$$\text{or} \quad \frac{1}{2} \cdot 5 \cdot t^2 + 5t(25 - 2t) + \frac{1}{2} \cdot 5 \cdot t^2 = 20 \times 25$$

$$\text{or} \quad t = 15 \text{ s}, 20 \text{ s}.$$

Here, $t = 20$ s is to be ruled out. Why? If $t = 20$ s, What will be the time of uniform motion?

Hence the time of uniform motion = 25 – 2 × 5 = 15 s.

Second Solution: Let us solve the problem with another variable. Let the velocity of the car at point B be v (see figure).

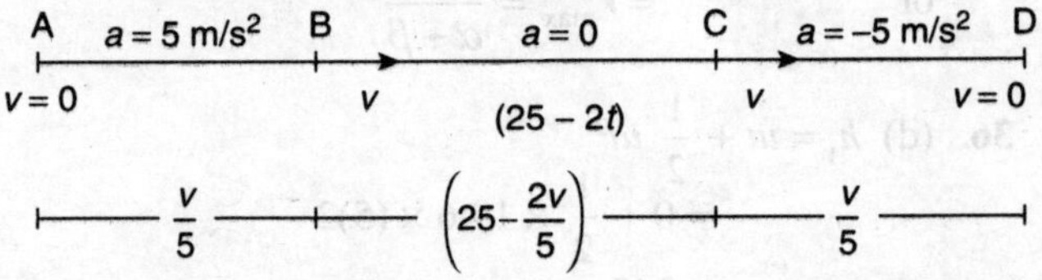

$$\text{Time from A to B} = \frac{v}{5} \qquad (\text{Using } v = u + at)$$

$$\text{Displacement AB} = \frac{v^2}{2 \times 5} \qquad (\text{Using } v^2 = u^2 + 2as)$$

$$\text{Time from C to D} = \frac{v}{5} \qquad (\text{Using } v = u + at)$$

$$\text{Displacement CD} = \frac{v^2}{2 \times 5} \qquad (\text{Using } v^2 = u^2 + 2as)$$

$$\text{Time from B to C} = \left(25 - \frac{2v}{5}\right)$$

$$\text{Displacement BC} = v\left(25 - \frac{2v}{5}\right)$$

Now, total displacement AB + BC + CD = 500 m

$$\text{or} \quad \frac{v^2}{2 \times 5} + v\left(25 - \frac{2v}{5}\right) + \frac{v^2}{2 \times 5} = 500$$

$$\text{or} \quad v = 25 \text{ m/s}.$$

Therefore, time of uniform motion

$$= 25 - \frac{2 \times 25}{5} = 15 \text{ s}.$$

Third Solution: It is handy to draw velocity–time graph for solving this problem. Let the car accelerate at 5 m/s^2 for time t. For the first t seconds, the velocity of the car increases linearly. And for the last t seconds, it decreases linearly. Velocity of the car remains constant for time $(25 - 2t)$ seconds.

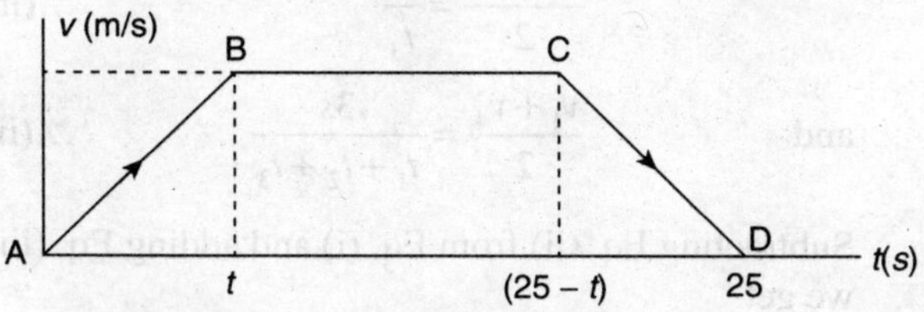

Slope of AB = –slope of CD = 5 m/s^2.

Total displacement = Area under the velocity–time graph = 500 m,

i.e., $\frac{1}{2}(t)(5t) + (25 - 2t)\cdot(5t) + \frac{1}{2}(t)(5t) = 500$

or $t = 5$ s, (20 *s* is ruled out, why?).

Time of uniform motion = $25 - 2t = 15$ s.

4. **Reasoning:** The shuttle is a large object, but we shall model it as a particle. The position of this particle is chosen at the bottom of the launch system because it is the bottom about which we have information. Let us assume that the acceleration is constant so that we can model the bottom of the launch system as a particle under constant acceleration and use the kinematic equations. If the acceleration is not truly constant during this time interval, what we shall calculate is the average acceleration during this interval.

Using the equations of motion and taking the initial velocity as zero (because the launch system begins from rest) and defining the initial position of the bottom of the launch system as the origin, we have

$$s = ut + \frac{1}{2}a + t^2$$

$$s = 0 + \frac{1}{2}a + t^2$$

$$a = \frac{2\times 75}{6^2} = 4.2 \text{ m/s}^2$$

b. We use the kinematic equation for velocity

$$v = u + at$$

$$v = 0 + 4.2 \times 6$$

$$= 25 \text{ m/s}$$

Notice that the acceleration of the shuttle in question is only about 0.4 *g*. As mentioned earlier, as the shuttle moves towards solid rocket booster separation, at about 120 seconds into the launch, the acceleration rises to about 3 *g*.

5. (i) OA is a straight-line graph between speed and time, it is sloping upward from O to A. Here OA represents uniform acceleration.

(ii) AB is a straight-line graph between speed and time, which is parallel to the time axis (X-axis). So, AB represents uniform speed. There is no acceleration from A to B.

(iii) BC is a straight-line graph between speed and time which is sloping downwards from B to C. Therefore, BC represents uniform retardation or negative acceleration.

(iv) Acceleration of the body, as we see from graph, is represented by line OA. So, the slope of velocity–time graph OA will give the acceleration of the body.

Thus, acceleration = slope of line OA = $\frac{\text{AD}}{\text{OD}}$

We have, AD = 6 m/s and OD = 4 s

So, acceleration = $\frac{6 \text{ m/s}}{4 \text{ s}} = 1.5 \text{ m/s}^2$

(v) The slope of line graph BC represents the retardation of the body.

So, retardation = Slope of line BC = $\frac{\text{BE}}{\text{EC}}$

we have, BE = 6 m/s, EC = 16 – 10 = 6 s

Retardation = $\frac{6 \text{ m/s}}{6 \text{ s}} = 1 \text{ m/s}^2$.

(vi) The distance travelled by the body is equal to the area enclosed between the speed–time graph and time axis.

Distance travelled from A to B = Area under the line AB and the time axis = Area of rectangle DABE = DA × DE

Here, DA = 6 m/s and DE = 10 – 4 = 6 s

Distance travelled from A to B = 6 × 6 = 36 m

6. (i) Position of the particle at $t = 2$ s, then

$$x(2) = 3 \times 2 - 4 \times (2)^2 + (2)^3$$

$$= 6 - 4 \times 4 + 8 = -2 \text{ m}$$

(ii) At, $t = 0$, $x(0) = 0$

$$x(4) = 3 \times 4 - 4 \times (4)^2 + (4)^3 = 12 \text{ m}$$

Displacement = (4) – (0) = 12 m

(iii) $v_{av} = \frac{x(4) - x(2)}{4 - 2} = \frac{12 - (-2)}{2} = 7 \text{ m/s}$

7. For train A, the initial velocity,

$$u = 54 \text{ kmh}^{-1} = 54 \times \frac{5}{18} = 15 \text{ ms}^{-1}$$

Final velocity $v = 0$ and time, $t = 5$ s.

For train B, $u = 36 \text{ kmh}^{-1} = 36 \times \frac{5}{18} = 10 \text{ ms}^{-1}$

$$v = 0, t = 10 \text{ s}$$

Speed–time graph for trains A and B are shown in the figure below:

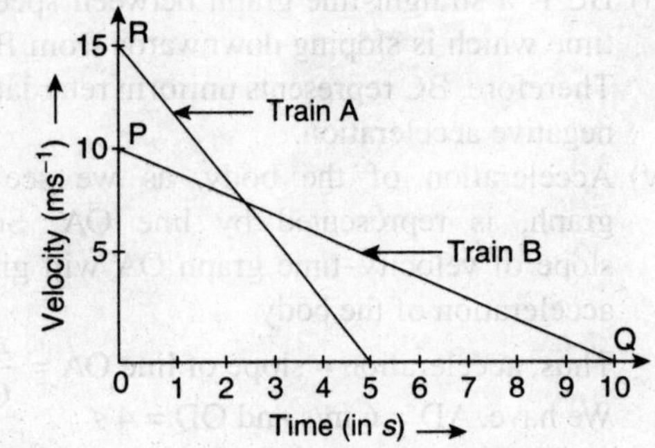

Distance travelled by train A = Area under straight-line graph

$$\text{RS} = \text{area of } \Delta\text{ORS} = \frac{1}{2} \times \text{OR} \times \text{OS}$$

$$= \frac{1}{2} \times 15 \times 5 = 37.5 \text{ m}$$

Distance travelled by train B = Area under PQ

$$= \frac{1}{2} \times \text{OP} \times \text{OQ}$$

$$= \frac{1}{2} \times 10 \times 10 = 50 \text{ m}$$

8. Let s be the successive equal distance and v_1, v_2, v_3 the initial velocities for the successive distances and v_4, the final velocity in the third distance. Since the acceleration is constant, the average velocities in the three intervals are

$$\frac{v_1 + v_2}{2}, \frac{v_2 + v_3}{2}, \frac{v_3 + v_4}{2} \text{ and } \frac{v_1 + v_4}{2}$$

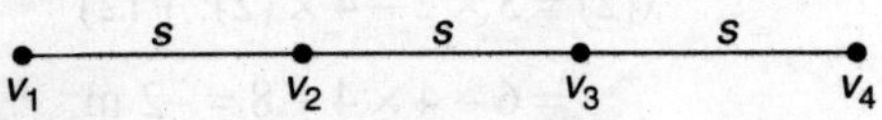

We know that average velocity

$$v_{av} = \frac{\text{distance}}{\text{time}}$$

Therefore,

$$\frac{v_1 + v_2}{2} = \frac{s}{t_1} \quad \text{...(i)}$$

$$\frac{v_2 + v_3}{2} = \frac{s}{t_2} \quad \text{...(ii)}$$

$$\frac{v_3 + v_4}{2} = \frac{s}{t_3} \quad \text{...(iii)}$$

and

$$\frac{v_1 + v_4}{2} = \frac{3s}{t_1 + t_2 + t_3} \quad \text{...(iv)}$$

Subtracting Eq. (ii) from Eq. (i) and adding Eq. (iii), we get

$$\frac{v_1 + v_4}{2} = s\left[\frac{1}{t_1} - \frac{1}{t_2} + \frac{1}{t_3}\right] \quad \text{...(v)}$$

Now from Eqs (iv) and (v), we get

$$\frac{1}{t_1} - \frac{1}{t_2} + \frac{1}{t_3} = \frac{3}{t_1 + t_2 + t_3}$$

9. Acceleration can be written as $a = 2 + 2 - t$ or $a = 4 - t$ for $t \le 2$ s and $a = 2 + t - 2$ or $a = t$ for $t \ge 2$ s. Therefore, acceleration–time graph of the particle will be as shown below:

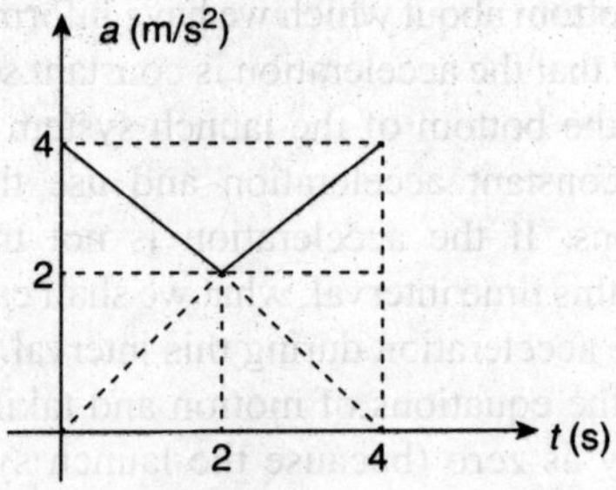

$$v_f - v_i = \text{Area under } (a\text{–}t) \text{ graph}$$

or $$v_f - v_i = (4 \times 4) - \frac{1}{2}(4)(2)$$

$$= 12 \text{ m/s}$$

CHAPTER 2

Force and Laws of Motion

LEARNING OBJECTIVES

At the end of this chapter, you will be able to learn about:

- Force and Newton's Laws of Motion
- Momentum and Impulse
- Conservation of Momentum
- Friction and Its types
- Laws of Friction
- Density and Relative density
- Pressure and Thrust
- Buoyancy and Archimedes' Principle
- Application of Newton's Laws of Motion

Introduction

A force is a physical quantity which tends to set a stationary body in motion or which tends to change the speed and direction of a moving body or which tends to change the shape of a body. Force is a vector quantity having magnitude as well as direction.

Balanced Forces

If the resultant of all the force acting on a body is zero, the forces are termed balanced forces. A body under the action of balanced forces does not change its state of rest or of uniform motion and it appears as if no force is acting on it.

Unbalanced Forces

If the resultant of all the forces acting on a body is not zero, the forces are called unbalanced forces. When unbalanced forces act on a body, they produce a change in its state of rest or of uniform motion.

Inertia

The inherent property of a body by virtue of which a body at rest tends to remain at rest and a body in motion tends to continue moving with same velocity is called inertia.

The mass of a body is a measure of its inertia.

Newton's Laws of Motion

First Law of Motion

A body at rest will remain at rest, and a body in motion will continue in motion in a straight line with uniform speed, unless it is compelled by an external force to change its state of rest or of uniform motion.

Second Law of Motion

The force acting on a body is directly proportional to the product of the mass of the body and the acceleration produced in the body by the action of the force, and it acts in the direction of the acceleration.

Suppose a force F acts on a body of mass m and produces an acceleration a in the body. Then, according to Newton's second law of motion

$$F = ma$$

In vector form,

$$\vec{F} = m\vec{a}$$

If $F = 0$, $a = 0$, as m cannot be zero.

Unit of Force

(a) The S.I. unit of force is Newton (N)

One Newton is that force which when applied on a body of mass 1 kg will produces in it an acceleration of 1 m/s^2.

i.e., 1 newton = 1 kilogram × 1 m/s^2

or 1N = 1 kg × 1 m/s^2

(b) The C G S unit of force is dyne (dyn)

One dyne is that force which when applied on a body of mass 1 g will produce in it an acceleration of 1 cm/s^2.

1N = 10^5 dyne

ILLUSTRATION 2.1

A body of mass 0.5 kg is resting on a frictionless surface. Calculate the distance travelled by it in 10s when a force of 100 N is applied on it.

Solution

$u = 0, F = 100$ N, $m = 0.5$ kg

$$F = ma$$

$$\therefore \quad 100 = 0.5\,a$$

$$\therefore \quad a = 2 \times 10^2 \text{ ms}^{-2}$$

$$S = ut + \frac{1}{2}at^2$$

$$= 0 + \frac{1}{2} \times (2\times10^2) \times (10)^2$$

$$\therefore \quad S = 10^4 = 10000 \text{ m} = 10 \text{ km}$$

Momentum

The amount of motion present in a moving object, is called the momentum of the object. It is represented by the symbol p. It is a vector quantity.

Momentum is measured as the product of mass and velocity.

$$p = mv$$

Units

S.I. unit of momentum is kg m/s. C.G.S unit of momentum is g cm/s.

Relation between force and momentum

Force applied on a body is equal to the rate of change of its momentum and this change of momentum takes place in direction of force.

If a force F applied to a moving body of mass m, changes its velocity from u to v in time t

$$F = \frac{mv - mu}{t}$$

ILLUSTRATION 2.2

A body of mass 250 g has a velocity of 8 ms^{-1}. Calculate its momentum.

Solution

$$m = 250 \text{ g} = \frac{250}{1000} \text{ kg}$$

$$v = 8 \text{ ms}^{-1}$$

Momentum $(P) = mv$

$$P = \frac{250}{1000} \times 8 = 2 \text{ kg ms}^{-1}$$

Impulse of a Force

The product of the magnitude of a force applied on a body and the time for which it is applied, is called impulse of the force. It is represented by the symbol I.

$$\text{Impulse} = \text{Force} \times \text{Time}$$

$$I = F \times t$$

Units

S.I. unit of impulse is Newton-second (Ns) and the C.G.S unit is dyne-second.

Impulse and Momentum

Impulse is equal to change in momentum

$$F \times t = P_2 - P_1$$

ILLUSTRATION 2.3

A football moving with a velocity of 20 ms^{-1} is stopped by a football player in 0.5 second. The mass of football is 5 kg. Calculate:

(a) Impulse

(b) Average retarding force

Solution

Mass of the ball $(m) = 0.5$ kg

Time $(t) = 0.5$ sec

Initial velocity $(u) = 20$ ms^{-1}

Final velocity $(v) = 0$ ms^{-1}

(a) Impulse $= mv - mu$

$$= 0.5 \times 0 - 0.5 \times 20$$

$$= -10.0 \text{ kg ms}^{-1}$$

Negative sign shows that impulsive force is a negative force.

(b) Retarding force $= \dfrac{\text{Impulse}}{\text{Time}}$

$$= \frac{-10}{0.5} = -20 \text{ N}$$

ILLUSTRATION 2.4

A disc of mass 10 g is kept floating horizontally by throwing 10 marbles/sec against it from below. If the mass of each marble is 5 g, calculate, stating the principle involved, the velocity with which the marbles are striking the disc. Assume that the marbles strike the disc normally and rebound downward with the same speed.

Solution

Let M = mass of disc = 10 g

m = mass of each marble = 5 g

n = number of marbles striking the disc/sec.

v = velocity with which the marbles strike the disc.

Then, weight of the disc acting downward = Mg

After striking the disc, the marbles rebound with the same speed so that the change in its velocity is $v-(-v)=2v$.

Hence, the change in momentum of each marble when it strikes the disc = $m\times 2v = 2mv$.

This change in momentum is imparted to the disc. Total momentum imparted per second to the disc is, therefore, $2\ mnv$.

Since the force is the rate of change of momentum, the force exerted by marbles on the disc in the upward direction is $2\ mnv$ and the disc will remain stationary if the net force acting on it is zero provided, it is initially at rest. Hence, for the disc to remain stationary, upward force on it = its weight

$$\Rightarrow \quad 2mnv = Mg$$

$$\Rightarrow \quad v = \frac{Mg}{2mn} = \frac{10\times 981}{2\times 5\times 10} = 98.1 \text{ cm/s}$$

$$\Rightarrow \quad v = 0.98 \text{ m/s}$$

Conservation of Momentum

The total momentum of a system remains constant if no external force is acting on it. In case of collision, total momentum before collision and after collision is equal.

ILLUSTRATION 2.5

A body A of mass (m_1) 100 kg collides with a body B of mass (m_2) 20 kg. Before collision, both bodies have the same velocity (v) 10 ms^{-1} in opposite direction. After the collision, body A moves with a velocity (v_1) 5 ms^{-1}. Find the velocity (v_2) of body B after collision.

Solution

Conservation of Momentum

$$m_1v - m_2v = m_1v_1 + m_2v_2$$

$$100\times 10 - 20\times 10 = 100\times 5 + 20\times v_2$$

$$\therefore \quad 300 = 20v_2$$

$$v_2 = 15 \text{ ms}^{-1}$$

Third Law of Motion

The law states that for every action there is equal and opposite reaction. Moreover, action and reaction act on different bodies.

ILLUSTRATION 2.6

A bullet of mass 10 g is fired with a rifle. The bullet takes 0.003s to move through its barrel and leaves it with a velocity of 300 ms^{-1}. What is the force exerted on the bullet by rifle? What is the force exerted on the rifle by the bullet?

Solution

$$m = 10 \text{ g} = 0.01 \text{ kg}$$

$$u = 0$$

$$v = 300 \text{ ms}^{-1}$$

$$t = 0.003 \text{ s}$$

F = Force and Laws of Motion

From impulse equation,

$$F\times t = mv - mu$$

$$\therefore \quad F = \frac{0.01\times(300-0)}{0.003} = 1000 \text{ N}$$

Force exerted on the bullet by the rifle = 1000 N (action)

Force exerted on the rifle by the bullet = –1000 N (reaction)

Friction

It is the force which opposes the motion of a body, when pushed or pulled on a horizontal surface.

Types

Static Friction

It is the friction acting on a body at rest.

It is the static friction which does not allow the body to move.

Dynamic Friction

It is the friction acting on a body in motion. It is slightly less than the limiting friction.

It is due to dynamic friction that the force is kept applied to make the body continue its motion.

Dynamic friction has two types:

(i) Sliding friction: It is the friction acting on a body in sliding motion.

(ii) Rolling friction: It is the friction acting on a body in rolling motion.

Rolling friction is less than the sliding friction.

Origin of Friction

The following two factors are responsible for friction.

1. Interlocking of the elevations and the depressions of the lower surface and bottom surface of the body.

2. Force of attraction between molecules in the lower surface and bottom surface of the body.

Dynamic friction is less than static friction.

Limiting friction (F_{max}) = P

Dynamic friction < P

Laws of limiting Friction

The maximum value of the static friction, is called limiting friction. It is equal to the minimum force required to move the body.

There are four laws of limiting friction.

1. **Law of direction:** Limiting friction acts tangential to the surfaces in contact, in backward direction (i.e. opposite to the direction of motion).
2. **Law of magnitude:** Limiting friction is directly proportional to the normal reaction acting on the body.
3. **Law of dependence:** Limiting friction depends upon the nature and material of the surfaces in contact.
4. **Law of independence:** Limiting friction is independent of the area and shape of the surfaces in contact.

Coefficient of Friction

The constant ratio of the limiting friction to normal reaction is called coefficient of friction. It is represented by the symbol μ.

$$\mu = \frac{\text{Limiting Friction}}{\text{Normal Reaction}} = \frac{F}{R}$$

or $$F = \mu R$$

ILLUSTRATION 2.7

A body is being pulled by force of 20 N. The normal reaction on the body is 40 N.

Find the coefficient of friction.

Solution

$$\mu = \frac{F}{R} = \frac{20}{40} = 0.5$$

ILLUSTRATION 2.8

A block of mass 2 kg is lying over a rough horizontal surface at rest with $\mu_s = 0.5$ and $\mu_K = 0.45$. A horizontal force F acts on block such that F increases from 5 N to 12 N continuously in a certain time and then decreases from 12 N to 5 N continuously in certain time. [$g = 10$ m/s²]

(a) Find the acceleration of the block when $F = 9.5$ N (increasing)

(b) Find the acceleration of the block when $F = 10.5$ N (increasing)

(c) Find the acceleration of the block when $F = 9.5$ N (decreasing).

Solution

Limiting static friction (L.S.F.)

$$= \mu_s N = 0.5 \times N = 0.5\ Mg = 10\text{ N} \because Mg = 20\text{ N}$$

$$f_K\ (\text{Kinetic friction})\ \mu_k \text{N} = 0.45\ Mg = 9\text{ N}$$

(a) When F is increasing until $F >$ L.S.F. = 10N, the block does not move at all. Thus, for $F \le 10$ N, the friction is static and adjustable and equals to the applied force; acceleration is zero.

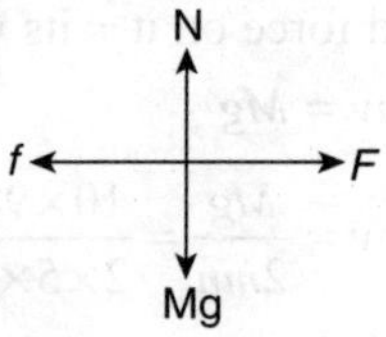

Free-body diagram (F.B.D.) of block

(b) When $F = 10.5$ N > L. S. F. (10N), the block moves and friction becomes kinetic $F = f_K$

$$a = \frac{F - f_k}{m} = \frac{10.5 - 9}{2} = \frac{1.5}{2} = 0.75\text{ m/s}^2$$

(c) Once the motion starts, friction becomes kinetic block will only be accelerating for $F > 9$ N, thus when force F = 9.5 N (decreasing) $F = f_K$

$$a = \frac{F - f_k}{m} = \frac{9.5\text{ N} - 9\text{ N}}{2} = 0.25\text{ m/s}^2.$$

Density

The mass per unit volume of a substance is called the density of the substance. It is represented by the symbol ρ (rho)

The S.I. unit is kilogram per cubic metre (kg/m³).

Relative Density or Specific Gravity

The ratio of the density of a substance to the density of water at 4°C, is called relative density (R.D.) or specific gravity (S.G.) of the substance. It is a pure number having no unit.

As the density of water at 4° C = 1g/cm³. The relative density of substance becomes numerically equal to its density in C.G.S units.

Thrust

The total force applied on a surface is called the thrust on the surface. It is represented by the symbol $\vec{F}$. Being a force, it is a vector quantity. It has the units of force.

Pressure

The force applied per unit area of a surface is called the pressure. It is represented by the symbol P. It is a scalar quantity.

The C.G.S unit of pressure is dyne per square cm (dyne/cm^2)

The S.I. unit of pressure is Newton per square metre (N/m^2) or pascal (Pa)

$$1 \text{ N/m}^2 = 10 \text{ dyn/cm}^2$$

$$\text{Thrust} = \text{Pressure} \times \text{Area}$$

$$F = PA$$

ILLUSTRATION 2.9

The mass of a rectangular block is 500 g. It has a base of length 10 cm and breadth 5 cm. Calculate in S.I. units the average pressure on its base. ($g = 10$ ms^{-2})

Solution

$$m = 500 \text{ g} = 0.5 \text{ kg}$$

$$\text{Force } (F) = W = mg \text{ (weight)}$$

$$= 0.5 \times 10 = 5 \text{ N}$$

$$\text{Area of the base } (A) = \text{length} \times \text{breadth}$$

$$\frac{5}{100} \times \frac{10}{100} = 0.005 \text{ m}^2$$

$$\text{Pressure } P = F/A$$

$$\therefore \quad P = \frac{5}{0.005}$$

$$P = 1000 \text{ Nm}^{-2} = 1000 \text{ Pa}$$

Fluid Pressure

A substance which can flow, is called a fluid. The term fluid includes both liquids and gases.

The pressure at any point in a fluid is defined as the normal force (or thrust) acting on unit area surrounding that point. Pressure is a scalar quantity.

The total thrust on any immersed plane surface is equal to the product of the area of the surface and the pressure on the surface, if the pressure is constant.

If however, the pressure over the immersed plane is not constant, (like on the wall of a tank full of water) total thrust on the surface is equal to the product of the pressure at centre of gravity (C.G.) of the surface and the area of the surface.

Pressure Due to Liquid Column

Let a vessel with vertical walls, having bottom area A have a liquid of density ρ up to a height h.

$$\text{Thrust at the bottom of vessel} = Ah\rho g$$

$$\text{Pressure at the bottom} = P = h\rho g$$

Thus, pressure at a point, due to the liquid, on the surface of the liquid is zero as height $h = 0$.

However, there is air above the liquid. The weight of the air gives rise to its own pressure, which is known as atmospheric pressure.

The total pressure at any point in a liquid is given as:

Total pressure = P = Pressure due to liquid [$h\rho g$] + Atmospheric pressure [P_A]

i.e., $$P = h\rho g + P_A$$

ILLUSTRATION 2.10

The pressure of water on the ground floor is 40000 Pa and on the first floor, it is 10000 Pa. Find the height of the first floor. Density of water 1000 kg m^{-3} and $g = 10$ ms^{-2}

Solution

Pressure on ground floor (P_1) = 40000 Pa

Pressure on first floor (P_2) = 10000 Pa

Difference of pressure ($P_2 - P_1$) = $h\rho g$ = 40000 – 10000 = 30000 Pa

$$P = h \times \rho \times g$$

$$30000 = h \times 1000 \times 10$$

$$\therefore \quad h = 3 \text{ m}$$

Buoyancy and Buoyant Force

When a body is immersed in a fluid (liquid or gas), it displaces the fluid whose volume is equal to the volume of the body immersed in the fluid. This displaced fluid exerts an upward force on the body. This tendency of the displaced fluid (exerting an upward force), is called buoyancy. The upward applied force is called the buoyant force. It is equal to the weight of the fluid displaced by the body.

The centre of gravity of the displaced fluid is called centre of buoyancy. It is represented by the letter B.

Archimedes' Principle

When a body is partially or wholly immersed in a fluid, it experiences an upthrust (or apparently loss in weight, which is equal to the weight of fluid displaced by the immersed part of the body.

Apparent weight of the block in the liquid = Resultant downward force on the block

= Its actual weight + Downward thrust on top − Upward thrust on bottom

i.e., $W' = W = Ah_1\rho g - Ah_2\rho g$

or $W - W' = Ah_2\rho g - Ah_1\rho g = Ah\rho g$

$W - W'$ = loss of weight of the body in the liquid

Ah = volume of body

= volume of liquid displaced by the body

$Ah\rho$ = mass of liquid displaced by the body

$Ah\rho g$ = weight of liquid displaced by the body

Thus,

Loss in weight of the body = Weight of liquid displaced by the body

ILLUSTRATION 2.11

The volume of a 500 g sealed tin is 350 cm³. What is the density of the packed tin ? Will it float or sink in water if it has density of 1 g/cm³? What will be the weight of water displaced by the tin?

Solution

$M = 500$ g

$V = 350$ cm³

Density of tin, ρ = ?

$$\rho = \frac{M}{V}$$

$$\therefore \quad \rho = \frac{500}{350} = 1.43\,\text{g/cm}^3$$

Density, $\rho = 1.43$ g/cm³

Since density of tin is more than density of water (1 g/cm³), it will sink in water.

Mass of water displaced = Volume of water displaced × Density of water = 350 × 1 = 350 g

$\therefore$ Weight = 0.35 kg × 10 ms⁻² = 3.5 N

ILLUSTRATION 2.12

A sinker weighs 25 gf in water. When weighed in water along with a cork, it weighs 16 gf. If the cork alone weighs 2.5 gf in air, find its R.D.

Solution

Weight of cork in air = 2.5 gf

Weight of sinker in water = 25 gf

(Weight of sinker + Weight of cork) in water = 16 gf

$\therefore$ Weight of cork in water = 16 gf − 25 gf = −9 gf

R.D. of floating body (cork)

$$= \frac{\text{Weight of solid in air}}{\text{Loss of weight of solid in water}}$$

R.D. of cork =

$$\frac{\text{Weight of cork in air}}{\text{Weight of cork in air} - \text{Weight of cork in water}}$$

$$\text{R.D. of cork} = \frac{2.5}{2.5-(-9)} = \frac{2.5}{11.5}$$

R.D. of cork = 0.217

ILLUSTRATION 2.13

A solid weighs 24 kgf in air, 20 kgf in water and 16 kgf in a certain liquid. Find the R.D. of (i) the solid (ii) the liquid

Solution

(i) R.D. of solid $= \dfrac{\text{Weight of solid in air}}{\text{Loss of weight in water}}$

$$= \frac{\text{Weight of solid in air}}{\text{Weight of solid in air} - \text{Weight of solid in water}}$$

$$= \frac{24}{24-20} = 6$$

R.D. of solid = 6

(ii) R.D. of liquid $= \dfrac{\text{Loss of weight of solid in liquid}}{\text{Loss of weight of solid in water}}$

$$= \frac{\text{Weight of solid in air} - \text{Weight of solid in liquid}}{\text{Weight of solid in air} - \text{Weight of solid in water}}$$

$$= \frac{24-16}{24-20} = \frac{8}{4}$$

R.D. of liquid = 2

Application of Newton's Laws

It is neither possible nor desirable to follow rigidly a prescribed set of rules in solving problems involving Newton's Laws. There are many ways of approaching a problem and the procedure we suggest is certainly not the only one. In fact, no cut-and-dry procedure can even substitute for an intelligent analytical thinking. Nevertheless, the following systematic method suggested will be helpful in getting started and we urge you master it even if you should later resort to short-cuts or a different approach.

Steps:

1. Divide the system into smaller sub-systems, each of which can be treated as point mass.
2. (i) Represent the body by a point or simply, as a symbol.

   (ii) Draw the **free-body diagram** – The representation of all the forces acting on a body is known as its free-body diagram, i.e., draw all the forces acting on the chosen object, for example, the body's weight, the forces applied by the other objects with which the chosen object is in contact and non-contact forces (in addition to the weight), such as electric and magnetic forces, if any, acting on the body. Remember that while applying Newton's second laws, we are concerned only with the external forces acting on the chosen object. The chosen body also exerts forces on its surrounding but such forces are irrelevant because they do not act on the chosen object. Indicate on the diagram the magnitude and direction of each force.

As an example, consider two blocks at rest on a table top.

The free-body diagram for block A is shown in figure (a). F_1 is the force exerted on block A by block B, and W_A is the force of gravity on A, called the weight. Similarly the free-body diagram for block B has been drawn in figure (b). W_B is the force of gravity on block B, and F_2 is the force exerted by A on B. There are no other physical interactions on B.

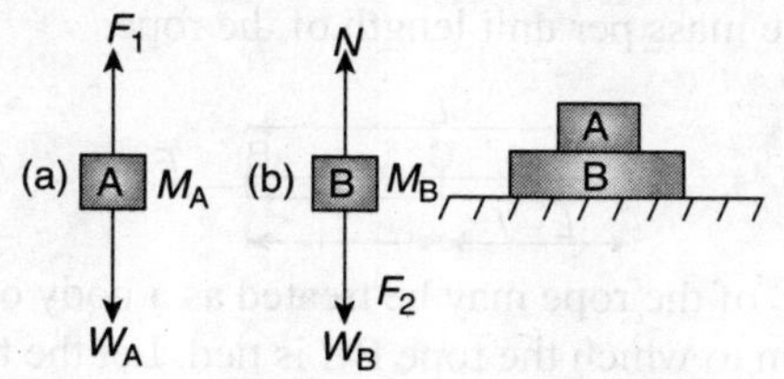

The following are some more examples of free body diagrams:

In these diagrams, the object of interest is isolated from its surroundings and the interactions between the object and the surroundings are represented in terms of forces.

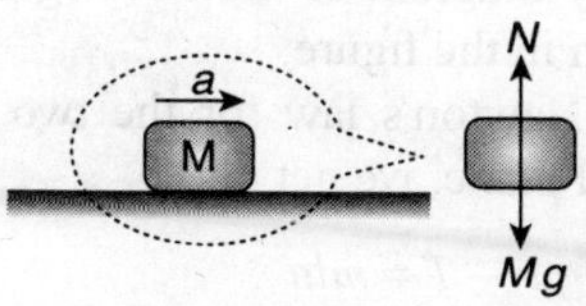

(a)

F.B.D. of a block placed on a horizontal surface. Two vertical forces act on the block: The earth pulls the block downward by *mg*, and the surface pushes the block upward by N.

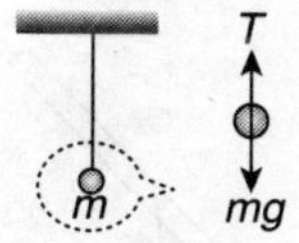

(b)

F.B.D. of a ball suspended by a string. Two vertical forces act on the ball: The earth pulls the ball downward by *mg* and the string pulls the ball upwards by *T*.

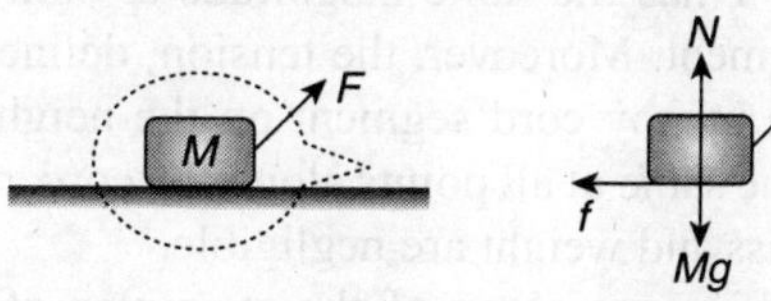

(c)

F.B.D. of a block placed on a rough surface being pulled by an external force. There are four forces acting on the block: the gravitational pull *mg*; the normal reaction *N*; the external force *F*; and the tangential force of friction *f*.

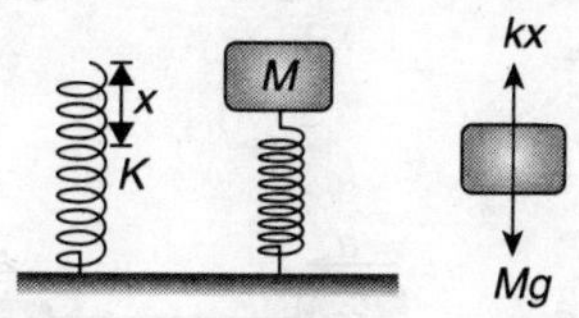

(d)

F.B.D. of a block supported by a spring of stiffness *k*. Two vertical forces act on the block: the gravitational pull, and the spring force *kx*.

Tension

Nature of the forces exerted by an ideal rope: An ideal rope is one which is perfectly flexible, inextensible and massless. No such rope or cord exists. But in many situations a real rope is very much close to the ideal. For a massless rope, the tension is the same at all points along the length of the rope irrespective of whether the rope is accelerated or in equilibrium.

Let us consider a short segment of a cord as shown in the figure. The tension on it from the neighboring parts on the cord at the two ends are labeled T_1 and T_2; its weight is *mg*. Applying Newton's second law to the small segment, we have

$$\sum \vec{F} = m\vec{a}$$

$$\vec{T_1} + \vec{T_2} + m\vec{g} = m\vec{a}$$

But if mass m is negligibly small, we may discard both mg and ma terms, and the above equation reduces to – $\vec{T}_1 = \vec{T}_2$, or in magnitude $T_1 = T_2$.

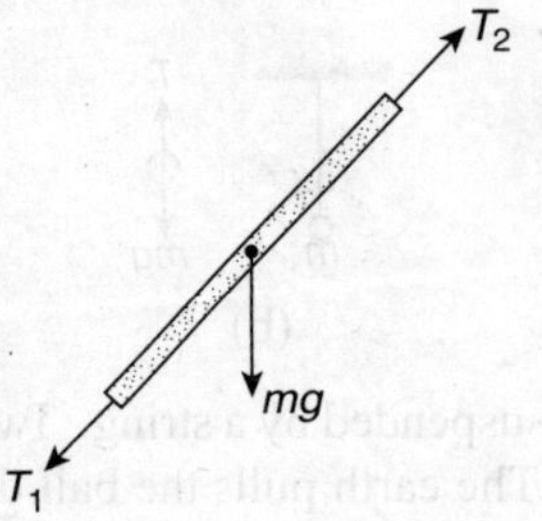

Then tension T has the same magnitude at both ends of the small segment. Moreover, the tension, defined as the force exerted by any cord segment on the neighbouring segment, is the same at all points along the cord, provided the cord's mass and weight are negligible.

The following are some of the examples of tension force when a body is connected by means of a string or a rope.

(i) If a string is ideal (inextensible and massless), then the magnitude of the accelerations of any number of masses connected through this string is always same.

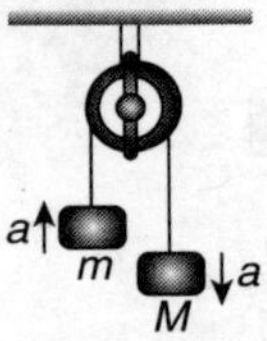

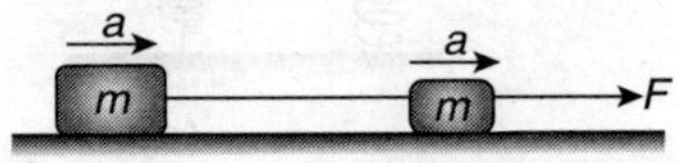

(ii) If a string is massless, the tension in it is same everywhere; on the other hand, if a string is not massless (i.e. its mass is not negligible), tension at different points of the string may be different.

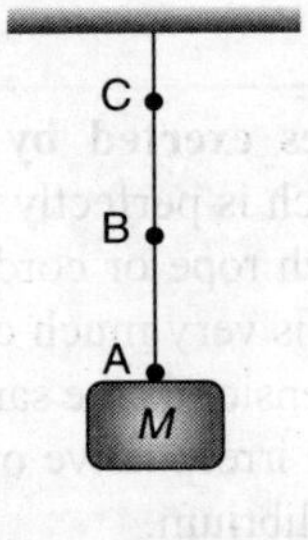

String is massless, i.e., $T_A = T_B = T_C$

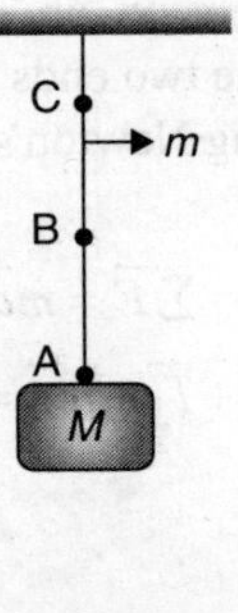

String is not massless i.e.

$$T_A \neq T_B \neq T_C \text{ and } T_A < T_B < T_C$$

(iii) If friction is present between the pulley and the string, tension is different on the two sides of the pulley But if the friction is absent between the pulley and the string, tension will be same on both sides of the pulley, provided the pulley is massless.

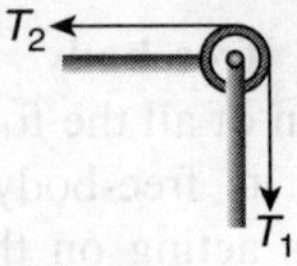

There is no friction between pulley and string, i.e., $T_1 = T_2$

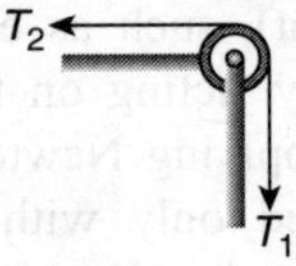

There is friction between pulley and string, i.e., $T_1 \neq T_2$

ILLUSTRATION 2.14

A heavy rope of length L, lying on a smooth horizontal surface, is pulled by a constant force F. What is the tension T in the rope at a distance l from the end where force F is applied?

Solution

Let AB be the rope of length L and C be the point such that CB equals l. The length of AC is $(L - l)$. Let m be the mass per unit length of the rope.

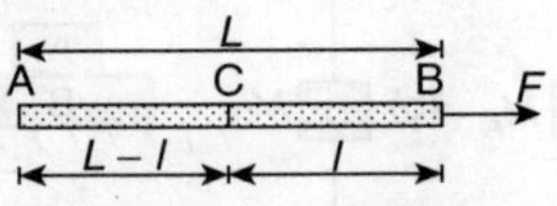

Part AC of the rope may be treated as a body of mass $(L - l)$ m to which the rope BC is tied. Let the tension between the rope AC and BC be T.

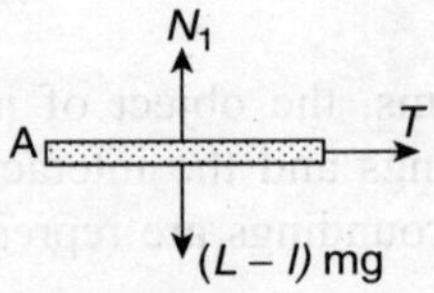

The free-body diagram of the two segments AC and CB are shown in the figure.

Applying Newton's law for the two segments in the horizontal plane, we get

$$F - T = mla \quad \text{...(i)}$$

$$T = m(L - l)a \quad \ldots\text{(ii)}$$

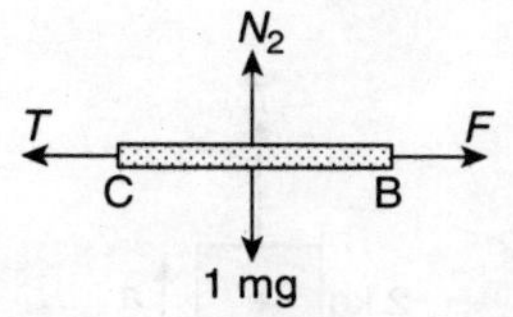

Eliminating *ma* from the two equations, we get

$$T = F\frac{(L-1)}{L}$$

ILLUSTRATION 2.15

A block of mass m_1 = 0.2 kg is suspended from the roof by a light string. Another block of mass m_2 = 0.3 kg suspended from the first block by another light string. Assuming that $g = 10\ \text{m/s}^2$, find the tensions in the two strings.

Solution

The situation is shown in the figure below. Since the strings are light, tensions at different points of the string will be equal. However, the tensions in the two strings will be different.

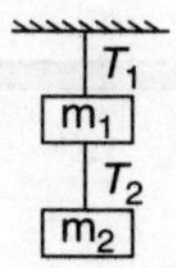

Considering the equilibrium of m_1, the different forces acting on it are:

(i) its weight m_1g downward

(ii) tensions T_1 upward and,

(iii) tension T_2 downward.

Since m_1 is in equilibrium, the net force acting on it must be zero, i.e.,

$$m_1g + T_2 - T_1 = 0$$

$$\Rightarrow \quad T_1 = m_1g + T_2$$

Similarly, considering the equilibrium of m_2, the different forces acting on it are:

(i) its weight m_2g downward and,

(ii) tension T_2 upward.

As m_2 is in equilibrium, the net force acting on it must be zero. Hence,

$$m_2g - T_2 = 0$$

$$\Rightarrow \quad T_2 = m_2g = 0.3\times 10$$

$$\Rightarrow \quad T_2 = 3\ \text{N}$$

Putting this value of T_2 in (1), we have,

$$T_1 = 0.2\times 10 + 3$$

$$\Rightarrow \quad T_1 = 5\ \text{N}$$

ILLUSTRATION 2.16

In the arrangement shown in the figure, the strings are light and inextensible. The surface over which the blocks are placed is smooth. Find:

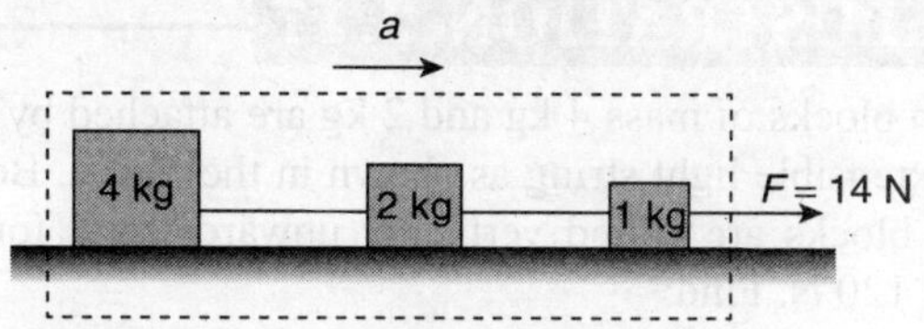

(a) the acceleration of each block,

(b) the tension in each string.

Solution

Let *a* be the acceleration of each block and T_1 and T_2 be the tensions, in the two strings as shown in figure.

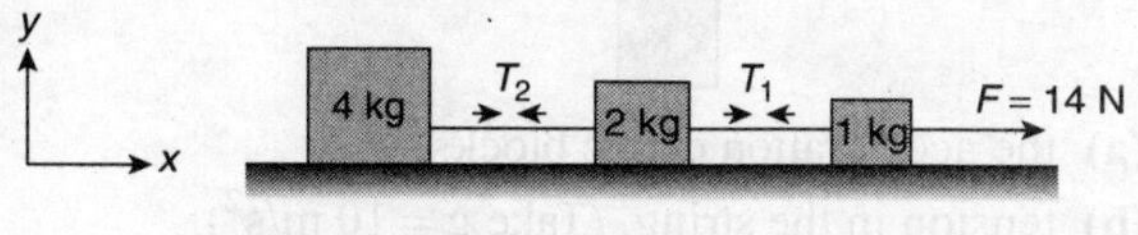

Taking the three blocks and the two strings as the system,

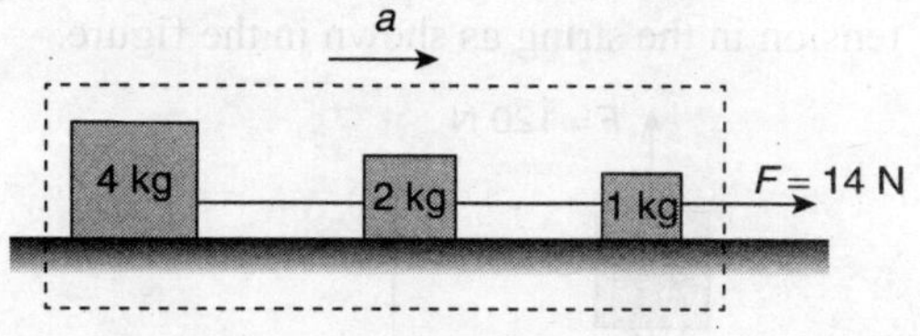

Using $\quad \Sigma F_x = ma_x$

or $\quad 14 = (4+2+1)a$

or $\quad a = \frac{14}{7} = 2\ \text{m/s}^2$

(b) Free-body diagram (showing the forces in x-direction only) of 4 kg block and 1 kg block are shown in the figure.

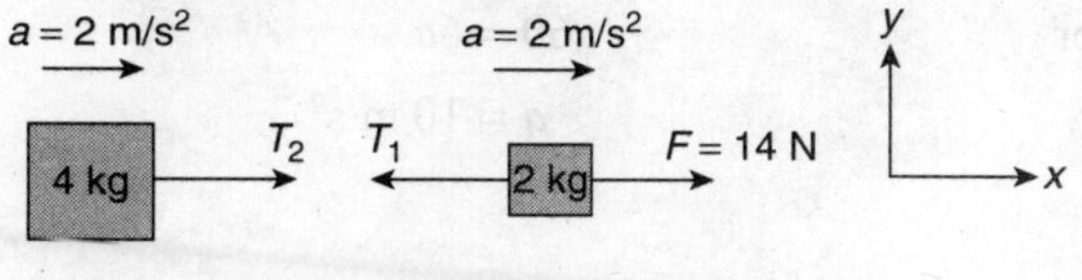

using $\quad \Sigma F_x = ma_x$

for 1 kg bock, $\quad F - T_1 = (1)(a)$

or $\quad 14 - T_1 = (1)(2) = 2$

$\therefore \quad T_1 = 14 - 2 = 12\,\text{N}$

For 4 kg block, $\quad T_2 = (4)(a)$

$\therefore \quad T_2 = (4)(2) = 8\,\text{N}$

ILLUSTRATION 2.17

Two blocks of mass 4 kg and 2 kg are attached by an inextensible light string as shown in the figure. Both the blocks are pulled vertically upwards by a force $F = 120$ N. Find:

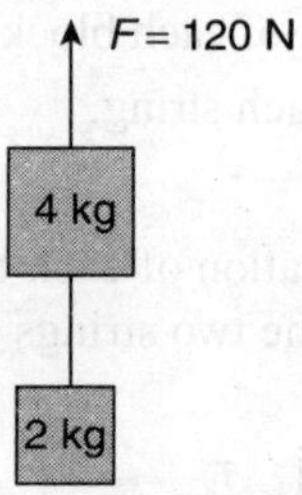

(a) the acceleration of the blocks,

(b) tension in the string. (Take $g = 10\ \text{m/s}^2$).

Solution

(a) Let a be the acceleration of the blocks and T the tension in the string as shown in the figure.

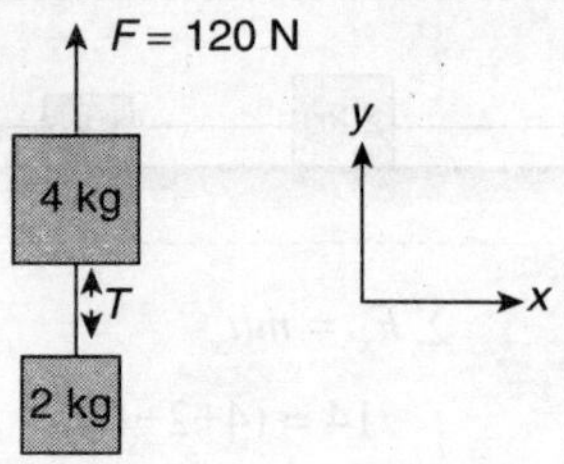

Taking the two blocks and the string as the system.

using $\quad \Sigma F_y = ma_y$

$$F - 4g - 2g = (4 + 2)a$$

or $\quad 120 - 40 - 20 = 6a$

or $\quad 60 = 6a$

$\therefore \quad a = 10\ \text{m/s}^2$

(b) Free-body diagram of 2 kg block is as shown in the figure.

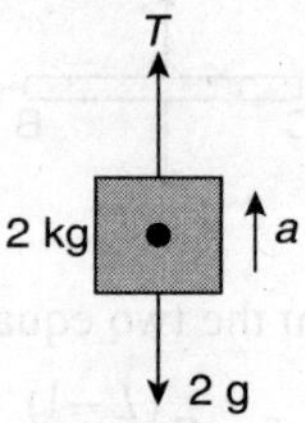

Using $\quad \Sigma F_y = ma_y \quad$ we get

$$T - 2g = 2a$$

or $\quad T - 20 = (2)\,(10)$

$\therefore \quad T = 40$ N

ILLUSTRATION 2.18

In the system shown in figure, all surfaces are smooth. The string is massless and inextensible. Find acceleration a of the system and tension T in the string. ($g = 10\ \text{m/s}^2$)

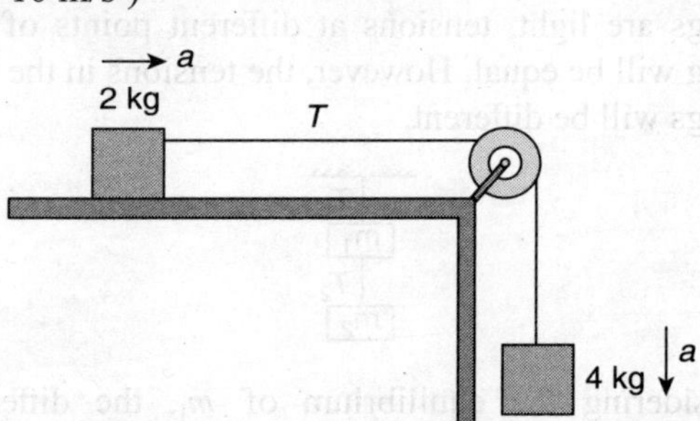

Solution

Here, the weight of 2 kg is perpendicular to motion (or a). Hence, it will not contribute in net pulling force. Only weight of the 4 kg block will be included.

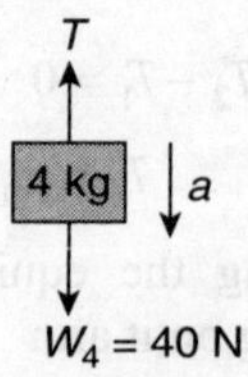

$$\therefore \quad a = \frac{\text{Net pulling force}}{\text{Toal mass}}$$

$$= \frac{(4)(10)}{(4+2)} = \frac{20}{3}\ \text{m/s}^2$$

For T, consider FBD of 4 kg block. Writing the equation of motion,

$$\therefore \quad 40 - T = 4a$$

$$T = 40 - 4a$$

$$= 40 - 4\left(\frac{20}{3}\right)$$

$$= \frac{40}{3}\text{ N}$$

POINTS TO REMEMBER

1. Newton's three laws of motion are:
   (i) Everybody continues in its state of rest or of uniform motion along a straight line unless an external force acts on it.
   (ii) The rate of change of momentum is directly proportional to the impressed force and takes place in the direction in which the force acts.
   (iii) To every action, there is an equal and opposite reaction.
2. The property of a body to maintain its state of rest or uniform motion along a straight line, unless an external force acts on it, is called inertia. Newton's first law is, therefore, also called the law of inertia.
3. The mass m of a body is a measure of its inertia. Larger the mass, more the time it takes to change its velocity by a given amount and vice-versa.
4. The momentum P of a body is given by the product of its mass m and velocity v and is, therefore, a vector quantity, i.e.,

   $$\vec{p} = m\vec{v}$$

   The unit of momentum in SI, is kg-m/s.
5. The principle of conservation of linear momentum states that the total linear momentum of a system in any given direction remains conserved if no external force acts on it.

   If u_1 and u_2 be the velocities of particles of masses m_1 and m_2 respectively along a given direction before collision, and v_1 and v_2 be their velocities along the same straight line after collision, then according to the principle of conservation of linear momentum,

   $$m_1u_2 + m_2u_2 = m_1v_1 + m_2v_2$$
6. Newton's second law gives the definition of force and a way to measure it. The force is an external agent that changes or tends to change the state of a body. If m be the mass of a body moving with an acceleration $\vec{a}$ under the action of a force $\vec{F}$ then,

   $$\vec{F} = m\vec{a}$$

   The units of force in CGS, FPS, and SI system are dyne, pound and newton respectively.
7. The impulse of a force $\vec{F}$ is defined as the product of the force and the time Δt during which it acts. Thus,

   Impulse of a force $= \vec{F}.\,\Delta t$

   The unit of impulse is Newton-second (N-s)
8. The second law also gives the principle of physical independence of forces. It states that if a number of forces acts on a body, then the effect produced by each force is independent of the effect produced by the other forces.
9. Action and reaction are nothing but forces. The first one to act is known as action and the other that comes into play as a result of the action is called reaction.
10. The forces of action-reaction appear simultaneously due to interaction between two bodies and reaction exists only so long as action is present.
11. The action and reaction always act on two different bodies and hence, in no case, can produce equilibrium.
12. Inertial frames are those which are either at rest or move with uniform velocity $\vec{v}$. Newton's laws are valid only in the inertial frames of reference.
13. Non-inertial frames are those which move with an acceleration or deceleration.
14. Newton's laws of motion are not valid if one is working from a non-inertial frame. If the frame is moving with respect to an inertial frame with acceleration $\vec{a}_0$, one must assume the existence of a pseudo-force $-m\vec{a}_0$ acting on the particle of mass m. Once this pseudo-force is included, one can use Newton's laws in their usual form.

15. A body shows a weight only because the force of reaction acts on it. If a body falls freely under gravity, its apparent weight during its free fall is zero because there is no force of reaction acting on it.

16. The apparent weight of a body in a moving lift is usually different from its actual weight. Hence, three cases may arise:

 (i) If the lift is moving downward with acceleration a, then the reaction of the floor of the lift on the body is

$$R = mg\left(1 - \frac{a}{g}\right)$$

$$\Rightarrow \quad R < mg$$

 that is, the apparent weight of the body is less than its actual weight mg.

 If $a = g$, it gives $R = 0$ which means that the apparent weight of a body falling freely under gravity is zero.

 (ii) If the lift is moving upward with acceleration a, then the reaction of the floor of the lift on the body is

$$R = mg\left(1 + \frac{a}{g}\right)$$

$$\Rightarrow \quad R > mg$$

 That is, the apparent weight of the body is greater than its actual weight mg.

 (iii) If the lift moves in either direction with uniform velocity, then $R = mg$, i.e., the apparent weight of the body is equal to its actual weight mg.

17. When several bodies are connected by unstretchable strings passing over several pulleys, the motion of the bodies takes place under certain constraints. The length of the string remains constant and the string is in state of tension. The tensions in different strings will be different.

 The tension in different parts of a string will be the same if the string is light and the pulley is frictionless; otherwise it will be different.

18. When the motion of body is to considered, all the forces acting on that body, and not the forces applied by that body, should be taken into account.

19. Whenever a spring is compressed or stretched by x, a force equal to $-Kx$ is developed in the spring that tends to bring the spring back to its normal length. The negative sign shows that the force developed is in a direction opposite to x. Here K is the force constant of the spring and is defined as the force developed in the spring per unit compression or elongation.

COMMON MISCONCEPTIONS

1. Newton's first law

Newton's first law does not say what happens for an object with zero net force, that is, multiple forces that cancel; it says what happens in the absence of a force. This subtle but important difference allows us to define force as that which causes a change in the motion. The description of an object under the effect of forces that balance is covered by Newton's second law.

2. Force is the cause of changes in motion

Be sure that you are clear on the role of force. Many times, students make the mistake of thinking that force is the cause of motion. But we can have motion in the absence of forces, as described in Newton's first law. Be sure to understand that force is the cause of changes in motion.

3. Newton's second law

The protocol for using Newton's second law is to add all of the forces on the object vectorally and then find the resultant acceleration. Do not find individual acceleration for each force and then add the accelerations vectorally.

4. Weight of an object

We are familiar with the common phrase, the weight of an object. Keep in mind that, despite this everyday-phrase, weight is not a property of an object. It is a measure of the gravitational force between the object and the Earth. Thus, weight is a property of a system of items – the object and the Earth. As the system changes, for example, by lifting the object well off the surface, the weight changes, due to the decreasing value of g. We could also move the object into a new system, consisting of the object and Mars, for example. Then the weight will be very different from that on Earth because of the different value of g on Mars.

5. Newton's third law

This is such as important and often misunderstood notion that it will be repeated here in a misconception. Newton's third law action and reaction forces act on different objects. Two forces acting on the same object, even if they are equal in magnitude and opposite in direction, cannot be an action–reaction pair.

6. Free-body diagrams

The most important step in solving a problem using Newton's laws is to draw a proper simplified pictorial representation – the free-body diagram. Be sure to draw only those forces that act on the object you are isolating. Be sure to draw all forces acting on the object, including any field forces, such as the gravitational force.

PRACTICE QUESTIONS

Very-short Answer Type Questions

1. If a body is not in its rest position, then the net external force acting on it cannot be zero. Is this statement true, or false?
2. If the net force acting on a body be zero, then will the body remain necessarily in rest position?
3. The distance travelled by a moving body is directly proportional to time. Is any external force acting on it.
4. A person sitting in the compartment of a train moving with uniform speed throws a ball in the upward direction. What path of the ball will appear to him? What to a person standing outside?
5. If force is acting on a moving body perpendicular to the direction of motion, then what will be its effect on the speed and direction of the body?
6. A thief jumps from the roof of a house with a box of weight W on his head. What will be the weight of the box as experienced by the thief during the jump?
7. A force is always required to move a body uniformly. Comment.
8. Calculate the mass of a body weighing 100 dyne, Take $g = 10$ ms^{-2}.
9. Why does a gun recoil when a bullet is fired?
10. How is impulse related to linear momentum?
11. Calculate the force acting on a body whose linear momentum changes by 20 kg ms^{-1} in 10 s.
12. A force of 5 N changes the velocity of a body from 10 ms^{-1} to 20 ms^{-1} in 5 sec. How much force is required to bring about the same change in 2 sec?

Short-answer Type Questions

1. A bird is sitting on the floor of a wire cage and the cage is in the hand of a boy. The bird starts flying in the cage. Will the boy experience any change in the weight of the cage?
2. When a ball is thrown upwards, its momentum first decreases and then increases. Is conservation of linear momentum violated in this process?
3. An athlete runs a certain distance before talking a long jump. Why?
4. What is meant by law of inertia?
5. What is the function of shockers in a scooter?

Descriptive Type Questions

1. Give some illustrative examples of Newton's third law of motion.
2. What is lubrication. How does it reduce friction? What are the different types of frictional forces?
3. What are the different kind of forces depending on position and motion?

NCERT Corner

1. There are three solids made up of aluminum, steel and wood, of the same shape and same volume. Which of them would have the highest inertia?
2. Two balls of the same size but of different materials, rubber and iron, are kept on the smooth floor of a moving train. The brakes are applied suddenly to stop the train. Will the balls start rolling? If so, in which direction? Will they move with the same speed? Give reasons for your answer.
3. Two identical bullets are fired one by a light rifle and another by a heavy rifle with the same force. Which rifle will hurt the shoulder more and why?
4. A horse continues to apply a force in order to move a cart with a constant speed. Explain why?
5. Suppose a ball of mass m is thrown vertically upward with an initial speed v, its speed decreases continuously till it becomes zero. Thereafter, the ball begins to fall downward and attains the speed v again before striking the ground. It implies that the

magnitude of initial and final momentums of the ball are same. Yet, it is not an example of conservation of momentum. Explain why.

6. Velocity versus time graph of a ball of mass 50 g rolling on a concrete floor is shown in the figure below. Calculate the acceleration and frictional force of the floor on the ball.

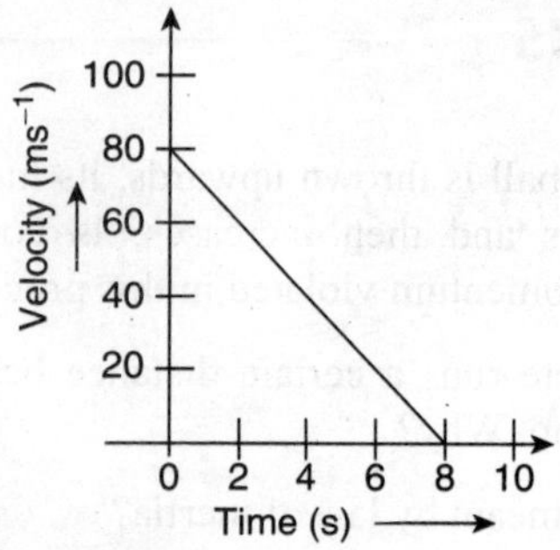

7. A truck of mass M is moved under a force F. If the truck is then loaded with an object equal to the mass of the truck and the driving force is halved, then how does the acceleration change?

8. Two friends on roller-skates are standing 5 m apart facing each other. One of them throws a ball of 2 kg towards the other, who catches it, How will this activity affect the position of the two? Explain your answer.

9. Water sprinkler used for grass lawns begins to rotate as soon as the water is supplied. Explain the principle on which it works.

10. Using second law of motion, derive the relation between force and acceleration. A bullet of 10 g strikes a sand-bag at a speed of 10^3 m s^{-1} and gets embedded after travelling 5 cm. Calculate:
(i) the resistive force exerted by the sand on the bullet
(ii) the time taken by the bullet to come to rest.

11. Derive the unit of force using the second law of motion. A force of 5 N produces an acceleration of 8 ms^{-2} on a mass m_1 and an acceleration of 24 ms^{-2} on a mass m_2. What acceleration would the same force provide if both the masses are tied together?

12. What is momentum? Write its SI unit. Interpret force in terms of momentum. Represent the following graphically:
(a) momentum versus velocity when mass is fixed
(b) momentum versus mass when velocity is constant.

Numericals

Level - I

1. A force acts for 8 seconds on a body of mass 10 kg. The force then stops acting and the body describes 80 m in next 5 s. Find the force.

2. A force of 0.04 newton acts upon a body. As a result, the speed of the body changes from 0.30 ms^{-1} to 0.10 ms^{-1}, in passing through a certain distance. Find the distance if the mass of the body is one kilogram.

3. A constant force acting on a body of mass 3.0 kg changes its speed from 2.0 m/s to 3.5 m/s in 25 s. The direction of motion of the body remains unchanged. What is the magnitude and direction of the force?

4. A truck starts from rest and rolls down a hill with constant acceleration. It travels a distance of 400 m in 20 s. Calculate the acceleration of the truck and force acting on it, if its mass is 7 metric ton.

5. A cricket ball of mass 150 gram moving with a speed of 12 ms^{-1} is hit by a bat so that the ball is turned back with a velocity of 20 ms^{-1}. Calculate the impulse received by the ball.

6. An automobile vehicle has a mass of 1500 kg. What must be the force between the vehicle and the road if the vehicle is to be stopped with a negative acceleration of 1.7 ms^{-2}?

7. A force of 9 N pulls a block of mass 4 kg through a rope of mass 0.5 kg. The block is resting on a frictionless surface. Calculate the force of reaction exerted by the block on the rope.

8. A uniform force acts on a body of mass 0.060 kg for 11 seconds and then ceases to act. During the next three seconds, the body travels a distance of 9.90 metre. Find the magnitude of the force.

9. Find the distance travelled by a body before coming to rest if it moving on the ground with a velocity of 10 m/sec and the coefficient of kinetic friction between the body and the ground is 0.2.

10. A body is rolling on the ground with a velocity of 1 m/s. After travelling a distance of 5 m, the body stops. Calculate the coefficient of friction.

11. A car is travelling at the rate of 70 km/h. Suddenly the brakes are applied causing all the tyres to skid. How far will the car travel before coming to rest? Given $\mu = 0.20$.

12. A force of 19.6 N, when applied parallel to the surface, just moves a body of mass 10 kg kept on a horizontal surface. If a 5 kg mass is kept on the first mass, what force should be applied parallel to the surface?

Level - II

1. A monkey is climbing on a rope that goes over a smooth light pulley and supports a mirror of equal mass at the other end at the monkey's level. Show that whatever force the monkey exerts on the rope, the monkey can never get away from his image seen in the mirror.

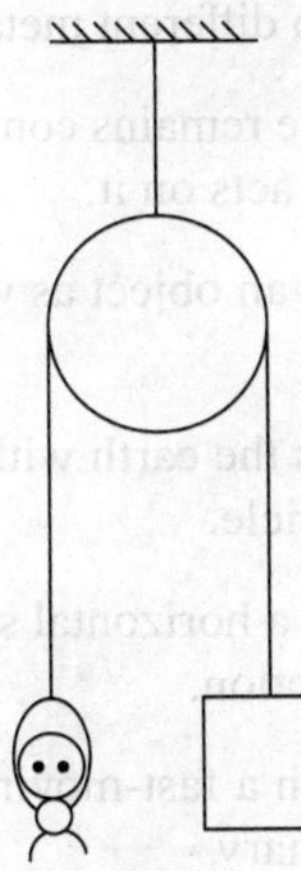

2. A mass A (500 g) is placed on a smooth table with a string attached to it. The string goes over a frictionless pulley and is connected to another mass B (200 g). At $t = 0$, the mass A is at a distance of 200 cm from the end and is moving with a speed of 50 cm/s towards the left. What will be its position and speed at $t = 1$s?

3. The total mass of an elevator, with an 80 kg man in it, is 1000 kg. This elevator moving upward with a speed of 8 m/s is brought to rest over a distance of 16 m. Calculate:
   (a) the tension T in the cable supporting elevator.
   (b) the force exerted on the man by the elevator floor.

4. A spring weighing machine inside a stationary lift reads 50 kg when a man stands on it. What would happen to the scale reading if the lift is moving upward with (i) constant velocity, and (ii) constant acceleration?

5. A particle of mass 0.40 kg moving initially with a constant speed of 10 m/s to the north is subject to a constant force of 8.0 N directed towards the south for 30 s. Take the instant the force is applied to be $t = 0$, the position of the particle at that time to be $x = 0$, and predict its position at $t = -5$ s, 25 s, 100 s.

6. A 20 kg box is pulled with uniform speed across a horizontal floor by a force directed 30° above horizontal. The coefficient of friction is 0.25. What is the magnitude of the force?

Fill in the Blanks

1. A dirty carpet can be cleaned by beating. This is in accordance with the ______ law of motion.
2. When a gun is fired, the bullet moves forward and the gun moves ______ with ______ having the same magnitude as that of the bullet.
3. If equal forces act on two bodies, one heavier than another, the heavier one will have ______ acceleration than that of the lighter one.
4. A man at rest in the middle of a pond on perfectly frictionless ice can get himself to the shore by taking the help of Newton's ______ law of motion.
5. Rate of change of momentum of a body is called ______.
6. For a rocket ejecting gases at a uniform speed, the thrust depends upon ______.
7. The force due to static friction is ______ than the force due to kinetic friction.
8. The force due to friction acts in an ______ to the direction of motion of the body.
9. Loss of weight of a body immersed in a liquid is equal to ______.
10. Swimming becomes possible because of ______ law of motion.
11. The action and reaction forces act on ______.
12. A heavier body having the same momentum will move with ______ speed compared to the speed of the lighter body.
13. If equal forces act on two bodies at rest, the heavier one will have ______ speed than the lighter one after 10 seconds.
14. The change of momentum of a body is called ______.
15. The total momentum of a system remains ______ provided ______ acts on the system.

16. Motor tyres have treads (corrugations) to ______ friction.

17. ______ friction has the least value.

18. A man lying over a wooden plank; then he stands over it. In the two cases he exerts same ______ but different ______.

19. A ______ spring pushes the objects attached to its ends.

20. The exerted force by the earth on a body is always in the ______ direction.

True or False

1. Rubbing reduces friction.
2. Dynamic friction is less than static friction.
3. Truck tyres are corrugated to increase friction and avoid sliding on the road.
4. Broad vehicle brakes exert more pressure.
5. On a block of iron immersed in water, force of buoyancy is more than the weight of the iron block.
6. Force is a vector quantity.
7. Mass is the inertia coefficient of a body.
8. The second law of motion measures force.
9. Gravitational thrust of a force have same value everywhere.
10. Slow action in catching a moving ball is less painful.
11. One can move over a frictionless ice surface by using the third law of motion.
12. Law of conservation of linear momentum is a consequence of the third law of motion.
13. Swimming can be explained using the third law of motion.
14. Too much smoothing reduces friction.
15. Sliding friction is less than rolling friction.
16. Flat top and pointed tip nails increase pressure.
17. Force of liquid displaced is called force of buoyancy.
18. Archimedes' principle can be applied to find percentage of two different metals in an alloy.
19. Speed of a particle remains constant. This means, no unbalanced force acts on it.
20. A spring can pull an object as well as it can push an object.
21. A particle attracts the earth with a force equal to the weight of the particle.
22. A ball moving on a horizontal surface stops because of the force of friction.
23. It is easier to catch a fast-moving ball with the arms kept nearly stationary.
24. A particle starts from rest under the action of a constant force. Here, the graph of distance versus time is a straight line.
25. Any pair of equal and opposite forces forms an action–reaction pair.

COMPETITION VISTA

Match the Columns

1. Match Column A with Column B.

Column A	Column B
(P) Displacement	(U) ms^{-1}
(Q) Speed	(V) newton
(R) Acceleration	(W) metre
(S) Force	(X) Ns
(T) Momentum	(Y) ms^{-2}

2. Match Column A with Column B.

Column A	Column B
(P) Dyne	(T) newton
(Q) Momentum	(U) CGS unit of force
(R) Limiting friction	(V) Mass × velocity
(S) Kg m s^{-2}	(W) Maximum static friction

3. Match Column A with Column B.

Column A	Column B
(P) Friction	(T) Newton
(Q) Coefficient of friction	(U) To reduce force of friction
(R) Unit of friction	(V) Opposition to the relative motion
(S) Lubricants	(W) Dimensionless and unit-less quantity

4. Match Column A with Column B.

Column A	Column B
(P) Weight	(T) N m^{-2}
(Q) Pressure	(U) Kg
(R) Impulse	(V) N s
(S) Mass	(W) Kg wt

Assertion-Reason Questions

In the following questions, a statement of Assertion (A) is followed by a statement of Reason (R). Mark the correct choice as:

(a) Both A and R are true and R is the correct explanation of A.
(b) Both A and R are true but R is NOT the correct explanation of A.
(c) A is true but R is false.
(d) A is false and R is true.

1. Assertion: No force is required to move a body uniformly along a straight line.
Reason: When the body is too light.

2. Assertion: Linear momentum of a body changes even when it is moving uniformly in a circle.
Reason: Force required to move a body uniformly along a straight line is zero.

3. Assertion: A rocket works on the principle of conservation of linear momentum.
Reason: Higher the velocity, smaller is the pressure and vice-versa.

4. Assertion: Friction is a self-adjusting force.
Reason: The magnitude of static friction is equal to the applied force and its direction is opposite to that of the applied force.

5. Assertion: A given force applied for a given time causes the same change in linear momentum of different bodies.
Reason: $F \times t = p_2 - p_1 =$ change in linear momentum = constant for all bodies.

6. Assertion: Forces of action and reaction do not cancel each other. Each produces its own effect.
Reason: Forces of action and reaction are equal and opposite.

7. Assertion: Force of friction increases when the surfaces in contact are too smooth.
Reason: Smoothness decreases friction.

8. Assertion: A table cloth can be pulled from a table without dislodging the dishes.
Reason: To every action, there is an equal and opposite reaction.

Comprehension Type Questions

1. A 8000 kg engine pulls a train of 5 wagons, each of 2000 kg along a horizontal track. If the engine exerts a force of 40000 N and the track offers a friction force of 5000 N, then calculate:
(i) the net acceleration force
(ii) the acceleration of the train and
(iii) the force of wagon 1 on wagon 2.

2. Two 100 g blocks hang at the ends of a light flexible cord passing over a small frictionless pulley. A 40 g block rests on the block on the right and is removed after 2 sec.
(a) How far will each block move in the first second after the 40 g block is removed?
(b) What was the tension in the cord before the 40 g block was removed?
(c) What was the tension in the cord supporting the pulley after the 40 g block was removed?

3. (i) Name the property of bodies to resist a change in their velocity.
(ii) What is the relationship between force and acceleration?
(iii) What name is given to the product of mass and velocity of a body?
(iv) Which physical quantity corresponds to the rate of change of momentum?
(v) Name the principle on which a rocket works.

4. The speed–time graph of a car of 1000 kg mass is given. Based on this, answer the following questions:

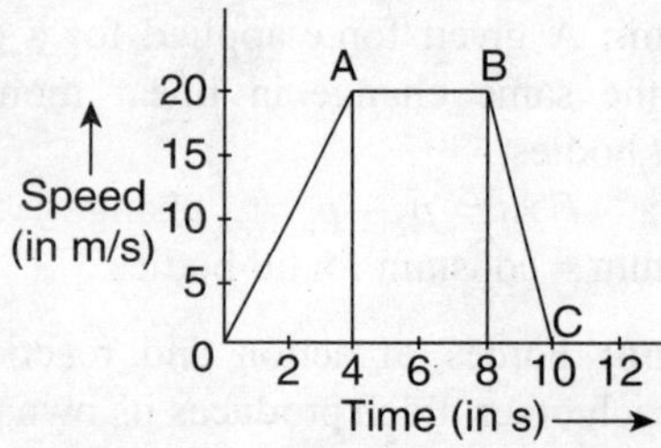

(i) When is the maximum force acting on car?
(ii) What is the retarding force acting on the car?
(iii) For how long is no force is acting on the car?
(iv) What is the velocity of the car after 4 s?
(v) Find the acceleration of the car during each of the first two intervals of four second each.

5. The motion of a body of mass 5 kg is shown in the velocity–time graph.

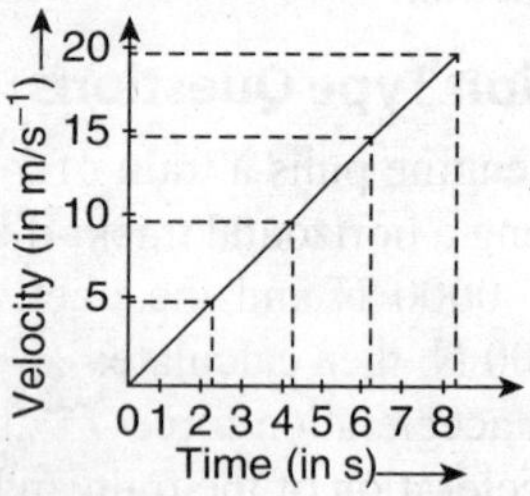

Find from the graph:
(i) its acceleration
(ii) the force acting on the body
(iii) the change in momentum of body in 2 s after start.

Objective Questions

One Option Correct

1. An unbalanced force acts on a body. Then the body:
(a) must remain at rest
(b) must move with uniform velocity
(c) must be accelerated
(d) must move along a circle.

2. If no force acts on a body, it will:
(a) change its shape
(b) move with an increased speed
(c) either remain at rest or move in a straight line
(d) break up.

3. By applying a force of 1 N, one can hold a body of approximate mass:
(a) 100 mg (b) 100 g
(c) 1 kg (d) 10 kg

4. The force of friction between two bodies is
(a) parallel to the contact surface
(b) perpendicular to the contact surface
(c) inclined at 30° to the contact surface
(d) inclined at 60° to the contact surface.

5. The speed of a falling body increases continuously. This is because:
(a) no force acts on it
(b) it is very light
(c) the air exerts a frictional force to it
(d) the earth attracts it.

6. When a bus starts, the passengers are pushed back. This is an example of:
(a) Newton's first law
(b) Newton's second law
(c) Newton's third law
(d) None of Newton's laws.

7. A force of a given magnitude acts on a body. The acceleration of the body depends on the
(a) mass of the body (b) volume of the body
(c) density of the body (d) shape of the body.

8. If a constant force acts on a body at rest, the distance moved by the body in time t will be proportional to ______.
(a) t (b) t^2
(c) t^3 (d) t^4

9. The momentum of a body of given mass is proportional to its ______.
(a) volume (b) shape
(c) speed (d) colour

10. The principle of conservation of linear momentum states that the linear momentum of a system
(a) cannot be changed
(b) cannot remain constant
(c) can be changed only if internal forces act
(d) can be changed only if external forces act.

11. Action–reaction forces
(a) act on the same body
(b) act on different bodies
(c) act along different lines
(d) act in the same direction.

12. Consider a porter standing on a platform with a suitcase which presses his head with a force of 200 N. If this force is taken to be action, then the reaction force is exerted by:
(a) the head on the suitcase
(b) the earth on the suitcase
(c) the earth on the porter
(d) the suitcase on the earth.

13. A force produces an acceleration of 0.5 m /s^2 in a body of mass 3.0 kg. If the same force acts on a body of mass 1.5 kg, the acceleration produced in it is:
(a) 3.0 m/s^2 (b) 1.0 m/s^2
(c) 5.0 m/s^2 (d) 7 m/s^2

14. A force produces an acceleration of 5.0 cm/s^2 when it acts on a body of mass 20 g. The force in newton is:
(a) 2×10^{-3} N (b) 4×10^{-3} N
(c) 1.0×10^{-3} N (d) 5×10^{-3} N

15. A force produces an acceleration of 2.0 m/s^2 in a body 'A' and 5.0 m/s^2 in another body 'B'. The ratio of the mass of 'A' to the mass of 'B' is:
(a) 2.5 (b) 3.5
(c) 5.5 (d) 4.5

16. A force acts on a particle of mass 200 g. The velocity of the particle changes from 15 m/s to 25 m/s in 2.5 s. Assuming the force to be constant, it magnitude is:
(a) 0.4 N (b) 0.6 N
(c) 0.8 N (d) 0.5 N

17. A force of 1.0 N acts on a body of mass 10 kg. The body covers 100 cm in 4 seconds moving along a straight line. The initial velocity is ______.
(a) 2 cm/s (b) 4 cm/s
(c) 6 cm/s (d) 5 cm/s

18. A force of 0.6 N acting on a particle increases its velocity from 5.0 m/s to 6.0 m/s in 2 sec. The mass of the particle is ______.
(a) 1.2 kg (b) 2.2 kg
(c) 4.2 kg (d) 6.2 kg

19. A force acting on a particle of mass 200 g displaces it through 400 cm in 2 seconds. The magnitude of the force, if the initial velocity of the particle is zero, is ______.
(a) 0.1 N (b) 0.3 N
(c) 0.4 N (d) 0.5 N

20. A body of mass 300 g, kept at rest, breaks into two parts due to internal forces. One part of mass 200 g is found to move at a velocity of 12 m/s towards east. The velocity of the other part is ______.
(a) 24 m/s towards west
(b) 14 m/s towards east
(c) 24 m/s towards north
(d) 54 m/s towards south

21. A force of 16 N is distributed uniformly on one surface of a cube of edge 8 cm. The pressure on this surface is ______.
(a) 3500 Pa (b) 2500 Pa
(c) 4500 Pa (d) 5500 Pa

22. Friction exerted by surfaces on each other when they are at rest is called ______.
(a) static friction (b) kinetic friction
(c) sliding friction (d) None of the above.

23. A body (without airspaces) of mass 150 gm and volume 250 cm^3 is put in water. The body will:
(a) float (b) sink
(c) partially sink (d) None of these.

24. A mass of 2 kg at rest travels for 4 seconds with an acceleration of 1.5 m /s^2. The gain of momentum of the body is:
(a) 5 kg m/s (b) 10 kg m/ s
(c) 12 kg m/s (d) 14 kg m/s

25. Two forces of 5 N and 3 N act on a body in opposite directions in a straight line as shown here. The resultant force is

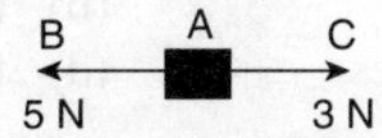

(a) 3 N in the direction of AB
(b) 2 N in the direction of AB
(c) 4 N in the direction of AC
(d) 5 N in the direction of AC.

26. A 6 kg mass 'A' moving with a velocity of 2 m/s collides with a 4 kg mass 'B' moving with a velocity of 1.5 m/s in the opposite direction in a straight line. If the two masses get stuck, then the velocity of the combination is ______.
(a) 0.4 m/s (b) 0.2 m/s
(c) 0.1 m/s (d) 0.6 m/s

27. Two bodies collide at the same temperature. Which of the following is conserved?
(a) Velocity (b) Momentum
(c) Kinetic energy (d) All of the above.

28. Impulse is equal to:
(a) the change of momentum
(b) the change of velocity
(c) the change of force
(d) the change of position.

29. In the S.I. System, momentum is expressed as ______.
(a) kgs (b) kgms^{-1}
(c) kgm^{-1}s^{-1} (d) kgms^{-2}

30. The property by which a body tends to retain its original state of rest or uniform motion in a straight line is called ______.
(a) impulse
(b) inertia
(c) instantaneous velocity
(d) momentum

31. 'Rate of change of momentum is proportional to the exerted force and takes place in the direction in which the force acts' is a statement of Newton's:
(a) first law of motion
(b) second law of motion
(c) third law of motion
(d) all the three laws of motion.

32. Which of the following are vector quantities?
(a) Force (b) Velocity
(c) Momentum (d) All of the above.

33. Due to an impulse, the change in momentum of a body is 1.8 kg ms^{-1}. If the duration of the impulse is 0.2 sec, the force produced in it is ____.
(a) 9 N (b) 8 N
(c) 7 N (d) 6 N

34. Action and reaction take place on ____.
(a) one body (b) two bodies
(c) three bodies (d) four bodies

35. Vehicles are provided with shock absorbers, which obey Newton's:
(a) first law of motion
(b) second law of motion
(c) third law of motion.
(d) All of the above.

36. A jet engine works on the principle of
(a) conservation of linear momentum
(b) conservation of kinetic energy
(c) conservation of angular momentum
(d) conservation of inertia.

37. A ball of mass 100 gm is moving with a velocity of 10 ms^{-1}. The force of the blow by the bat acts for 0.01 seconds. The average force exerted on the ball by the bat is ____.
(a) 100 N (b) 200 N
(c) 300 N (d) 400 N

38. Passengers standing in a bus are thrown outwards when the bus takes a sudden turn. This happens due to:
(a) outwards pull on them
(b) inertia
(c) change in momentum
(d) change in acceleration.

39. Ratio of force to acceleration measures
(a) velocity (b) impulse
(c) momentum (d) mass.

40. When we kick a stone, we get hurt. Due to which of the following properties of the stone does it happen?
(a) Inertia (b) Velocity
(c) Reaction (d) Momentum

41. A body is at rest on the surface of earth. Which of the following statements is correct?
(a) Only weight of the body acts on it.
(b) Only upward force acts on the body.
(c) Frictional force acts on the body.
(d) Net upward force is equal to the net downward force.

42. A book is lying on a table. What is the angle between the action of the book on the table and the weight of the book?
(a) 0° (b) 45°
(c) 90° (d) 180°

43. In which of the following cases is the net force not equal to zero?
(a) A kite skillfully held stationary in the sky
(b) A ball falling freely from a height
(c) An aeroplane rising upwards at an angle of 45° with the horizontal with a constant speed
(d) A cork floating on the surface of water.

44. An electric fan is placed on a stationary boat and air is blown with it on the sail of the boat. Which of the following statements is correct?
(a) The boat will start moving with uniform speed.
(b) The boat will be uniformly accelerated in the direction of the flow of air.
(c) The boat will be uniformly accelerated opposite to the direction of flow of air.
(d) The boat will remain stationary as before.

45. When the total force acting on the system of particles is zero the linear momentum of the system of the particles is ____.
(a) positive
(b) negative
(c) either positive or negative
(d) zero

46. Let E, G and N represent the magnitudes of electromagnetic, gravitational and nuclear forces between two electrons at a given separation Then,
(a) N > E > G (b) E > N > G
(c) G > N > E (d) E > G > N

47. A 60 kg man pushes a 40 kg man by a force of 60 N. Then the 40 kg man has pushed the other man with a force of ____.
(a) 40 N (b) 0
(c) 60 N (d) 20 N

48. A body of weight W_1 is suspended from the ceiling of a room through a chain of weight W_2. The ceiling pulls the chain by a force of:

(a) W_1 (b) W_2

(c) $W_1 + W_2$ (d) $\frac{W_1 + W_2}{2}$

49. When a horse pulls a cart, the force that helps the cart to move forward is the forcer exerted by:
(a) the cart on the horse
(b) the ground on the horse
(c) the ground on the cart
(d) the horse on the ground.

50. A car accelerates on a horizontal road due to the force exerted by:
(a) the engine of the car (b) the driver of the car
(c) the earth (d) the road.

51. While walking on ice, one should take small steps to avoid slipping. This is because smaller steps ensure ________.
(a) larger friction
(b) smaller friction
(c) larger normal force
(d) smaller normal force

52. Newton's first law of motion gives the concept of:
(a) energy (b) work
(c) momentum (d) inertia.

53. Inertia of motion depends on:
(a) mass (b) velocity
(c) acceleration (d) displacement.

54. A man drawing a bucket of water falls back when the rope snaps. This is because of:
(a) law of gravitation
(b) Newton's I law of motion
(c) Newton's II law of motion
(d) Newton's III law of motion.

55. Impulse has the same unit as that of:
(a) momentum (b) force
(c) torque (d) couple.

56. If there is an option for the soldiers to use rifles of different weights but with bullets of a fixed weight, they would prefer:
(a) light guns, because handling them is easy
(b) heavy guns so that they can be held firmly
(c) heavy guns because they have less recoil.
(d) light guns so that they can be carried easily.

57. A rider on the horse falls forward when the horse stops suddenly. This is due to:
(a) inertia of rest (b) inertia of the horse
(c) inertia of the rider (d) loss of balance.

58. A goalkeeper in a game of football pulls his hands backwards after holding the ball shot at the goal. This enables the goal keeper to
(a) exert larger force on the ball
(b) reduce the force exerted by the ball on hands
(c) increase the rate of change of momentum
(d) decrease the rate of change of momentum.

59. The inertia of an object tends to cause the object
(a) to increase its speed
(b) to decrease its speed
(c) to resist any change in its state of motion .
(d) to decelerate due to friction.

60. Two bodies of equal masses move with uniform velocities v and $4v$ respectively. Find the ratio of their K.E.
(a) 1 : 4 (b) 1 : 16
(c) 1 : 9 (d) 1 : 1

61. A water tanker filled up to $\frac{2}{3}$ of its height is moving with a uniform speed. On sudden application of the brake, the water in the tank would _____.
(a) move backward (b) move forward
(c) be unaffected (d) rise upwards

62. A boy is standing in a lift. The force on the floor of the lift due to the weight of the boy becomes zero
(a) when the lift moves up with an acceleration of 9.8 ms^{-2}.
(b) when the lift moves down with an acceleration 9.8 ms^{-2}
(c) when the lift is not moving.
(d) None of these.

63. A passenger in a moving train tosses a coin which falls behind him. It means that motion of the train is ______.
(a) accelerated
(b) uniform
(c) retarded
(d) along circular tracks

64. An object of mass 2 kg is sliding with a constant velocity of 4 ms^{-1} on a frictionless horizontal table. The force required to keep the object moving with the same velocity is ______.
(a) 32 N (b) 0 N
(c) 2 N (d) 8 N

65. A bullet of mass m moving with a speed v strikes a wooden block of mass M and gets embedded in it. The speed of this embedded block will be:

(a) $\left(\sqrt{\frac{M}{M+m}}\right)v$ (b) $\left(\sqrt{\frac{m}{M+m}}\right)v$

(c) $\left(\frac{m}{M+m}\right)v$ (d) $\left(\frac{m+M}{Mm}\right)v$

66. If we know the magnitude and direction of force exerted on a body of given mass, Newton's II law of motion permits us to calculate its:
(a) speed (b) momentum
(c) position (d) acceleration

67. A body whose momentum is constant must always
(a) be accelerated
(b) be decelerated
(c) be moving in a circle
(d) be having constant velocity.

68. If a bomb at rest explodes and splits into two equal fragments, these fragments will move in:
(a) one direction and with the same velocity
(b) opposite directions with different velocities
(c) opposite directions with unequal speeds
(d) opposite directions with equal speed.

69. A lift is descending with a constant velocity v. A man in the lift drops a coin. The coin experiences acceleration towards the floor equal to:
(a) $g + v$ (b) $g - v$
(c) g (d) zero.

70. If a book is pressed against a wall using a horizontal force you can avoid its falling down. The frictional force along the wall in the upward direction acts, if the book moves, down. Suppose a block of weight W is held against a vertical wall by applying a horizontal force F, the minimum value of F needed to hold the block would be _______.
(a) less than W
(b) equal to W
(c) greater than W
(d) It cannot be determined.

71. A body of mass 20 g moving horizontally with a velocity 100 cm s^{-1} strikes a pendulum bob of mass 20 g and both of them move together. Estimate the height to which the system moves up. (Assume g = 10 m/s^2)
(a) 1.25 cm (b) 12.5 cm
(c) 0.125 cm (d) 125 cm

The following table shows the data recorded for a cyclist. Study it carefully and answer the questions from 72 to 74.

Time in (hrs)	Distance (km)
0	0
1	0.5
2	2
3	4.5
4	8
5	12.5
6	18
7	24.5
8	32.0

72. The cyclist's speed is/can _______.
(a) uniform (b) non-uniform
(c) be both. (d) not be determined.

73. The cyclist moves with:
(a) uniform speed
(b) non-uniform speed
(c) constant acceleration
(d) variable acceleration.

74. The graph between t and v should be a:
(a) straight line parallel to X-axis
(b) straight line inclined to X-axis and originating from origin
(c) straight lien originating from Y-axis and inclined to X-axis
(d) straight line parallel to Y-axis.

Following is the data recorded for a car in a race. Study it carefully and answer the questions from 75 to 76.

Time in (hrs)	Distance (km)
0	0
1	1
2	8
3	27
4	64
5	125
6	216
7	343

75. The motion is:
(a) uniform (b) non-uniform
(c) Both. (d) None.

76. The acceleration is:
(a) uniform (b) non-uniform
(c) Both. (d) None.

77. A number of forces acting on a body do not cause any change in its state of rest or of uniform motion. The forces are ______.
(a) parallel (b) unbalanced
(c) balanced (d) inclined

78. Balanced forces ______ a body.
(a) move (b) accelerate
(c) retard (d) deform

79. Newton's first law of motion is
(a) qualitative
(b) quantitative
(c) both qualitative and quantitative
(d) neither qualitative nor quantitative.

80. S.I. unit of force is ______.
(a) metre/sec (b) sec/metre
(c) dyne (d) newton

81. Inertia measures the amount of ______ in a body.
(a) mass (b) motion
(c) velocity (d) acceleration

82. Force measures rate of change of ______ of a body.
(a) mass (b) inertia
(c) velocity (d) momentum

83. Forces of action and reaction are
(a) equal in same direction
(b) equal and opposite
(c) unequal in same direction
(d) unequal and opposite.

84. Due to action and reaction forces involved in the interaction of two bodies in a system, the total momentum of the system
(a) remains constant (b) becomes zero
(c) become less (d) becomes more.

85. Friction has
(a) advantages only
(b) disadvantages only
(c) both advantages as well as disadvantages
(d) neither advantages nor disadvantages.

86. Loss of weight of a body immersed in a liquid is
(a) equal to the weight of liquid displaced
(b) more than the weight of liquid displaced
(c) less than the weight of liquid displaced
(d) zero.

87. For moving a body from rest or stopping a body in motion, we need ______.
(a) force (b) mass
(c) direction (d) time

88. A number of forces acting on a body change the velocity of the body. The forces are ______.
(a) parallel (b) unbalanced
(c) balanced (d) inclined

89. The first law of motion defines ______.
(a) inertia
(b) force
(c) both inertia and force
(d) neither inertia nor force

90. Newton's second law of motion is
(a) qualitative
(b) quantitative
(c) both qualitative and quantitative
(d) neither qualitative nor quantitative.

91. C.G.S. unit of force is ______.
(a) metre/sec (b) sec/metre
(c) dyne (d) newton

92. Momentum measures amount of ______ in a body.
(a) mass (b) motion
(c) velocity (d) acceleration

93. Impulse measures change of ______ of a body.
(a) mass (b) inertia
(c) momentum (d) velocity

94. Forces of action and reaction act
(a) one after the other on same body
(b) simultaneously on same body
(c) one after the other on different bodies
(d) simultaneously on different bodies.

95. The force of friction
(a) has to be applied
(b) has unlimited value
(c) is self-acting and fixed
(d) is self-acting and self-adjusting.

96. A rectangular iron block is kept over a table with different forces touching the table. In different cases, the block exerts
(a) same thrust and same pressure
(b) same thrust and different pressure
(c) different thrust and same pressure
(d) different thrust and different pressure.

97. Which one of the following statements regarding Newton's first law of motion is correct?
(a) It is an independent statement.
(b) It defines an inertial frame of reference.
(c) It was first enunciated by Galileo.
(d) None of these.

98. Physical independence of force is a consequence of Newton's
(a) third law of motion
(b) second law of motion
(c) first law of motion
(d) All of these laws.

99. A person is sitting in a lift accelerating upwards. The measured weight of the person will be
(a) less than actual weight
(b) equal to actual weight
(c) more than actual weight.
(d) None of these.

100. Which is a non-central force?
(a) Electrostatic force
(b) Nuclear force
(c) Gravitational force
(d) None of these.

101. In the arrangement shown in the figure, the strings are light and inextensible. The surface over which blocks are placed is smooth. What is the acceleration of each block?

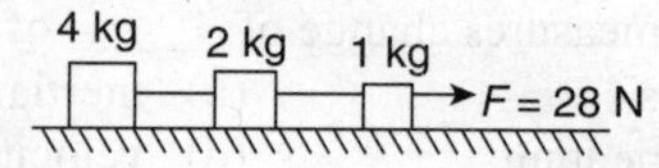

(a) 8 m/s^2 (b) 4 m/s^2
(c) 2 m/s^2 (d) 14 m/s^2

102. A body is executing uniform circular motion. At any instant its velocity and acceleration vector will be
(a) along the same direction
(b) in opposite directions
(c) normal to each other
(d) not related to each other.

103. The spring balance inside a lift suspends an object. As the lift begins to ascend, the reading indicated by the spring balance will
(a) increase
(b) decrease
(c) remain unchanged
(d) depend on the speed of ascend.

104. According to the third law of motion, action and reaction
(a) always act on the same body
(b) always act on different bodies in opposite directions·
(c) have the same magnitude and direction
(d) act on either body at normal to each other.

105. Which of the following statement is not correct for an object moving along a straight path in an accelerated motion?
(a) Its speed keeps changing.
(b) Its velocity always changes.
(c) It always goes away from the earth.
(d) A force is always acting on it.

106. A block moving on a surface with velocity 20 m/s comes to rest because of surface friction over a distance of 40 m. Taking $g = 10 \text{ m/s}^2$, the coefficient of dynamic friction is ______.
(a) 0.5 (b) 0.3
(c) 0.2 (d) 0.1

107. A pulley has two different arrangements (i) and (ii) as shown. Neglecting the masses of the rope and the pulley, the ratio of the acceleration of the mass M in the arrangement (i) to that in the arrangement (ii) is ______.

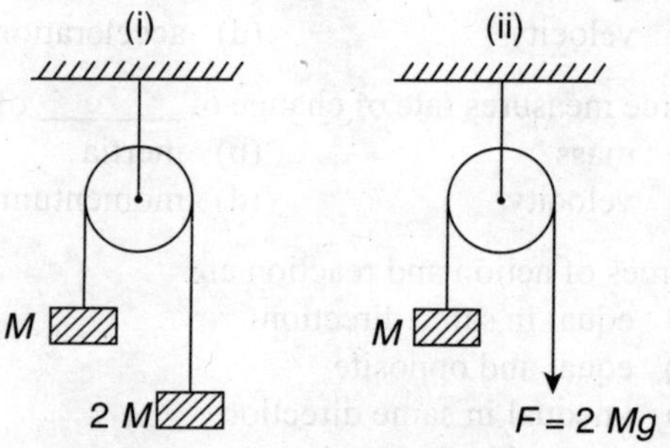

(a) 3 : 1 (b) 2 : 1
(c) 1 : 2 (d) 1 : 3

108. A 600 kg rocket is set for a vertical firing. If the exhaust speed is 1000 ms^{-1}, the mass of the gas ejected per second to supply the thrust needed to overcome the weight of rocket is ______.
(a) 117.6 kgs^{-1} (b) 58.6 kgs^{-1}
(c) 6 kgs^{-1} (d) 76.4 kgs^{-1}

109. A force of 6 N acts on a body at rest and of mass 1 kg. During this time, the body attains a velocity of 30 m/s. The time for which the force acts on the body is ______.

(a) 7 seconds (b) 5 seconds
(c) 10 seconds (d) 8 seconds

110. Pulling is easier than pushing because:
(a) Force of friction is in opposite direction.
(b) When we pull a roller, the vertical component of the pulling force acts in the direction of weight.
(c) It is possible in the case of roller only.
(d) The vertical component of the pulling forces acts in the opposite direction of weight.

111. A 1.5 kg ball drops vertically on a floor hitting with a speed of 25 m/s. It rebounds with an initial speed of 15 m/s. If the ball was in contact for only 0.03 second, the force exerted on the floor by the ball is ______.
(a) 2000 N (b) 3000 N
(c) 3500 N (d) 4000 N

112. Two blocks of masses M and m are connected by a string passing over a pulley as shown in the figure. The downward acceleration of the block with mass m is ______.

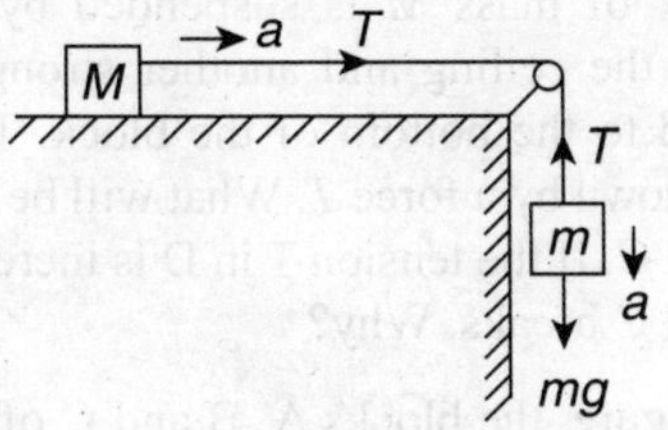

(a) $\frac{M}{(m+M)g}$ (b) $\frac{mg}{(m+M)}$

(c) $\frac{(m+M)}{mg}$ (d) $\frac{(m+M)}{mg}$

113. You are on a frictionless horizontal plane. How can you get off it so that no horizontal force is exerted by pushing against the surface?
(a) By jumping
(b) By spitting and sneezing
(c) By rolling your body on the surface
(d) By running on the plane.

114. A body is moving along a rough horizontal surface with an initial velocity 6 m/s. If the body comes to rest after travelling a distance 9 m, then the coefficient of sliding friction will be ______.
(a) 0.4 (b) 0.2
(c) 0.6 (d) 0.8

115. While launching a rocket of mass 2×10^4 kg, a force of 5×10^5 N is applied for 20 second. Calculate the velocity attained by the rocket at the end of 20 second.
(a) 300 ms^{-1} (b) 320 ms^{-1}
(c) 500 ms^{-1} (d) 100 ms^{-1}

More Than One Option Correct

1. Which of the following forces obey inverse square law?
(a) Nuclear forces
(b) Weak force
(c) Gravitational force
(d) Electromagnetic forces

2. When a force is applied to a moving body,
(a) both momentum and KE change
(b) neither momentum nor KE changes
(c) KE changes, but not the momentum
(d) momentum changes, but not the KE.

3. When a bicycles is in motion, the force of friction exerted by the ground on the two wheels is such that it acts
(a) in backward direction on both the wheels
(b) in the forward direction on both wheels
(c) in backward direction on front wheel and in forward direction on rear wheel
(d) in forward direction on front wheel and in backward direction on rear wheel.

4. In which of the following cases, net force is zero?
(a) A drop of rain falling down with terminal velocity
(b) a cork of mass 5 gram floating on water
(c) A car moving with constant speed on a rough road
(d) A charged particle moving parallel to magnetic field.

5. Action and reaction:
(a) act on two different objects
(b) have opposite direction
(c) have equal magnitude
(d) have zero resultant.

6. If the tension in the cable supporting an elevator is equal to the weight of the elevator, the elevator may be
(a) going up with uniform speed
(b) going down with uniform speed
(c) going up with increasing speed
(d) going down with increasing speed.

Scholar Zone

1. In the following example, try to identify the number of times the velocity of the ball changes. A football player kicks football towards his team. Another player kicks the ball to the goal. The goalkeeper of opposite team collects the ball and kicks it towards a player of his own team. Also identify the agent supplying the force in each case.

2. Two blocks of masses 2.9 kg and 1.9 kg are suspended from a rigid support S by two inextensible wires, each of length 1 m, as shown in the figure.

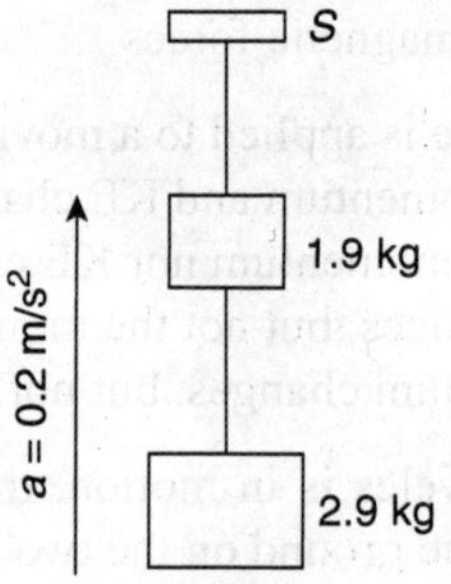

The upper wire has negligible mass and the lower wire has a uniform mass of 0.2 kg/m. The whole system of blocks, wires and support have an upward acceleration of 0.2 m/s^2.
   (i) Find the tension at the mid-point of the lower wire.
   (ii) Find the tension at the mid-point of the upper wire.

3. A force-time graph for a linear motion of a body is shown in the figure. Find the change in linear momentum between 0 s and 7 s.

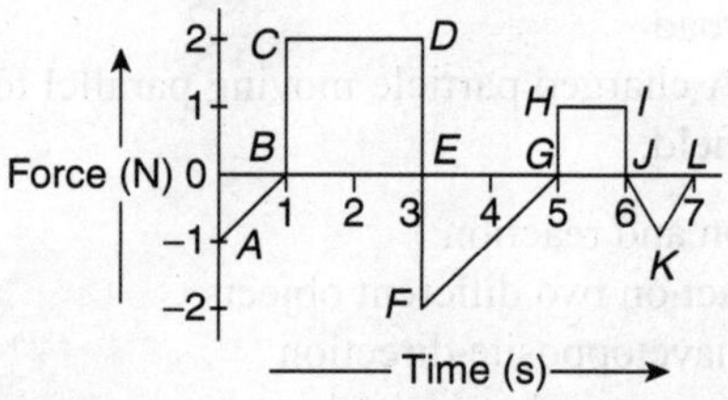

4. In the absence of air friction, it is claimed that all objects fall with the same acceleration. A heavier object is pulled to the Earth with more force than a light object. Why does the heavier object not fall faster?

5. Suppose you are talking by interplanetary telephone to your friend, who lives on the Moon. He tells you that he has just won a piece of gold weighing one newton in a contest. Excitedly, you tell him that you entered the Earth version of the same contest and also won a newton of gold! Who is richer?

6. In a tug of war, three men pull on a rope to the left at A and three men pull to the right at B with forces of equal magnitude. Now a weight of 5 kg is hung vertically from the centre of the rope.
   (a) Can the men get the rope AB to be horizontal? If so, determine the magnitude of the forces required at A and B to do this. If not, explain.

7. A large man and a small boy stand facing each other on frictionless ice. They put their hands together and push toward each other, so that they move away from each other. Who exerts the larger force? Who experiences the large acceleration? Who moves away with higher velocity? Who moves over a longer distance while their hands are in contact?

8. A block of mass M is suspended by a light cord C from the ceiling and another stronger cord D is attached to the bottom of the block. The cord D is pulled down by a force T. What will be the tension in the cord C. If the tension T in D is increased steadily, the cord C breaks. Why?

9. In the figure, the blocks A, B and C of mass m each have acceleration a_1, a_2, and a_3 respectively. F_1 and F_2 are external forces of magnitude 2 mg and mg respectively. Then find out the acceleration a_1, a_2, a_3 in increasing order.

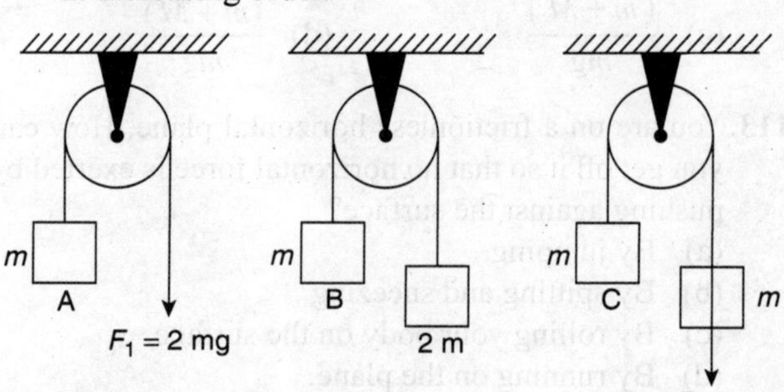

Answer Keys

Objective Questions

One Option Correct

1. (c) **2.** (c) **3.** (b) **4.** (a) **5.** (d) **6.** (a) **7.** (a) **8.** (b) **9.** (c) **10.** (a)
11. (b) **12.** (c) **13.** (b) **14.** (c) **15.** (a) **16.** (c) **17.** (d) **18.** (a) **19.** (c) **20.** (a)
21. (b) **22.** (a) **23.** (a) **24.** (c) **25.** (b) **26.** (d) **27.** (b) **28.** (a) **29.** (b) **30.** (b)
31. (b) **32.** (d) **33.** (a) **34.** (b) **35.** (c) **36.** (a) **37.** (b) **38.** (b) **39.** (d) **40.** (c)
41. (c) **42.** (a) **43.** (b) **44.** (d) **45.** (d) **46.** (d) **47.** (c) **48.** (c) **49.** (b) **50.** (d)
51. (d) **52.** (d) **53.** (a) **54.** (d) **55.** (a) **56.** (c) **57.** (c) **58.** (b) **59.** (c) **60.** (b)
61. (b) **62.** (b) **63.** (a) **64.** (b) **65.** (c) **66.** (d) **67.** (d) **68.** (d) **69.** (c) **70.** (a)
71. (a) **72.** (b) **73.** (c) **74.** (b) **75.** (b) **76.** (b) **77.** (c) **78.** (d) **79.** (a) **80.** (d)
81. (a) **82.** (d) **83.** (b) **84.** (a) **85.** (c) **86.** (a) **87.** (a) **88.** (b) **89.** (c) **90.** (b)
91. (c) **92.** (b) **93.** (c) **94.** (d) **95.** (d) **96.** (b) **97.** (c) **98.** (b) **99.** (c) **100.** (b)
101. (b) **102.** (c) **103.** (a) **104.** (b) **105.** (c) **106.** (a) **107.** (d) **108.** (c) **109.** (b) **110.** (d)
111. (a) **112.** (b) **113.** (b) **114.** (b) **115.** (c)

More Than One Option Correct

1. (c) and (d) **2.** (a) and (d) **2.** (a) and (c) **4.** (a), (b), (c) and (d) **5.** (a), (b) and (c)
6. (a) and (b)

Hints and Solutions

Very-short Answer Type Questions

1. False. It is possible that the body may be moving with a constant velocity and net external force acting on the body is zero.

2. No, the body may be moving uniformly along a straight line.

3. When $s \propto t$, acceleration = 0. Therefore, no external force is acting on the body.

4. Vertically downwards; parabolic to a person standing outside.

5. No change in speed, but change in direction is possible.

6. Weight of the box during jump ,

$$W' = m(g-a) = m(g-g)$$
$$= \text{zero.}$$

7. The statement is wrong when we assume that there is no force of friction and air resistance, etc. In that case, force is required only for producing acceleration in the body.

8. From $W = mg$,

$$m = \frac{W}{g}$$

$$= \frac{100 \text{ dyne}}{1000 \text{ cm s}^{-2}} = 10^{-1}\text{g}$$

9. The recoiling of a gun is accounted for by the principle of conservation of linear momentum. The total linear momentum of the gun and the bullet on firing is equal to their total linear momentum before firing, which is zero (as both are at rest initially).

10. Impulse = Change in linear momentum of the body.

11. Force = Rate of change of linear momentum

$$= \frac{20}{10} = 2 \text{ N}$$

12. Initial velocity, $u = 10$ m/s

Final velocity, $v = 20$ m/s

Time taken, $t = 5$ sec

Use formula, $v = u + at$

$$20 = 10 + 5a \Rightarrow a = 2 \text{ m/s}^2$$

We know from Newton's 2nd law of motion,

$$F = ma$$
$$5 = m \times 2 \Rightarrow m = 2.5 \text{ kg}$$

Now we have to find acceleration when the velocity changes from 10m/s to 20m/s in just 2 sec.

so, $a = (20 - 10)/2 = 5 \text{ m/s}^2$

Now force , $F' = ma = 2.5 \times 5 = 12.5$ N

Short-answer Type Questions

1. In a wire cage, air inside is in free contact with atmospheric air. Therefore, when the bird starts flying inside the cage the weight of bird is no more experienced. Hence, the cage will appear lighter than before.
2. No, the momentum conservation principle is not violated. This is because vector sum of linear momenta of the ball and earth remains constant.
3. An athlete runs a certain distance before talking a long jump. This is because, the velocity acquired by running is added to the velocity of the athlete at the time of jump. Hence he can jump over a longer distance.
4. In the absence of an external force, a body cannot change, on its own, its state of rest or the state of uniform motion along a straight line. This statement is called Galileo's Law of Inertia.
5. When the scooter move over an uneven road, impulsive forces are exerted by the road. The function of shocker is to increase the time of impulse. This would reduce the force/jerk experienced by the rider of the vehicle.

Descriptive Type Questions

1. Following are few examples of Newton's third law of motion:
   (i) Rotating water fountains in the gardens are the examples of Newton's third law of motion.
   (ii) When a horse pulls a cart, the force exerted by horse on cart is always equal and opposite to the force exerted by cart on the horse. The cart moves forward because the action and reaction though equal and opposite but they act on different bodies.
   (iii) When a man jumps on the shore from a boat, the boat pushes backward due to reaction force.
   (iv) When a bullet is fired from the gun, the bullet moves forward and gun pushes backward due to reaction force.
   (v) When a ball falls on earth, the earth exerts a force on the ball. Due to this force the ball suffers acceleration equal to g. The ball also exerts same amount of force ($m \times g$, m being the mass of the body) on the earth but it does not appear to move due to its larger mass.
   (vi) When a man walks on a surface, the man pushes the surface backward while the surface pushes him forward.
2. When we put lubricant (like oil or grease) between the surfaces in contact, then this is known as lubrication. The lubricant forms a thin layer between the two surfaces and hence avoids direct contact. In fact, by lubrication, the dry friction is converted into fluid friction, which of course, is comparatively lesser in magnitude. In heavy machines, continuous supply of oil or some other suitable lubricant is provided to reduce the friction as well as to protect the moving part from overheating. Flow of compressed air is also used as lubricant. It reduces the friction between moving parts by acting as an elastic cushion and carry away the heat generated. It has the advantage of preventing dust and dirt from collecting over the moving part.
   There are three types of frictional forces:
   (a) Static frictional force,
   (b) Kinetic frictional force, and
   (c) Rolling frictional force.
3. All the forces, depending on their motion and position, can be divided in following four categories.
   (i) **Constant force:** If the direction and magnitude of a force is constant it is said to be a constant force. Mechanical forces of constant magnitude are produced by machines. Tension in a string can be a constant force.
   (ii) **Time-dependent force:** If a force depends on time it is said to be time dependent, *e.g.*, in case of impulse or motion of a charged particle in an alternating electric field, force is time dependent.
   (iii) **Position-dependent force:** If force acting on a body depends on the position of body, it is said to be position-dependent force. Gravitational force between two bodies $(= Gm_1m_2/r^2)$ or force between two charged particles $[= q_1q_2/(4\pi\varepsilon_0 r^2)]$ are examples or position-dependent forces.
   (iv) **Velocity-dependent force:** If force acting on a body depends on the velocity, it is said to be velocity-dependent. In physics, in the opinion of the author, only two forces, viz., viscous force $(= 6\pi\eta rv)$ and force on a charged particle in a magnetic field $(qvB\sin\theta)$ are velocity-dependent.

NCERT Corner

1. Steel. As the mass is a measure of inertia, the ball of same shape and size, having more mass than the other balls will have highest inertia. Since steel has greatest density and greatest mass, it has highest inertia.

2. Yes. the balls will start rolling in the direction in which the train was moving. Due to the application of the brakes, the train comes to rest but due to inertia the balls try to remain in motion; therefore, they begin to roll. Since the masses of the balls are not the same, the inertial forces are not same on both the balls. Thus, the balls will move with different speeds.

3. From the light rifle, according to law of conservation of momentum or explanation by Newton's laws of motion.

4. The force applied by the horse balances the force of friction.

5. Law of conservation of momentum is applicable to isolated system (no external force is applied). In this case, the change in velocity is due to the gravitational force of earth.

6. Acceleration $= a = \dfrac{v-u}{t} = -\dfrac{80}{8}\,\text{ms}^{-2} = -10\,\text{ms}^{-2}$

$$\text{Force} = ma = \frac{50}{1000}\times 10 = 0.5\,\text{N}$$

7. Calculate using $F = ma$
Acceleration becomes one-fourth of the original.

8. Separation between them will increase. Initially the momentum of both of them are zero as they are at rest. In order to conserve the momentum the one who throws the ball would move backward. The second will experience a net force after catching the ball and therefore will move backwards, that is, in the direction of the force.

9. The working of rotation of sprinkler is based on third law of motion. As the water comes out of the nozzle of the sprinkler, an equal and opposite reaction force comes into play. So the sprinkler starts rotating.

10. (i) $m = 10\,\text{g} = \dfrac{10}{1000}\,\text{kg}$

$$u = 10^3\ \text{m/s}$$
$$v = 0$$
$$s = \frac{5}{100}\,\text{m}$$
$$v^2 - u^2 = 2as$$
$$0 - (10^3)^2 = 2.a.\frac{5}{100}$$
$$a = \frac{-1000\times 1000}{2\times 5}\times 100$$
$$= -10^7\ \text{m s}^{-2}$$
$$F = m.a = \frac{10}{1000}\times 10^7 = 10^5\ \text{N}$$

11. $F = ma = \text{kg m s}^{-2}$
This unit is also called newton. Its symbol is N.

$$m_1 = \frac{F}{a_1} = \frac{5}{8}\ \text{kg},$$
$$m_2 = \frac{F}{a_2} = \frac{5}{24}\ \text{kg},$$
$$M = \left(\frac{5}{8} + \frac{5}{24}\right)\text{kg} = \left(\frac{5}{6}\right)\text{kg}$$

Acceleration produced in M

$$a = \frac{F}{M} = \frac{5}{5/6} = 6\,\text{ms}^{-2}$$

12. Momentum = Mass × Velocity
SI unit of momentum is kg/m s^{-1}
Force = Rate of change in momentum

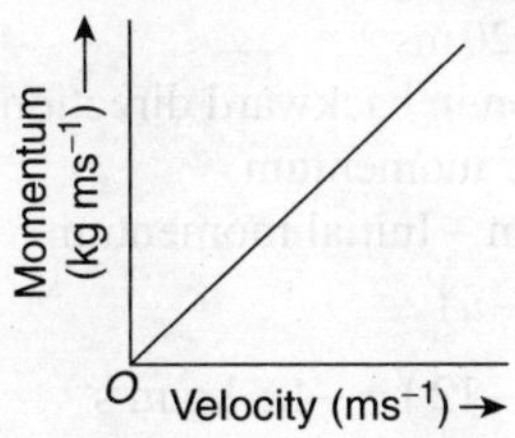

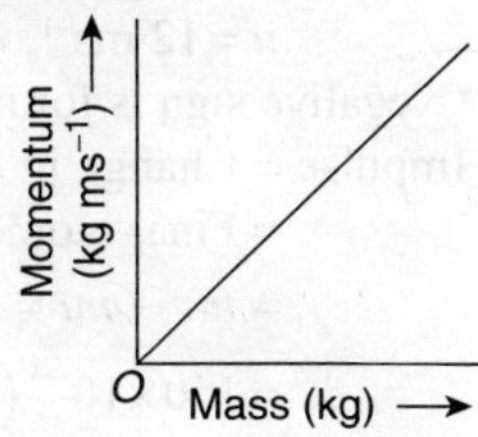

Numericals

Level – I

1. Here, $m = 10$ kg; $t = 8$ sec
After the force stops acting, the body moves with a uniform velocity given by

$$v = \frac{\text{Distance travelled}}{\text{Time taken}} = \frac{80}{5} = 16\ \text{ms}^{-1}$$

Initial velocity of the body, $u = 0$

From $v = u + at$

$$16 = 0 + a\times 8 \text{ or } a = 2\ \text{ms}^{-2}$$

As $F = ma$

$$F = 10\times 2 = 20\ \text{N}$$

2. Here, $F = -0.04$ N,
$u = 0.30\ \text{ms}^{-1}$, $v = 0.10\ \text{ms}^{-1}$
$m = 1$ kg, $s = ?$

From $F = ma$

$$-0.04 = 1\times a$$

or $a = -0.04\ \text{ms}^{-2}$

Again, from $v^2 = u^2 + 2as$

∴ $(0.10)^2 = (0.30)^2 + 2\times(-0.04)\,s$

or $0.01 = 0.09 - 0.08s$

or $0.08s = 0.09 - 0.01 = 0.08$

or $s = 1$ m

3. Here, $m = 3.0$ kg, $u = 2.0$ ms^{-1}
$v = 3.5$ ms^{-1}, $t = 25$ s, F = ?

$$F = ma = \frac{m(v-u)}{t}$$

$$= \frac{3.0(3.5-2.0)}{25} = 0.18 \text{ N}$$

The force is along the direction of motion.

4. 2 m/s^2; 14000 N
Here, $u = 0$, $s = 400$ m, $t = 20$ s, $a = ?$
F = ?, $m = 7$ metric ton = 7000 kg

From $s = ut + \frac{1}{2}at^2$

$$400 = 0 + \frac{1}{2} \times a(20)^2 = 200a$$

$$a = 400/200 = 2 \text{ m/s}$$

As F = $m\,a$ ∴ F = 7000 × 2 = 14000 N

5. –4.8 kg ms^{-1}
Here, $m = 150$ g $= 150 \times 10^{-3}$ kg.
$u = 12$ ms^{-1}, $v = -20$ ms^{-1}
(Negative sign is for motion in backward direction).
Impulse = Change in linear momentum
= Final momentum – Initial momentum

$$= mv - mu = m(v-u)$$

$$= 150 \times 10^{-3}(-20-12) = -4.8 \text{ kg m s}^{-1}$$

6. Mass, $m = 1500$ kg, acceleration, $a = 1.7$ ms^{-2}
From Newton's second law of motion

$$F = ma = 1500 \times (-1.7) = -2550 \text{ N}$$

7. Here the block is at rest. Hence the reaction of the surface and weight of the block balance each other.
Total mass pulled by the applied force

$$= (4 + 0.5) = 4.5 \text{ kg}$$

$$\therefore \text{Acceleration} = \frac{9 \text{ N}}{4.5 \text{ kg}} = 2 \text{ m/s}^2$$

Force of reaction = 4 kg×2 m/s^2= 8 N

8. $s = ut$ or $9.90 = u \times 3$
∴ $u = 3.30$ m/s
Thus during the time the force acts, it changes the velocity of the body from 0 to 3.30 m/sec in 11 seconds.
Using $v = u + at$

$$3.30 = 0 + a \times 11 \text{ or } a = 0.30 \text{ m/sec}$$

Now $F = ma = 0.060 \times 0.30$

$$= 18 \times 10^{-3} \text{ newton}$$

9. $a = \frac{F}{m} = \frac{\mu R}{m} = \frac{\mu mg}{m} = \mu g$

$$= 0.2 \times 9.8 = 1.96 \text{ m/sec}^2$$

Now $v^2 - u^2 = 2as$

$$\therefore \quad s = \frac{v^2 - u^2}{2a}$$

$$= \frac{0-(10)^2}{2 \times (-1.96)} = 25.5 \text{ metre}$$

10. Let a be the acceleration of the body, then

$$v^2 = u^2 + 2as$$

$$0 = 1 + 2a \times 5$$

or $a = -\left(\frac{1}{10}\right) = 0.1 \text{ m/s}^2$

Hence, there is a retardation equal to 0.1 m/s^2.
Now $F = \text{mass} \times (0.1) = 0.1$ m
Normal reaction
R = weight of the body = mg
$= m \times 9.8$ newton

$$\therefore \quad \mu = \frac{F}{R} = \frac{m \times 0.1}{m \times 9.8} = \frac{1}{98} = 0.0102$$

11. Given that

$$u = 70 \text{ km/h} = \{70 \times 10^3/(3600)\} \text{ m/s}$$

$$= (175/9) \text{ m/s}$$

Final velocity = 0
Now,

$$\mu = \frac{F}{R} = \frac{m(-a)}{mg} = \frac{-a}{g} \text{ or } -a = \mu \times g$$

∴ Retardation = $0.2 \times 9.8 = 1.96$ m/s^2
Using $v^2 = u^2 + 2as$, we have

$$0 = \left(\frac{175}{9}\right)^2 + 2(-1.96)s$$

Solving we get, $s = 96.45$ m

12. $F = \mu mg$
or $19.6 = \mu \times 10 \times 9.8$
∴ $\mu = 0.2$
With 5 kg $F = \mu(m_1 + m_2)g$

$$= 0.2(10+5) \times 9.8$$

$$= 29.4 \text{ newton}$$

Level – II

1. Let m be the mass of both the monkey and the mirror and T, the uniform tension in the rope.
When both are stationary, $T = mg$.

Let the monkey pull the rope by a force T' and in turn, is pulled up by the rope by an equal force T'. This tension T' will be the same throughout the rope. Then, on each side of the pulley, the net force in the upward direction is ($T' - mg$), which gives equal acceleration to both the monkey and the mirror in the same direction as their masses are equal.

Hence, the monkey can never get away from his image seen in the mirror.

2. The situation is shown in the figure below. Let u be the speed of A towards left initially and S be its distance from the end.

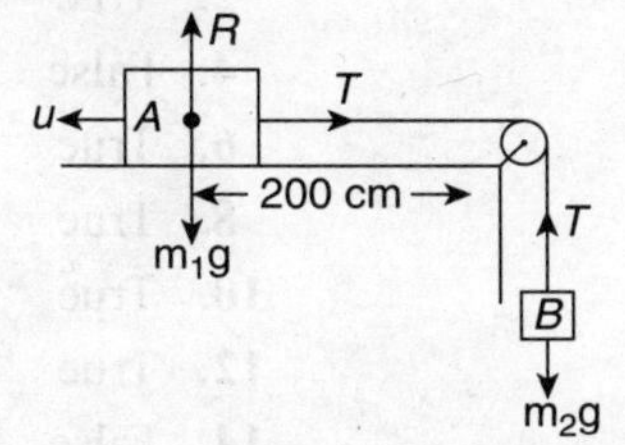

Since A and B are connected by an unstretchable string, their acceleration will be the same. Let a be the acceleration of A towards right and that of B downward.

The different forces acting on A are:

(i) its weight m_1 g downward,

(ii) reaction R of the table upward, and

(iii) tension T of the string towards right.

Since A is in vertical equilibrium,

$$R = m_1 g \quad (1)$$

The equation of motion of A along the horizontal towards right is

$$T = m_1 a \quad (2)$$

The forces acting on B are:

(i) its weight $m_2 g$ downward and

(ii) the tension T of the string upward.

Its equation of motion is

$$m_2 g - T = m_2 a \quad (3)$$

Adding (2) and (3), we get

$$m_2 g = (m_1 + m_2)a$$

$$\Rightarrow \quad a = \frac{m_2}{m_1 + m_2} \cdot g = \frac{200}{500 + 200} \times 980$$

$$\Rightarrow \quad a = 280\,\text{cm/s}^2$$

If the velocity of A after $t = 1$ s be v, then

$$v = u + at = 50 \times 1 - 280 \times 1$$

$$\Rightarrow \quad v = -230\text{ cm/s} = -2.3\text{ m/s}$$

The negative value of v indicates that the body is moving towards right after 1s.

If x be the displacement of A towards left in 1 s. then

$$x = ut - \frac{1}{2}at^2 = 50 \times 1 - \frac{1}{2} \times 280 \times 1^2$$

$$\Rightarrow \quad x = -90\,\text{cm} = -0.9\,\text{m}$$

The negative value of x shows that the body has moved towards right by a distance of 0.9 m.

Hence, the distance of A from the end after 1 s is 200 cm – 90 cm = 110 cm = 1.1 m.

3. Acceleration a of the elevator is given by the formula.

$$v^2 - u^2 = 2as \text{ or } 0 - 64 = 2 \times a \times 16$$

$$a = -2\text{ m/s}^2$$

(a) Let T be the tension in the cable. If M be the total mass of the elevator, then

$$T - Mg = Ma \text{ or } T = M(g + a)$$

$$T = 1000 \times (9.8 - 2) = 7800\text{ N}$$

(b) Let R be the force exerted on the man by the elevator floor. Then,

$$R - mg = ma \text{ or } R = m(g + a)$$

$$\therefore \quad R = 80 \times (9.8 - 2) = 624\text{ N}$$

4. (i) The reading of machine will be 50 kg wt. In this case, there will be no reaction, therefore,

Apparent weight = Actual weight

(ii) In this case when the lift is moving upward with constant acceleration, the reaction R (= ma) acts downward. Hence,

$$\text{Apparent weight } W' = W + R$$

$$= m(g + a)\text{ newton}$$

$$= \frac{mg[1 + (a/g)]}{g}\text{ kg}$$

$$= \left(50 + \frac{50a}{g}\right)\text{kg wt}$$

5. Here, $m = 0.40$kg,

$u = 10$ m/s due north

$F = -8.0$ N

(minus sign for opposite direction of force)

$$a = \frac{F}{m} = \frac{-8.0}{0.40} = -20\,\text{ms}^{-2}$$

For $\quad 0 \le t \le 30$ s

(i) At $t = -5$s

$$x = ut = 10 \times (-5) = -50\,\text{m}$$

(ii) At $t = 25\,\text{s}, x = ut + \frac{1}{2}at^2$

$$= 10 \times 25 + \frac{1}{2}(-20)(25)^2 = -6000\,\text{m}$$

(iii) At $t = 100\,\text{s}$, the problem is divided into two parts. Up to 30 s, there is retarding force.

$\therefore$ From $x_1 = ut + \frac{1}{2}at^2$

$$x_1 = 10 \times 30 + \frac{1}{2}(-20)(30)^2 = -8700\ \text{m}$$

At $t = 30s$, $v = u + at$

$$= 10 - 20 \times 30 = -590\,\text{m/s},$$

$\therefore$ For motion from 30 s to 100 s

$$x_2 = vt = -590 \times 70$$

$$= -41300\,\text{m}.$$

$$\therefore x = x_1 + x_2 = -8700 - 41300 = -50000\ \text{m}$$

$$= -50\ \text{km}.$$

6. The situation is shown in the figure. Since the body is moving with uniform speed,

$$f = F\cos 30°$$

Normal reaction,

$$R = mg - F\sin 30°$$

We know that $f = \mu R$

$\therefore$ $F\cos 30° = \mu\ (\text{mg} - F\sin 30°)$

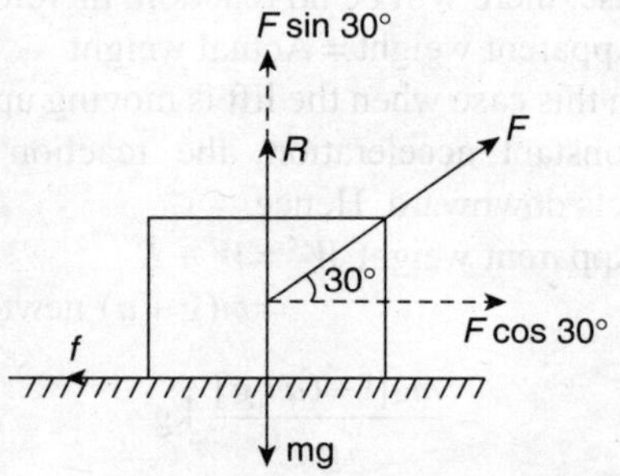

$$F\left(\cos 30° + \mu \sin 30°\right) = \mu\ \text{mg}$$

or $\quad F = \dfrac{\mu\ \text{mg}}{\cos 30° + \mu \sin 30°}$

$$= \frac{0.25 \times 20 \times 9.8}{0.866 + 0.25 \times (1/2)} = 49.45\ \text{newton}$$

Fill in the Blanks

1. first
2. backward, momentum
3. smaller
4. third
5. force
6. the rate of fuel consumption
7. more
8. opposite direction
9. the weight of the liquid displaced
10. third
11. two different bodies
12. lesser
13. lesser
14. impulse
15. constant, no external force
16. increase
17. rolling
18. thrust, pressure
19. compressed
20. downward

True or False

1. True
2. True
3. True
4. False
5. False
6. True
7. True
8. True
9. False
10. True
11. True
12. True
13. True
14. False
15. False
16. True
17. True
18. True
19. True
20. True
21. True
22. True
23. False
24. False
25. False

Match the Columns

1. (P) – (W), (Q) – (U), (R) – (Y), (S) – (V), (T) – (X).
2. (P) – (U), (Q) – (V), (R) – (W), (S) – (T).
3. (P) – (V), (Q) – (W), (R) – (T), (R) – (U).
4. (P) – (W), (Q) – (T), (R) – (V), (S) – (U).

Assertion-Reason Questions

1. (c) This is as per Newton's first law and on the assumption that no force is opposing the motion of the body.
2. (b) $\vec{p} = m\vec{v}$. In uniform motion along a circle, $\vec{v}$ is changing because of change in direction.
3. (b) Assertion and reason are both correct. But the reason refers to Bernoulli's theorem, which has nothing to do with the assertion.
4. (d) Static friction alone is a self-adjusting force and not all types of friction.
5. (a) Both the assertion and reason are true as per Newton's second law and reason is correct explanation of the assertion.

6. (b) Both the assertion and reason are true, but reason is not a correct explanation of the assertion. In fact, forces of action and reaction act on different bodies. That is why they do not cancel out.

7. (b) Assertion and reason, both are true and the latter is a correct explanation of the former.

8. (b) Both assertion and reason are true, but reason is not correct explanation of assertion.

Comprehension Type Questions

1. (i) Net force = Force exerted by engine – Friction force
[Here, frictional force is subtracted because it opposes the motion]
$= 40000 - 5000 = 35000 = 3.5\times10^4$ N
(ii) From Newton's second law of motion
Accelerating force = Mass of the train × Acceleration of train

$$\Rightarrow \quad a = \frac{F}{m}$$

Mass of train = Mass of engine + Mass of all wagons = 8000 + 5 × 2000
= 8000 + 10000 = 18000 kg

$$\therefore \text{Acceleration} = \frac{35000}{18000} = \frac{35}{18} = 1.95 \text{ ms}^{-2}$$

(iii) Force of wagon 1 on wagon 2 = Mass of the engine × Acceleration
= 8000 × 1.95 = 15600 N

2. (a) When two block masses of masses m_1 and m_2 are hanging from the ends of a string passing over a fixed pulley, their acceleration is:

$$a = \frac{(m_2 - m_1)g}{(m_2 + m_1)}$$

and the tension is: $T = \dfrac{2m_1 m_2 g}{m_1 + m_2}$

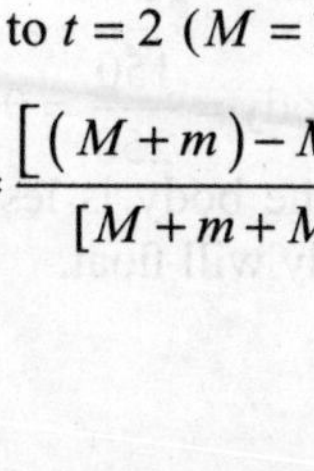

From $t = 0$ to $t = 2$ ($M = 100$ g, $m = 40$ g),

$$a = \frac{[(M+m) - M]g}{[M+m+M]} = \frac{g}{6}$$

$$v = u + at = 0 + (g/6)(2) = 980/3 \text{ cm/s}$$

From $t = 2$s to $t = 3$s (In first second after 40 g block is removed). Both blocks move with a constant velocity.

$$s = \frac{9010}{3}\times1 = 326.6 \text{ cm}$$

(b) Hence at $t = 2s$, the left block is moving up with velocity 980/3 cm/s and the right block is moving down with velocity 980/3 cm/s. Tension in the string during this interval is:

$$T = \frac{2M(m+M)g}{m+2M}$$

$$= \frac{2\times0.1\times1.4\times9.8}{0.24} = 1.143 \text{ N}$$

(c) Tension in the string = Weight of each block.

$$T = Mg = 0.98 \text{ N and } T' = 2T = 1.96 \text{ N}$$

3. (i) The property of bodies to resist a change in their velocity is inertia.
(ii) The relation between force and accelerations is given by $F = m \times a$.
(iii) The product of mass and velocity of a body is momentum.
(iv) The physical quantity that corresponds to the rate of change of momentum is force.
(v) The principle on which a rocket works is Newton's third law of motion.

4. (i) $a = \dfrac{20}{4} = 5\,\text{m/s}^2$ from $t = 0$ to $t = 4$ s
Thus, force is maximum in this period.

$$\therefore\ a = \frac{0-20}{10-8} = -10\,\text{m/s}^2$$

(iii) From $t = 4$ s to $t = 8$ s, no force acts.
(iv) $v = 20$ m/s from 4 s to 8 s.
(v) a from 0 to 4 s = 5 m/s²
a from 4 to 8 s = 0 m/s²

5. (i) $\text{Acceleration} = \dfrac{\text{Change in velocity}}{\text{Time}}$

Let us choose an interval of 2s.

Then $a = \dfrac{5}{2} = \dfrac{10}{4} = \dfrac{15}{6} = 2.5$ m/s²

(ii) The force acting on the body is given by,
$F = ma = 5\times2.5\times12.5$ N
(iii) $u = 0$, $v = 5$ m/s, $t = 2$ s
Thus, change in momentum = $mv - mu$
$= 5\times5 - 5\times0 = 25$ kg-m/s

Objective Questions

One Option Correct

1. (c) Unbalanced means there is a net force
$F = ma$

2. (c) $F = mg$

$\Rightarrow \quad m = \frac{F}{G} = \frac{1}{10}$
$= 0.1$ kg
$= 100$ gm

5. (d) Gravitational force

6. (a) Inertia of rest

7. (a) $F = ma$
Greater the mass, lesser the acceleration

8. (b) $S = \frac{1}{2}at^2$
$S \propto t^2$

9. (c) Momentum = Mass × Velocity

10. (a) Momentum is conserved in absence of net force

12. (c) Because it is the earth which is exerting force on the porter.

13. (b) 1st case: $F = ma$
$= 3 \times 0.5$
$= 1.5$ N
2st case: $1.5 = 1.5a$
or $\quad a = 1 m/s^2$

14. (c) $F = ma$
$= \frac{20}{100} \times \frac{5}{100}$
$= 1 \times 10^{-3}$ N

15. (a) $F = m_A \times 2$
$F = m_B \times 5$
$\frac{m_A}{m_B} = \frac{5}{2} = 2.5$

16. (c) $F = ma$
$= m\frac{(v-n)}{t}$
$= \frac{200}{100} \times \frac{(2.5-1.5)}{2.5}$
$= \frac{1}{5} \times \frac{10}{2.5}$
$= 0.8$ N

17. (d) $a = \frac{F}{m} = \frac{1}{10}$
$= 0.1$ m/s²

$S = ut + \frac{1}{2}at^2$

$1 = (u \times 4) + (\frac{1}{2}0.1 \times (4)^2)$
$= 4u + 0.8$
$u = 0.2/4 = 0.05$ m $= 5$ cm

18. (a) $a = \frac{v-u}{t}$
$a = \frac{6-5}{2}$
$= 0.5$ m/s²
$F = ma$
$0.6 = m \times 0.5$
$\Rightarrow \quad m = 1.2$ kg

19. (c) $S = \frac{1}{2}at^2$
$4 = \frac{1}{2}a \times (2)^2$
$= 2a$
$\Rightarrow \quad a = 2$ m/s²
$F = \frac{200}{1000} \times 2$
$= 0.4$ N

20. (a) Applying conservation of momentum
Momentum before explosion = 0
Momentum after explosion = $m_1v_1 + m_2v_2$
$m_1v_1 + m_2v_2 = 0$
$200 \times 12 + 100v_2 = 0$
$v_2 = -\frac{200 \times 12}{100}$
$= -24$ m/s

21. (b) Pressure $P = \frac{F}{A}$
$P = \frac{16}{8 \times 10^{-2} \times 8 \times 10^{-2}}$
$= 2500$ Pa

23. (a) Density of body $= \frac{150}{250} = 0.6$ g/cm³
The density of the body is less than the density of water, so the body will float.

24. (c) $v = u + at$

$= 0 + 1.5 \times 4$

$= 6$ m/s

Momentum $= mv$

$= 2 \times 6$

$= 12$ kg m/s

25. (b) $5 - 3 = 2$ N

Towards greater force

26. (d) Let v be the velocity of combined mass. Then,

$mv = m_1 v_1 + m_2 v_2$

$10v = 6 \times 2 - 4 \times 1.5$

$= 6$

$v = \frac{10}{6} = 0.6$ m/s

28. (a) Impulse $= mv_2 - mv_1$

29. (b) Momentum $= mv =$ kg m/s

33. (a) $F \times t =$ Change in momentum

$F \times 0.2 = 1.8$

$F = 9$ N

39. (d) $m = \frac{F}{a}$

47. (c) Action = Reaction

48. (c) Total action force $= w_1 + w_2$

Reaction force $= w_1 + w_2$

60. (b) $\frac{1}{2}mv^2 : \frac{1}{2}m(4v)^2$

1:16

61. (b) Inertia of motion

64. (b) Surface is frictionless

65. (c) $(M + m)V = mv$

$V = \frac{mv}{M + m}$

66. (d) Acceleration $= \frac{F}{m}$

67. (d) $P = mv$

68. (d) Law of conservation of momentum

70. (c) $F = \mu R = \mu W$

Since μ is less than 1,

Therefore $F < W$.

71. (a) Law of conservation of momentum and conservation of energy

$m_1 v_1 = (m_1 + m_2)V$

$20 \times 100 = (20 + 20)\ V$

$\Rightarrow \quad V = 50$ cm/s

Now, $\frac{1}{2}(m_1 + v_1)v^2 = (m_1 + m_2)\,gh$

$\Rightarrow \quad h = \frac{v^2}{2g} = \frac{50 \times 50}{2 \times 980}$

$= 1.25$ cm

86. (a) Archimedes' principle

97. (c) Newton's law is an independent statement, whose idea was given by Galileo.

98. (b) First law states that the body will continue in its state of rest or uniform motion if no net force is acting on it.

99. (c) Reaction by surface in lift is the apparent weight of person. When a person of mass m is sitting in a lift, then actual weight of the person is mg. This acts on a weighing machine which offers a reaction R, given by the reading of the weighing machine. This reaction exerted by the surface of contact on the person is apparent weight of the person.

When lift is accelerating upwards at the rate of a, then reaction

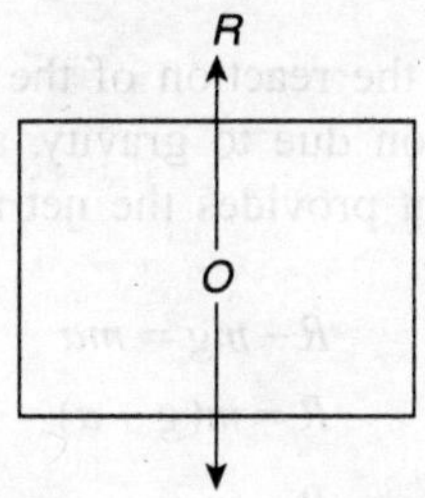

$R - mg = ma$

$\therefore \quad R = m(g + a)$

Hence, apparent weight is greater than actual weight.

100. (b) A force, which always acts toward or away from a fixed point, and whose magnitude depends only on the distance from that point is called a 'central force'. This point is called as 'centre of force'.

101. (b) Let a be the acceleration of each block and T_1 and T_2 be the tensions in the two strings as shown in the figure.

Taking the three blocks and the two strings as the system and

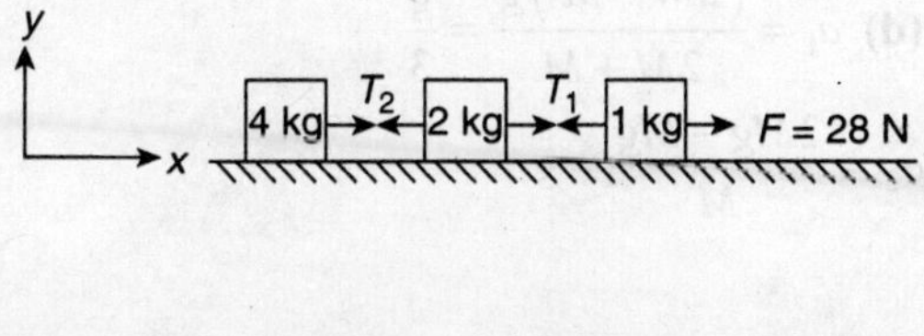

using $\quad \Sigma F_x = ma_x$

or $\quad 28 = (4+2+1)a$

Or $\quad a = \frac{28}{7} = 4 \text{ m/s}^2.$

102. (c) An object moving in uniform circular motion is moving around the perimeter of the circle with a constant speed. While the speed of object is constant, its velocity is changing. Velocity, being a vector quantity, has a constant magnitude but a changing direction. The direction is always tangent to the circle and as the object turns the circle, the tangent line always points in a new direction. Also, when the object is moving in circular motion, it is acted upon by a centripetal force directed towards the centre; hence acceleration is perpendicular to velocity.

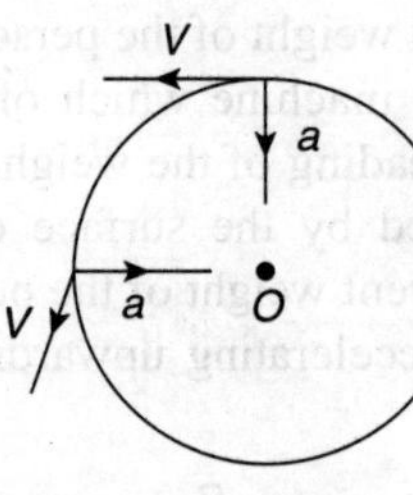

103. (a) Let R be the reaction of the base of lift and g the acceleration due to gravity, acting downwards. Their resultant provides the net acceleration to the lift. Therefore

$$R - mg = ma$$
$$R = m(g+a)$$

Also, $\quad R = mg'$

$$g' = g + a$$

Therefore, net acceleration increases and hence, the reading indicated by the spring balance will increase.

106. (a) $v^2 = u^2 - 2as$

$$(0)^2 = (20)^2 - 2a \times 40$$
$$80a = 400$$

$\Rightarrow \quad a = 5 \text{ m/s}^2$

$\therefore$ Coefficient of dynamic friction

$$\mu = \frac{a}{g} = \frac{5}{10} = \frac{1}{2} = 0.5$$

107. (d) $a_1 = \frac{(2M - M)g}{2M + M} = \frac{g}{3}$

$$a_2 = \frac{2Mg - Mg}{M} = g$$

$$\frac{a_1}{a_2} = \frac{1}{3}$$

$\Rightarrow \quad a_1 : a_2 = 1:3$

109. (b) Force (F) = 6 N;
Initial velocity (u) = 0;
Mass (m) = 1 kg and final velocity (v) = 30 m/s

Therefore acceleration (a) $= \frac{F}{m} = \frac{6}{1} = 6 \text{ m/s}^2$ and final velocity (v) $= 30 = u + at = 0 + 6 \times t$ or $t = 5$ seconds.

110. (d) The vertical component of the pulling force acts in the opposite direction of weight.

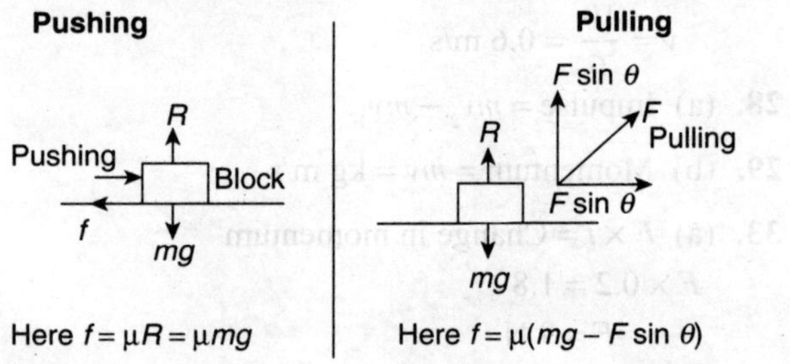

111. (a) $m = 1.5$ kg, $u = 25$ m/sec, $v = 15$ m/sec, $t = 0.03$ sec

$$F = \frac{m(v-u)}{t} = \frac{1.5[15-(-25)]}{0.03}$$

$$F = \frac{1.5 \times 40}{0.03} = 2000 \text{ N}$$

112. (b) $mg - T = ma$
$T = ma$
From equations (i) and (ii),
mg = $(m + M)a$

$$a = \frac{mg}{m+M}$$

113. (b) By spitting or sneezing we get a momentum in opposite direction which will help us in getting off the horizontal plane. In all other cases we will slip on ice as there is no friction.

114. (b) Given: $u = 6$ m/s; $s = 9$ m; $v = 0$
Using $v^2 - u^2 = 2as$

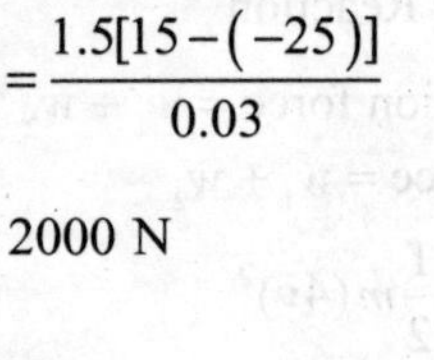

FBD

$$\Rightarrow \quad a = \frac{u^2}{2s} = \frac{6^2}{2\times 9}$$

$$= -2 \text{ m/s}^2$$

$$\because \quad f = \mu mg = ma$$

$$\Rightarrow \quad \mu = \frac{a}{g} = \frac{2}{10} = 0.2 \ (g = 10 \text{ m/s}^2)$$

115. (c) Mass, $m = 2 \times 10^4$ kg; Force, $F = 5 \times 10^5$ N

Time, $t = 20$; Initial velocity, $u = 0$, $v = ?$

We know that *impulse, Ft = change of momentum*

$$5 \times 19^5 \times 20 = 2 \times 10^4 [v - 0]$$

$$v = \frac{5\times 10^5 \times 20}{2\times 10^4} \text{ ms}^{-1} = 500 \text{ ms}^{-1}$$

More Than One Option Correct

1. Gravitational forces and electromagnetic forces obey inverse square law.
2. Force applied on a moving body may change magnitude of velocity. Therefore, both momentum and KE. will change. If force applied on a moving body changes only the direction of motion, momentum changes, but not the KE.
3. When pedaling is being done, force of friction is in backward direction on the front wheel and in forward direction on the rear wheel. However, when pedaling is stopped, force of friction on both the wheels is in the backward direction.
4. In all the four cases, net force = 0.
5. Forces of action and reaction are equal in magnitude, opposite in direction; and act on two different bodies.
6. $T = W$ indicates net force $f = 0$. Therefore, $a = 0$. Speed of the elevator must be uniform, whether it moves up or down.

Scholar Zone

1. There are a number of times at which the velocity of the ball changes which are as given below:
   (i) When the first player kicks the ball towards another player of his team, then the velocity of ball will change because the first player applies some force on the ball
   (ii) When another player kicks the ball towards the goal, then the velocity of ball will change; here again the force is applied on the ball.
   (iii) When the goalkeeper of the opposite team collects the ball, then the velocity of ball will change and it becomes zero. Here the goalkeeper applies some force on the ball to stop.
   (iv) When the goalkeeper kicks the ball towards his own team, the velocity of the ball changes because the goalkeeper applies some force on the ball.

2. Let m = mass per unit length of the lower wire $= 0.2$ kg

   $m_1 = 2.9$ kg

   $m_2 = 1.9$ kg

   l = length of each wire = 1 m

   Since the upper wire is of negligible mass, the tension T_1 will be the same throughout this wire. The tension in the lower wire will, however, be different at different points. Let T_2, be the tension at the middle point A of the wire.

   Let us divide the entire system into two systems, P and Q.

   The system P consists of mass m_1, the upper wire and the portion of the lower wire above the point A. The system Q consists of the portion of the lower wire below the point A and the block of mass m_2.

   Applying Newton's second law to system P, we get

$$T_1 - T_2 - m_1 g - m \times \frac{l}{2} g = \left(m_1 + m\frac{l}{2} \right) a$$

$$\Rightarrow \quad T_1 - T_2 = \left(m_1 + \frac{ml}{2} \right)(g + a)$$

$$\Rightarrow \quad T_1 - T_2 = \left(2.9 + 0.2 \times \frac{1}{2} \right)(9.8 + 0.2)$$

$$\Rightarrow \quad T_1 - T_2 = 30\,\text{N} \qquad (1)$$

   Applying Newton's second law to system Q, we get

$$T_2 - \left(\frac{ml}{2} \times g \right) - m_2 g = \left(m_2 + \frac{ml}{2} \right) a$$

$$\Rightarrow \quad T_2 = \left(m_2 + \frac{ml}{2} \right)(g + a)$$

$$\Rightarrow \quad T_2 = \left(1.9 + 0.2 \times \frac{1}{2} \right)(9.8 + 0.2)$$

$$\Rightarrow \quad T_2 = 20\,\text{N} \qquad (2)$$

   From (1) and (2),

$$T_1 = 50\,\text{N}$$

3. $\Delta p = F.t$

Now, $F.\,t$ = Area under the curve

Area under the curve =

$$\left(\frac{1}{2}\times1\times1\right)+(2\times2)+\left(\frac{1}{2}\times2\times2\right)+(1\times1)+\left(\frac{1}{2}\times1\times1\right)$$

$F.\,t = 2$ N-s

$\therefore \Delta p = 2$ N-s

4. It is true that the heavier object is pulled with a larger force. The strength of the force is determined by the gravitational mass of the object. The resistance to the force and, therefore, to the change in motion of the object, is represented by the inertial mass. Thus, if an object has twice as much mass as another, it is pulled to the Earth with twice the force, but it also exhibits twice the resistance to having its motion changed. These factors cancel, so that the change in motion and the acceleration is the same for all objects, regardless of mass.

5. Because the value of g is smaller on the Moon than on the Earth, more mass of gold would be required to represent 1 newton of weight on the Moon. Thus, your friend on the Moon is richer roughly by a factor of 6.

6. Infinite tension is required to straighten the rope. So, it is impossible to straighten the rope.

7. According to Newton's third law, the force of the man on the boy and the force of the boy on the man are reaction forces, so they must be equal in magnitude. Both individuals experience the same magnitude of force. Hence the boy, with the smaller mass, experiences acceleration over the same time interval. The larger acceleration of the boy results in his moving away from the interaction with a larger velocity. Again, because of the larger acceleration, the boy moves through a larger distance during the interval the hands are in contact.

8. The tension in the cord C will be $(T + mg)$. When a sudden jerk is given to the cord D, it immediately experiences tension while the cord C tends to remain at rest due to its inertia. So the tension produced in D is not transmitted to C. Hence, D breaks.

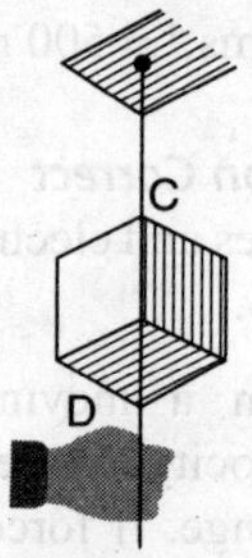

When a steady jerk is given to the cord, the tension produced in it is transmitted to C. Now the tension C is equal to tension D plus the weight Mg. Hence the cord C breaks

9. In the first case, the acceleration of block is g; in the second case it is $g/3$ and in the third case it is $g/2$.

$$a_2 < a_3 < a_1$$

CHAPTER

3 Work, Power and Energy

LEARNING OBJECTIVES

At the end of this chapter, you will be able to:

- Define work and work done by a force
- Describe energy and its kinds
- Define conservative and non-conservative forces
- Formulate work–energy theorem
- Define elastic and inelastic collision in one dimension

INTRODUCTION

Energy, in its different forms, is present in the universe. Every physical process in the universe involves energy and energy transfers. Therefore, energy is an extremely important concept to understand. Unfortunately, despite its importance the concept is bit abstract. However, we do have experiences of energy in our daily life. It could be better understood qualitatively in terms of energy transformation.

Examples of Energy Transformation

1. Heat energy is converted into kinetic energy by the steam engine.
2. If a compressed spring is released, it is able to return to its original state by converting the stored energy (PE) into energy of motion (KE).
3. At a thermal power station, the chemical energy of the coal is transformed into heat energy which is further changed into electrical energy.
4. The solar energy of sunlight is converted to chemical energy in food by green plants by the process of photosynthesis.
5. Nuclear fusion on the surface of the Sun is the source of solar heat energy and light energy.

In daily life, the term 'work' is applied to any activity requiring the exertion of muscular or mental effort. In Physics, the term is applied in a much more specific sense, which requires the application of a force to a body and displacement of that body.

Let d be the distance moved by a body along a straight line, while acted upon by a constant force of magnitude F in the same direction as the motion. The work W done by the force is

$$W = F \times d$$

However, the constant force acting on the body may not act in the direction in which the body moves. In Fig. 3.1, the constant force F makes an angle θ with the displacement.

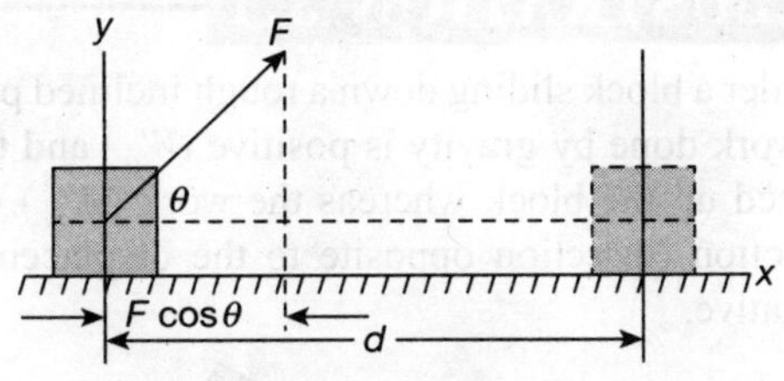

Figure 3.1

The work W done by this force when its point of application undergoes a displacement d is defined as the product of the magnitude of displacement and the component of the force in the direction of the displacement.

The component of F in the direction of d is $F \cos \theta$. Then

$$W = F \cos \theta \, d$$

An alternative interpretation of the above expression is that $d \cos \theta$ is the component of displacement in the direction of F.

Thus, work may also be defined as the product of magnitude of force and displacement in the direction of force.

Work is a scalar quantity. However, it is an algebraic quantity, that is, it can be positive or negative. When the component of force is in the same direction as the displacement, ($\theta = 0$), the work W is positive. If the force is at right angles ($\theta = 90°$) to the displacement it has no component in the direction of the displacement and work is zero. When it is opposite to the displacement, the work W is negative ($\theta = 180°$).

If θ is acute, work is positive.

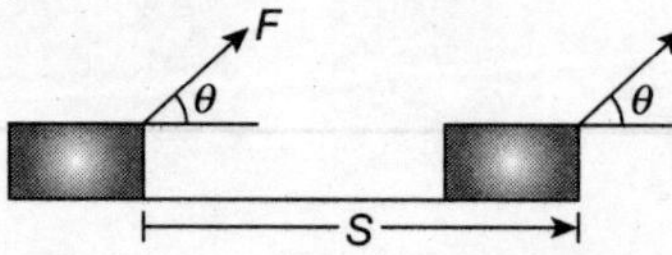

If $\theta = 90°$, force and displacement vectors are mutually perpendicular and $W = 0$.

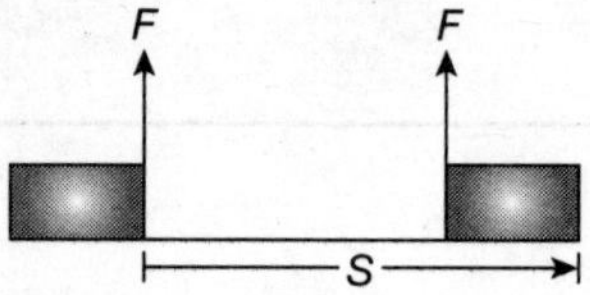

If θ is obtuse, work is negative.

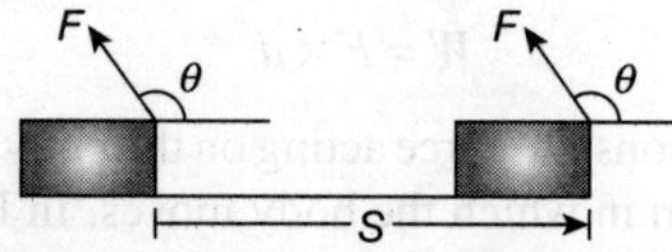

ILLUSTRATION 3.1

Consider a block sliding down a rough inclined plane. The work done by gravity is positive (W_g) and tends to speed up the block whereas the work (W_{f_r}) done by friction (direction opposite to the displacement) is negative.

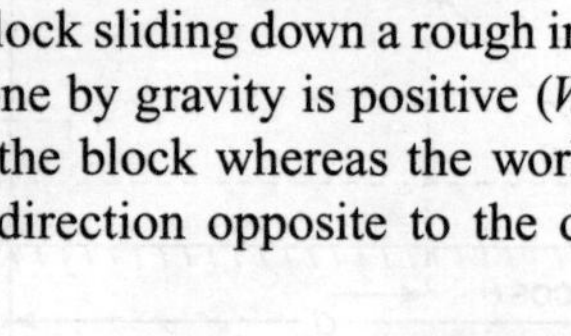

Solution

$$W_g = mg.S\cos\phi = mg\ S\sin\theta \text{ is positive.}$$

$$W_{fr} = -\mu_K mg\cos\theta\ S \text{ is negative.}$$

The third force acting on the block is the normal reaction N which does zero work since its direction is perpendicular to the displacement.

The SI unit of work is joule (abbreviated as J) and is defined as the work done by a force of 1 N to produce a displacement of 1 m in the same direction.

$\therefore$ 1 J = 1 N.m

Conclusion

(i) When a body is lifted, the work done by the lifting force is positive.

(ii) The work of gravitational force on a body being lifted is negative.

(iii) When a body slides on a stationary surface, the work of the frictional force exerted on the body is negative.

(iv) When gas is compressed in a cylinder, the work of the compressing force is positive.

(v) The work of the normal force exerted on a body by a surface on which it moves is zero.

(vi) The work of the centripetal force on a body moving in a circle is also zero.

(vii) In order to find the total work done on a body by several forces, we first identify the force and find the work done by each force. The algebraic sum gives the total work done.

ILLUSTRATION 3.2

A block of mass $m = 2$ kg is pulled by a force $F = 40$ N upwards through a height $h = 2$ m. Find the work done on the block by the applied force F and weight mg. ($g = 10$ m/s²)

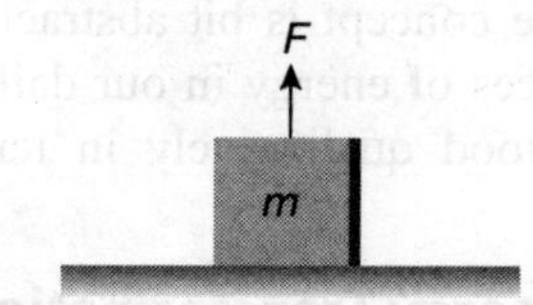

Solution

Weight mg = (2) (10) = 20 N

Work done by the applied force $W_F = F\,h\cos 0°$.

As the angle between force and displacement is 0°, $\cos 0° = 1$

or $\quad W_F = (40)\,(2)\,(1) = 80$ J

Similarly, work done by its weight

$$W_{mg} = (mg)\,(h)\cos 180°$$

or $\quad W_{mg} = (20)\,(2)\,(-1) = -40$ J

ILLUSTRATION 3.3

A man pushes a roller with a force of 50 N through a distance of 20 m. Calculate the work done if the handle of the roller is inclined at an angle of 60° with the ground.

Solution

Here, $F = 50$ N, $s = 20$ m, $\theta = 60°$, $W = ?$

As $W = F s \cos\theta$

$\therefore$ $W = (50)(20)\cos 60°$

$= 1000 \times \frac{1}{2} = 500$ J

Conservative and Non-conservative Forces

In mechanics, we encounter two types of forces: conservative and non-conservative force.

Conservative Force

A force is said to be conservative if the work by the force in moving a mass through any closed path is zero. Alternatively, if the net work done by the force in moving a mass between two points is path-independent, i.e., the work done depends only on the location of two points and not on the path followed, the force may be termed conservative. Conservative forces are non-dissipative and can store work (as energy)

Examples of conservative forces are Gravitational force, Electrostatic force, etc.

Non-conservative Force

The forces which do not fulfill the above-mentioned conditions are non-conservative. For these the energy is not stored but dissipated during displacement. Examples of non-conservative forces are friction, viscous forces etc.

Mechanical Energy: The following are the two kinds of mechanical energy.

(i) Kinetic Energy: Energy possessed by a body because of its motion is known as kinetic energy and it is equal to $\frac{1}{2}mv^2$.

Work–Energy Theorem

The work–energy theorem is a powerful tool that connects an object's speed with its position and is independent of complicacy of motion.

Let us consider the case of an object moving along a straight line as shown below in the figure.

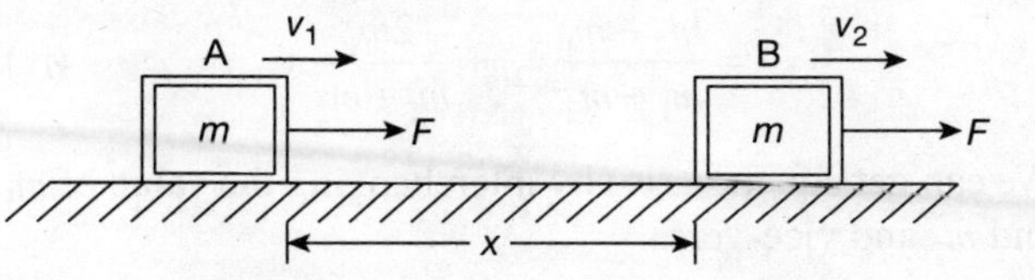

At position A the kinetic energy of the object is $\frac{1}{2}mv_1^2$. A constant force F acts on its direction of motion and produces an acceleration a, which increases its velocity. Let us say that after a displacement x, its velocity becomes v_2 and its kinetic energy $\frac{1}{2}mv_2^2$. Its energy is increased. It is due to the work done by the external force F on the object, which we can write as

$$\frac{1}{2}mv_1^2 + Fx = \frac{1}{2}mv_2^2$$

If, in the above case, force F acts in the direction opposite to the motion of the body, the energy of object decreases as work is done by the object in overcoming the opposition of the force F, so we write our equation as

$$\frac{1}{2}mv_1^2 - Fx = \frac{1}{2}mv_2^2$$

Thus, if force is acting in the direction of motion, it does position work on the object and its energy increases. If it is acting in opposition to motion of the object, obviously the object will do work on it and its energy decreases. We can define a general statement about the work associated and the energy of an object as

Initial kinetic energy of the object + Work done on it − Work done by it = Final kinetic energy

(ii) Potential Energy: When work is done on a system, it gains energy. If the system stores this energy in such a way that it can be subsequently recovered, the system is said to possess potential energy. Examples of potential energy are gravitational, elastic (in stretched or expanded spring), and electrostatic (interaction between charges) potential energies.

ILLUSTRATION 3.4

A ball of mass m is projected from a height h_1 with speed v_1 in the vertically upward direction. If its speed reduces to v_2 at height h_2, find the work done on the ball by air resistance.

Solution

Work done by gravity is

$$W_g = -mg(h_2 - h_1) \quad [\because \text{ weight and displacement are opposite}]$$

Let the work done by air resistance be W_{air}.

From work–energy principle,

$$W_g + W_{air} = \frac{1}{2}mv_2^2 - \frac{1}{2}mv_1^2$$

or, $-mg(h_2 - h_1) + W_{air} = \frac{1}{2}m(v_2^2 - v_1^2)$

$$\therefore \quad W_{air} = mg(h_2 - h_1) + \frac{1}{2}m(v_2^2 - v_1^2)$$

$$= mg(h_2 - h_1) - \frac{1}{2}m(v_1^2 - v_2^2).$$

Note: In the absence of air resistance, you will get the common equation of motion under gravity, which is $v_1^2 - v_2^2 = 2gh = 2g(h_2 - h_1)$.

ILLUSTRATION 3.5

A particle of mass 0.5 kg travels in a straight line with velocity $v = ax^{3/2}$, where $a = 5\,m^{-1/2}s^{-1}$. What is the work done by the net force during its displacement from $x = 0$ to $x = 2$ m?

Solution

$m = 0.5$ kg

$v = ax^{3/2}$, $a = 5\ m^{-1/2}s^{-1}$, $W = ?$

Initial velocity at $x = 0$, $v_1 = a \times 0 = 0$

Final velocity at $x = 2$, $v_2 = a\,2^{3/2} = 5 \times 2^{3/2}$

Work done = Increase in KE

$$= \frac{1}{2}m\,(v_2^2 - v_1^2)$$

$$W = \frac{1}{2} \times 0.5[(5 \times 2^{3/2})^2 - 0]$$

$$= 50 \text{ J}$$

Elastic and Inelastic Collisions in One Dimension

Collisions: A collision is said to take place when either two bodies physically collide against each other or the path of one body is changed by the influence of the other body.

When the path of one body is changed by the influence of some force of other body

The result of collision:

(i) The momentum and kinetic energy of the interacting bodies many change

(ii) The interacting forces are internal (action–reaction); therefore, total momentum as well total energy is conserved.

We can broadly classify the collision in two categories:

1. Elastic

2. Inelastic

Elastic Collision

A collision is said to be an elastic collision if both the kinetic energy and momentum are conserved in the collision.

During collision, the bodies are deformed but in elastic collision, they regain their original shape completely. In elastic collision, the interacting forces are conservative in nature.

Inelastic Collision

A collision is said to be an inelastic collision if the kinetic energy is not conserved in the collision. However, the momentum is conserved.

The loss in kinetic energy appears in the form of heat, sound and light energy. Here, the forces are non-conservative. Most of the collisions between macroscopic bodies are inelastic.

Perfectly Inelastic or Plastic Collision

If the two bodies stick together after the collision, the collision is said to be perfectly inelastic.

One-dimensional Elastic Collision (Head-on Collision)

One-dimensional elastic collision is that elastic collision in which the colliding bodies move along the same straight-line path before and after the collision.

Consider two bodies of masses m_1 and m_2 respectively with velocities $\vec{v}_{1i}$ and $\vec{v}_{2i}$ moving along the same straight line in the same direction and $|\vec{v}_{1i}| > |\vec{v}_{2i}|$.

Since they are moving in the same direction along the same line and the velocity of one body is greater than that of the other, they will collide after some time.

Let $\vec{v}_{1i}$ and $\vec{v}_{2i}$ be velocities of bodies respectively after the collision.

Applying the law of conservation of momentum,

Total momentum before collision = Total momentum after collision

$$\therefore \quad m_1 v_{1i} + m_2 v_{2i} = m_1 v_{1f} + m_2 v_{2f} \quad \text{(i)}$$

Kinetic energy before collision = Kinetic energy after collision

$$\frac{1}{2}m_1 {v_{1i}}^2 + \frac{1}{2}m_2 {v_{2i}}^2 = \frac{1}{2}m_1 {v_{1f}}^2 + \frac{1}{2}m_2 {v_{2f}}^2 \quad \text{(ii)}$$

From equations (i) and (ii) we get

$$v_{1f} = \frac{m_1 - m_2}{m_1 + m_2}v_{1i} + \frac{2m_2}{m_1 + m_2}v_{2i} \quad \text{(iii)}$$

$$v_{2f} = \frac{m_2 - m_1}{m_1 + m_2}v_{2i} + \frac{2m_1}{m_1 + m_2}v_{1i} \quad \text{(iv)}$$

We can get equation (iv) by interchanging the value of m_1 and m_2 and vice-versa.

Special cases

Case (1): Equal masses, $m_1 = m_2$

$$v_{1f} = \frac{m_1 - m_2}{m_1 + m_2} v_{1i} + \frac{2m_1}{m_1 + m_2} v_{2i} = v_{2i}$$

We find that the velocity of body A after collision is equal to the velocity of body B before collision.

$$v_{2f} = \frac{m_2 - m_1}{m_1 + m_2} v_{2i} + \frac{2m_1}{m_1} v_{1i}$$

Putting $m_2 = m_1$

$$v_{2f} = v_{1i}$$

The velocity of body (2) after collision is equal to the velocity of body (1) before collision.

To summarize, in a one-dimensional elastic collision of two bodies of equal masses, they interchange their velocities

Case (2): When the body (2) is initially at rest, i.e., $v_{2i} = 0$

$$v_{1f} = \frac{m_1 - m_2}{m_1 + m_2} v_{1i}$$

$$v_{2f} = \frac{2m_1}{m_1 + m_2} v_{1i}$$

Further, three sub-cases arise when the body (2) is at rest.

Sub-case (1): $m_2 << m_1$, i.e., the mass of body (2) is negligible as compared to the mass of the body (1).

$$v_{1f} \approx v_{1i}$$

So, the velocity of the massive body (1) after collision is nearly equal to its velocity before collision.

$$v_{2f} = 2v_{1i}$$

Sub-case (2): The velocity of the light body (2) after collision becomes approximately twice that of the massive body (1) before collision

Sub-case (3): $m_1 << m_2$, i.e., when light body (1) collides with a massive body (2).

$$v_{1f} \approx -v_{1i}$$

$$v_{2f} \approx 0$$

The velocity of the massive body after collision is very small.

ILLUSTRATION 3.6

A body of mass 2 kg makes an elastic collision with another body at rest and continues to move in the original direction with a speed equal to one-third of its original speed. Find the mass of the second body.

Solution

Here, $m_1 = 2\,\text{kg},\ v_2 = 0, v_1 = \frac{u_1}{3}$,

$$m_2 = ?$$

In a perfectly elastic collision in one dimension,

$$v_1 = \frac{(m_1 - m_2)u_1}{m_1 + m_2} + \frac{2m_2u_1}{m_1 + m_2}$$

$$\frac{u_1}{3} = \frac{(2 - m_2)}{2 + m_2} u_1 + 0$$

$$2 + m_2 = 6 - 3m_2$$

$$4m_2 = 4, m_2 = 1\text{kg}.$$

ILLUSTRATION 3.7

A bullet of mass 0.012 kg and horizontal speed 70 ms^{-1} strikes a block of wood of mass 0.4 kg and instantly comes to rest with respect to the block. The block is suspended from the ceiling by thin wire. Calculate the height to which the block rises. Also, estimate the amount of heat produced in the block.

Solution

Here, $m_1 = 0.012\,\text{kg}, u_1 = 70\,\text{m/s}$

$$m_2 = 0.4\,\text{kg}, u_2 = 0$$

As the bullet comes to rest with respect to the block, the two behave as one body. Let v be the velocity acquired by the combined mass.

Applying the principle of conservation of linear momentum,

$$(m_1 + m_2)v = m_1u_1 + m_2u_2 = m_1u_1$$

$$v = \frac{m_1u_1}{m_1 + m_2}$$

$$v = \frac{0.012 \times 70}{0.012 + 0.4} = \frac{0.84}{0.412} = 2.04\,\text{ms}^{-1}$$

The following table gives an idea of energy associate with some important phenomena.

S. No.	Phenomenon	Energy (J)
1	Energy required to break a bond in DNA	$= 10^{-20}$
2	Energy spent in turning a page	$= 10^{-3}$
3	Work done by human heart/beat	$= 0.5$

S. No.	Phenomenon	Energy (J)
4	Daily food intake of a human adult	$= 10^7$
5	Energy released in burning 1 litre of gasoline	$= 3 \times 10^7$
6	KE of a jet aircraft	$= 10^9$
7	Energy released in burning 1000 kg of coal	$= 3 \times 10^{10}$
8	Energy release of 15 megaton fusion bomb	$= 10^7$
9	Annual solar energy incident on earth	$= 5 \times 10^{24}$
10	Rotational energy of earth	$= 10^{29}$

Energy: The capacity to do work is known as energy.

Various forms of energy

Internal energy, kinetic energy, potential energy, chemical energy, thermal energy, electrical energy etc.

Kinetic energy

It is the energy possessed by a body by virtue of its motion.

$$KE = \frac{1}{2}mv^2$$

Work done by a force acting on a body is equal to the change in KE of the body.

$$W = \frac{1}{2}mv^2 - \frac{1}{2}mu^2$$

Potential energy

It is the energy possessed by a body by virtue of its position or distortion.

$$PE = m.g.h.$$

POINTS TO REMEMBER

1. The capacity of a body to do work is known as energy. Its units are the same as that of work.

   In dynamics, we are concerned with mechanical energy, which includes kinetic and potential energies.
2. The work–energy theorem states that the work done on a particle by the resultant force is equal to the change in its kinetic energy, i.e.,

   Work done = Change in kinetic energy
3. A force is called conservative if the work done by it during a complete cycle of a system is zero; otherwise, it is called non-conservative. Gravitational force, Coulomb force, force by a spring, etc. are a few examples of conservative forces. Frictional force and viscous force are examples of non-conservative forces.
4. Einstein's mass energy equivalence states that a mass m is equivalent to an energy E and the two are related by

   $$E = mc^2$$

   where $c = 3\times10^8$ m/s is the speed of light in vacuum.
5. Energy can neither be created nor destroyed. It can only be converted from one form to another.

COMMON MISCONCEPTIONS

1. **Work is done by … on …**

   It is important to identify the interaction of the system with the environment. When discussing work, try always to use the phrase, "the work done by … on. …" After "by", insert the part of the environment that is interacting directly with the system. After "on", insert the system. For example, "the work done by the hammer on the nail' identifies the nail as the system and the force from the hammer represents the interaction with the environment.
2. **Conditions for the work–kinetic energy theorem**

   Always remember these special conditions for the work–kinetic energy theorem. We will see many situations in which other changes occur in the system besides its speed, and there are other interactions with the environment besides work. The work–kinetic energy theorem is important, but limited in its application. It is not a general principle.
3. **The work–kinetic energy theorem: speed, not velocity**

   Note that the work–kinetic energy theorem states that work is related to a change in the speed of an object, not a change in velocity. For example, if an object is in uniform circular motion, the speed is constant. So, even though the velocity is changing, no work is done by the force causing the circular motion.

This is another reason why the energy approach is so attractive – we can handle many problems with only scalar calculations.

4. Heat is not a form of energy

The word heat is one of the most misused words in our popular language. Note that heat is a method of transferring energy, it is not a form of storing energy. Thus, phrases such as "heat content", "the heat of the summer," and "the heat escaped" all represent incorrect uses of this word.

PRACTICE QUESTIONS

Very-short Answer Type Questions

1. What is the work done in holding a 10 kg suitcase while waiting for a bus for 5 minutes?
2. Is it practically possible to have situations where $(E-U)<0$? (E is total energy, and U is potential energy.)
3. Can kinetic energy of a system be increased or decreased without applying any external force on the system?
4. Give some examples where a force does work on a body but fails to change its kinetic energy.
5. Does the sun do any work on earth, when earth revolves around the sun in a perfectly circular orbit?
6. Is work done by a non-conservative force always negative? Comment.
7. Is linear momentum of a system always conserved?
8. Is it possible to have a collision in which the whole of kinetic energy is lost?
9. What should be the angle between the force and the displacement for maximum and minimum work?
10. Can a body have energy without momentum?
11. In which type of collision, the momentum is conserved – elastic or inelastic? What about kinetic energy?
12. In which motion, momentum changes but kinetic energy does not?
13. A shot fired from a cannon explodes in the air. What will be the changes in momentum and kinetic energy?
14. The momentum of an object is doubled. How does its kinetic energy change?
15. Does the work done in moving a body depend on how fast or how slow the body is moved?
16. What type of energy is stored in the spring of a watch?
17. Is it possible to exert a force which does work on a body without changing its kinetic energy? If so, give example.
18. Define 1 joule (J).
19. Give two examples of objects having potential as well as kinetic energy.
20. An object is thrown vertically upwards. What is the kinetic energy of the body at the highest point of its motion?
21. A body of mass m falls freely from a height h. What will be its kinetic energy just before it touches the ground?
22. A dry cell converts one form of energy into another. Name the forms of energy.
23. A freely falling object eventually stops on reaching the ground. What happens to its kinetic energy on reaching the ground?
24. A mass of 10 kg is at a point A on a table. It is moved to a point B. If the line joining A and B is horizontal, what is the work done on the object by the gravitational force? Explain your answer.
25. The potential energy of a freely falling object decreases progressively. Does this violate the law of conservation of energy? Why?
26. Does the transfer of energy take place when you push a huge rock with all your might and fail to move it? Where is the energy you spend doing?
27. What is the work done by the force of gravity on a satellite moving round the earth? Justify your answer.

Short-answer Type Questions

1. Explain the reason why the potential energy in the following cases increases or decreases:
   (a) A spring is compressed
   (b) A spring is stretched

(c) Two dis-similar charges are brought near each other
(d) A body is taken away against the gravitational forces.

2. A meteorite burns in the atmosphere before it reaches earth's surface. What happens to its momentum?

3. A body is thrown upwards by applying a force *F* in the upward direction. The body is also simultaneously acted upon by the force due to gravity. Find out:
(i) The work done by applied force.
(ii) The work done by force of gravity.

4. Why is the work done on a body moving along a circular path zero?

5. What would have a greater effect on the kinetic energy possessed by an object – doubling its mass or doubling its velocity? Explain.

6. Derive an expression for the energy possessed by a body of mass *m* kept at a height *h* from the ground.

7. A ball of mass *m* moving with velocity *v* strikes head-on elastically with a number of balls of same mass at rest in a line. Only one ball from other side moves with same velocity. Explain why not two balls move simultaneously, each with velocity $v/2$?

8. If two objects collide and one is initially at rest,
(a) is it possible for both to be at rest after collision?
(b) is it possible for anyone to be at rest after collision?

9. Can a body have momentum without energy?

10. The momentum of a body increased by 50%. What is the percentage change in its KE?

11. An object thrown at a certain angle to the ground moves in a curved path and falls back to the ground. The initial and the final points of the path of the object lie on the same horizontal line. What is the work done by the force of gravity on the object?

12. A battery lights a bulb. Describe the energy changes involved in the process.

13. A trolley of mass 300 kg carrying a sand bag of 25 kg is moving uniformly with a speed of 27 km/h on a frictionless track After a while, sand starts leaking out of a hole on the trolley's floor at the rate of 0.05 kg s^{-1}. What is the speed of the trolley after the entire sand bad is empty?

14. A certain household has consumed 250 units of energy during a month. How much energy is this in joules?

15. A freely falling object eventually stops on reaching the ground. What happens to its kinetic energy?

16. Is it possible that a body be in accelerated motion under a force acting on the body, without work being done by the force? Give example.

17. Mountain roads rarely go straight up the slope, but wind up gradually. Why?

18. The velocity of an aeroplane is doubled. (a) What will happen to the momentum? Will the momentum remain conserved? (b) What will happen to the KE? Will the energy remain conserved?

19. What are the various energy transformations that occur when you are riding a bicycle?

20. What is power? What is the SI unit of power? Define it.

Descriptive Type Questions

1. Explain the meaning of kinetic energy with examples. Obtain an expression for kinetic energy of a body moving uniformly?

2. State and derive work–energy theorem.

3. What are the different forms of energy? Explain.

4. Energy remains constant in the case of freely falling bodies. Prove this statement.

5. Illustrate the law of conservation of energy by discussing the energy changes in an oscillating pendulum. Is the stopping of the pendulum, eventually, a violation of the law of conservation of energy?

NCERT Corner

1. Define watt. Express kilowatt in terms of joule per second. A 150 kg car engine develops 500 W for each kg. What force does it exert in moving a car at a speed of 20 m s^{-1}?

2. Is it possible that an object is in a state of accelerated motion due to external force acting on it, without work being done by the force. Explain it with an example.

3. A girl having mass of 35 kg sits on a trolley of mass 5 kg. The trolley is given an initial velocity of 4 ms^{-1}

by applying force. The trolley comes to rest after traversing a distance of 16 m. (a) How much work is done on the trolley? (b) How much work is done by the girl?

4. Four men lift a 250 kg box to a height of 1 m and hold it without raising or lowering it. (a) How much work is done by the men in lifting the box? (b) How much work do they do in just holding it? (c) Why do they get tired while holding it? ($g = 10$ m s^{-2})

5. Can any object have momentum even if its mechanical energy is zero? Explain.

6. The power of a motor pump is 2 kW. How much water can the pump can raise to a height of 10 m in a minute? (Given $g = 10$ m s^{-2})

7. A light and a heavy object have the same momentum. Find out the ratio of their kinetic energies. Which one has a larger kinetic energy?

8. The velocity of a body moving in a straight line is increased by applying a constant force *F*, for some distance in the direction of the motion. Prove that the increase in the kinetic energy of the body is equal to the work done by the force on the body.

9. Shivansh can run with a speed of 8 m s^{-1} against a frictional force of 10 N, and Pradeep can move with a speed of 3 m s^{-1} against the frictional force of 25 N. Who is more powerful and why?

10. A ball is dropped from a height of 10 m. If the energy of the ball reduces by 40% after striking the ground, how high can the ball bounce back? ($g = 10$ m s^{-2})

11. What is power? How do you differentiate kilowatt from kilowatt hour? The Jog Falls in Karnataka state are nearly 20 m high. 2000 tonnes of water fall from it in a minute. Calculate the equivalent power if all this energy can be utilized? ($g = 10$m s^{-2})

12. The weight of a person on a planet A is about half that on the earth. He can jump up to 0.4 m high on the surface of the earth. How high he can jump on the planet A?

13. Compare the power at which each of the following is moving upward against the force of gravity? ($g = 10$m s^{-2})

(i) a butterfly of mass 1.0 g that flies upward at a rate of 0.5 m s^{-1}

(ii) a 250 g squirrel climbing up on a tree at a rate of 0.5 m s^{-1}.

14. A boy is moving on a straight road against a frictional force of 0.5 N. After travelling a distance of 1.5 km he forgot the correct path at a roundabout of radius 100 m. However, he moves on the circular path for one-and-a-half cycles and then he moves up to 2.0 km. Calculate the work done by him.

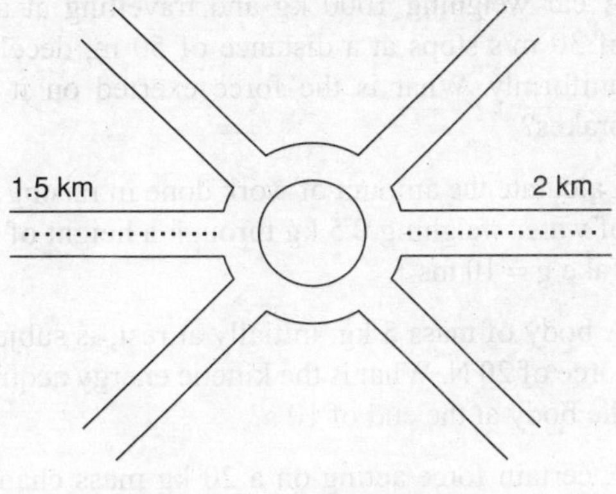

15. Can any object have mechanical energy even if its momentum is zero? Explain.

16. If an electric iron of 1200 W is used for 30 minutes every day, find the electric energy consumed in the month of April.

17. How is power related to the speed at which a body can be lifted? How many kilograms will a man working at the power of 100 W, be able to lift at constant speed at 1 ms^{-1} vertically? ($g = 10$ m s^{-2})

18. A rocket is moving up with a velocity *v*. If the velocity of this rocket is suddenly tripled, what will be the ratio of the two kinetic energies?

19. An automobile engine propels a 1000 kg car (A) along a levelled road at a speed of 36 km h^{-1}. Find the power if the opposing frictional force is 100 N. Now, suppose after travelling a distance of 200 m, this car collides with another stationary car (B) of same mass and comes to rest. Let its engine also stop at the same time. Now car (B) starts moving on the same level road without getting its engine started. Find the speed of the car (B) just after the collision.

Numericals

Level - I

1. A boy pulls a bucket of water of mass 5 kg from a well which is 10 m deep in 10 s. Calculate the power used by him. ($g = 10$ m/s^2)

2. A girl weighing 500 N climbs 40 m vertically when walking up the stairs in a building. How much work does she do against gravity? Enumerate the energy changes here.

3. Calculate the kinetic energy of a body of mass 2 kg moving with a velocity of 0.1 metre per second.

4. Two bodies of equal masses move with uniform velocities v and $3v$ respectively. Find the ratio of their kinetic energies.

5. A car weighing 1000 kg and travelling at a speed of 30 m/s stops at a distance of 50 m, decelerating uniformly. What is the force exerted on it by the brakes?

6. Calculate the amount of work done in raising a glass of water weighing 0.5 kg through a height of 20 cm. Take $g = 10\ ms^{-2}$.

7. A body of mass 5 kg, initially at rest, is subject to a force of 20 N. What is the kinetic energy acquired by the body at the end of 10 s?

8. A certain force acting on a 20 kg mass changes its velocity from $5ms^{-1}$ to $2ms^{-1}$. Calculate the work done by the force.

9. An object of mass 40 kg is raised to a height of 5 m above the ground. What is its potential energy? If the object is allowed to fall, find its kinetic energy when it is half-way down.

10. An electric heater is rated 1500 W. How much energy does it use in 10 hours?

Level – II

1. An automobile is moving at 100 km/hour and is exerting a tractive force of 400 kg. wt. What H.P. must the engine develop if 20% of the power developed is wasted?

2. A rubber ball falls on a floor from a height of 19.6 m. Calculate the velocity with which it strikes the ground. To what height will the ball rebound if it loses 25% of its energy on striking the ground?

3. The human heart discharges 75 ml of blood at each beat against a pressure of 0.1 m of Hg. Calculate the power of the heart assuming that the pulse frequency is 80 beats per minute. Given, density of mercury = $13.6 \times 10^3\ kg/m^3$.

4. A shot travelling at the rate of $100\ ms^{-1}$ is just able to pierce a plank 4 cm thick. What velocity is required to just pierce a plank 9 cm thick?

5. How much mass is converted into energy per day in Tarapur nuclear power plant operated at 10^6 kW?

6. The bob of a pendulum is released from a horizontal position A as shown in figure. If the length of the pendulum is 15 m, what is the speed with which the bob arrives at the lowermost point B, given that it dissipates 5% of its initial energy against air resistance?

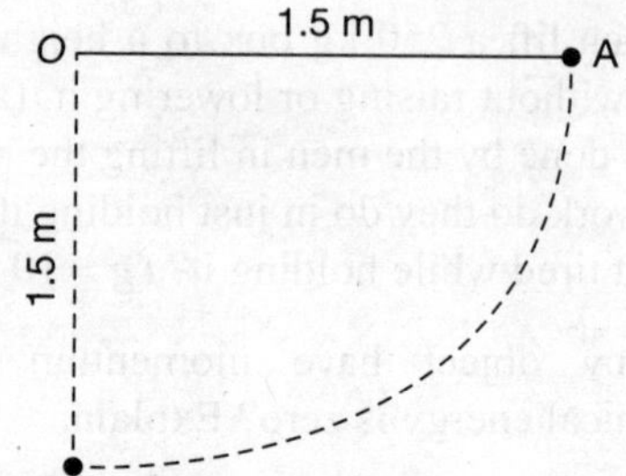

7. A light body and a heavy body have the same kinetic energy. Which one has greater linear momentum?

8. A bob of mass 0.3 kg falls from the ceiling of an elevator moving down with a uniform speed of $7\ ms^{-1}$. It hits the floor of the elevator (length of the elevator = 3m) and does not rebound. What is the heat produced by the impact? Would your answer be different if the elevator were stationary?

9. A gardener moves a lawn roller through a distance of 100 m with a force of 50 N. Calculate his wages, if he is to be paid 10 paise for doing 50 J of work. It is given that the applied force is inclined at 60° to the direction of motion.

Fill in the Blanks

1. A 1kg mass has a kinetic energy of 1 joule when its speed is ________.

2. A photoelectric cell converts ________ into ________ energy.

3. $F \times s = \frac{1}{2}$ ________.

4. Newton-meter is the unit of ________.

5. When a body falls a vertical distance its ________ energy gets converted into ________ energy.

6. When non-zero force and displacement are observed with zero work done, the angle between the force and displacement is ________.

True or False

1. Power is the rate of doing work.

2. If the mass of an object is doubled, its potential energy is increased four times.

3. Watt is the SI unit of energy.

3. (d)

4. (c)

5. (c)

6. (e)

7. (d)

8. (a)

9. (a)

10. (a)

Comprehension Type Questions

1. (a) As we know work done is given by the area under F–x curve, we have from the figure

$$W=\frac{1}{2}\times3\times30+\frac{1}{2}\times1\times50+20\times2+\frac{1}{2}\times2\times20$$
$$=130\,\text{J}$$

(b) Gain in kinetic energy of the block = 130 J

(c) $\frac{1}{2}mv^2=130$

or $v=\sqrt{65}$ m/s = 8.062 m/s

(d) As the force ceases to act, the speed remains constant, i.e., $\sqrt{65}$ m/s or 8.062 m/s.

(e) Distance = Speed × Time

$$=\sqrt{65}\times5$$
$$=5\sqrt{65}=5\times8.062=40.31\,\text{app.}$$

2. Here, m = 2 kg, u = 0, F = 7 N;

$\mu=0.1.\text{W}=?\,t=10$ s

Acceleration produced by applied force

$$a_1=\text{F}/m=7/2=3.5\text{ m/s}^2$$

Force of friction, $F_f=\mu\text{R}=\mu mg$

$=0.1\times2\times9.8=1.96$ N

Retardation produced by friction,

$$a_2=\frac{-F_f}{m}=-\frac{1.96}{2}=-0.98\,\text{ms}^{-2}$$

Net acceleration with which body moves,

$$a=a_1+a_2=3.5-0.98=2.52\text{ ms}^{-2}$$

Distance moved by the body in 10 second, from

$$s=ut+\frac{1}{2}at^2$$

$$s=0+\frac{1}{2}\times2.52\times(10)^2=126\text{ m.}$$

(a) Work done by the applied force

$=\text{F}\times s$

$\text{W}_1=7\times126=882$ J.

(b) Work done by the force of friction

$\text{W}_2=-F_f\times s=-1.96\times126$

$=-246.9$ J

(c) Work done by the net force

W_3 = Net force × Distance

$$=\left(\text{F}-F_f\right)s=(7-1.96)126=635\text{ J.}$$

(d) From $v=u+at$

$$v=0+2.52\times10=25.2\text{ ms}^{-1}.$$

$\therefore$ Final KE $=\frac{1}{2}mv^2$

$$=\frac{1}{2}\times2\times(25.2)^2=635\text{J.}$$

Initial KE $=\frac{1}{2}mu^2=0.$

$\therefore$ Change in KE = 635 – 0 = 635 J

This shows that change in KE of the body is equal to work done by the net force on the body.

3. Here, $m=10\,\text{kg}$, $h=0.5\,m$, $n=1000$

(a) Work done against gravitational force

$W=n(mgh)$

$=1000\times(10\times9.8\times0.5)=49000$ J.

(b) Mechanical energy supplied by 1 kg of fat

$$=3.8\times10^7\times\frac{20}{100}$$

$$=0.76\times10^7\text{ J/kg}$$

$\therefore$ Fat used up by the dieter

$$=\frac{1\text{kg}}{0.76\times10^7}\times49000$$

$$=6.45\times10^{-3}\text{ kg}$$

4. (a) Volume of wind flowing/sec = A v

Mass of wind flowing/sec = A $v\rho$

Mass of air passing in t sec = A $v\rho t$

(b) KE of air $=\frac{1}{2}mv^2=\frac{1}{2}(\text{A}\,v\rho t)v^2$

$$=\frac{1}{2}\text{A}v^3\rho t$$

(c) Electrical energy produced

$$=\left(\frac{25}{100}\right)\times\text{KE of air}$$

$$=\frac{1}{4}\times\frac{1}{2}\text{A}v^3\rho t$$

Electric power $=\frac{1}{8}\frac{\text{A}v^3\rho t}{t}=\frac{1}{8}\text{A}v^3\rho$

$$P = \frac{1}{8} \times 30 \times (10)^3 \times 1.2$$

$$= 4500 \text{ watt} = 4.5 \text{ kW}$$

5. (a) J
   (b) When $\cos\theta$ is negative, i.e., $\theta > 90°$.
   (c) No, because point of application of force does not move.
   (d) Work done is zero because $\theta = 90°$.
   (e) Work is a scalar quantity but at the same time it is an algebraic quantity.

Objective Questions

One Option Correct

1. (b) $W = Fd\cos\theta$

$0 = Fd\cos\theta$

$\Rightarrow \cos\theta = 0$

$\theta = 90°$

2. (b) In circular path, the angle between F and d is 90°, so no work is done.

4. (b) PE is converted into KE

PE = mgh

$= 2 \times 9.8 \times 1$

$= 19.6$ J

KE = 19.6 J

5. (b) Energy has the same unit as Work

$W = P \times t$

= Watt second

6. (b) $\text{KE} = \frac{1}{2}mv^2$

$\text{KE}_A = \frac{1}{2}m_A v^2$

$\text{KE}_B = \frac{1}{2}m_B v^2$

$\text{KE}_A : \text{KE}_B = m_A : m_B$

$= 2:1$

8. (c) $\text{KE} = \frac{1}{2}mv^2$

$50 = \frac{1}{2} \times 1 \times v^2$

$v^2 = 100$

$v = 10$ m/s

10. (a) $W = Fd$

$= 6 \times 2 = 12$ J

$E_A = Fd\,\text{Cos}\theta$

$= 50 \times 10 \times 5\,\text{Cos}30°$

$= 2500 \times \frac{\sqrt{3}}{2}$

$E_B = \frac{1}{2}mv^2$

$= \frac{1}{2} \times 25 \times (2)^2$

$= 50$ J

13. (a) In a circular path, the force and displacement are always perpendicular to each other; therefore work done is Zero

21. (d) Watt and Horse power are the units of power.

26. (d) Total mass = (50 + 20) = 70 kg
Total height = 20 × 0.25 = 5m
∴ Work done = $mgh = 70 \times 9.8 \times 5 = 3430$ J

50. (c) According to work–energy theorem
W = change of the kinetic energy

$$FS\cos\theta = \frac{1}{2}mv^2 - \frac{1}{2}mu^2$$

Substituting the given values, we get

$$20 \times 4 \times \cos\theta = 40 - 0$$

$$\cos\theta = \frac{40}{80} = \frac{1}{2}$$

$$\theta = \cos^{-1}\left(\frac{1}{2}\right) = 60°$$

51. (c) There is no displacement.

52. (a) Let d_s be the distance travelled by the vehicle before it stops.
Here, final velocity $v = 0$, initial velocity $= u$
Using equation of motion $v^2 = u^2 + 2aS$
∴ $0^2 = u^2 + 2ad_s$

or Stopping distance, $d_s = \frac{u^2}{2a}$

54. (d) $W = FS\cos\theta = 10 \times 4 \times \cos 60° = 20$ joule

55. (b) Work × Displacement
If force and displacement both are doubled, then work would be increased four times.

56. (c) When a force of constant magnitude which is perpendicular to the velocity of a particle acts on the particle, work done is zero and hence changes in kinetic energy is zero.

59. (a) Maximum height reached by the particle

$$H_{max} = \frac{u^2}{2g} = \frac{(5)^2}{2 \times 10} = 1.25 \text{ m}$$

Work done $= \vec{F}.\vec{d} = Fd\cos\theta$

$= mg\times(H_{max})\times\cos(180°)$

$= 0.1\times10\times1.25\times(-1) = 1.25$ J

60. (c) When the ball is released from the top of the tower, then ratio of distances covered by the ball in first, second and third second

$h_1 : h_2 : h_3 = 1:3:5:$ [Because $h_n \propto$]

$\therefore$ Ratio of work done mgh_I: mgh_{II}: mgh_{III} = 1: 3: 5

61. (a) As the surface is smooth, work done against friction is zero. Also, the displacement and force of gravity are perpendicular; so, work done against gravity is zero.

62. (a) KE acquired by the body = Work done on the body.

KE $= \frac{1}{2}mv^2 = Fs$, i.e., it does not depend upon the mass of the body although velocity depends upon the mass $v^2 \propto$ [if F and s are constant].

63. (d) $W = Fd \text{ Cos } \theta$

$25 = 5\times10\times\text{Cos }\theta$

$\Rightarrow \text{Cos}\theta = \frac{1}{2}$

or $\theta = 60$

64. (a) The potential energy does not change because the car is moving on a level road.

$PE = mgh$

h remains same, therefore PE remains same

65. (d) $W = F\times h$

$= 10\times2$

$= 20\ J$

66. (b) $W = mg\sin\theta\times s$

$= 2\times10^3\times\sin15°\times10$

$= 5.17$ kJ

67. (c) $W = Fs\cos\theta \Rightarrow \cos\theta = \frac{W}{Fs} = \frac{25}{50} = \frac{1}{2} \Rightarrow \theta = 60°$

More Than One Option Correct

2. The minimum speed required to complete the circular motion is $(5gl)^{\frac{1}{2}} = 5.42\frac{\text{m}}{\text{s}}$, which is more than the imparted speed. So the particle will not be able to attain the maximum height due to the initial speed, which is $\frac{v^2}{2\text{g}} = 90$m. The string will slack before it reaches the 90 m height above the initial point.

3. Inside a system of particles, the conservative forces cause a decrease in potential energy of the system when work done by these forces is positive, which imparts kinetic energy in particles of the system.

4. When the direction of friction (static or kinetic) on a body is in the direction of motion of the point where friction is acting then work done by the friction will be positive in a given frame of reference. In a conservative force field, internal forces of system can increase the kinetic energy of system by doing work and by accounting of work done by pseudo-force on the particles of a system. We can use work–energy theorem for analysis of dynamics of the body in non-inertial reference frames.

5. The work done by a force on an object in any reference frame is zero only if the displacement of point of application of force in the direction of force is zero.

7. As KE is proportional to time, we can say speed of the body is proportional to $\sqrt{t}$. Then as acceleration $a = dv/dt$, we get acceleration is inversely proportional to $\sqrt{t}$. Hence options (b) and (c) are correct.

Scholar Zone

1. The work done is

$W = FS$, as $\theta = 0°$

$\Rightarrow W = 2.50\times2.50$

$\Rightarrow W = 6.25$ J

The acceleration of the particle is

$$a = \frac{F}{m} = \frac{2.50\text{ N}}{15\text{ g}}$$

$$\Rightarrow a = \frac{2.50\text{ N}}{0.015\text{ kg}} = 166.67\text{ m/s}^2$$

If t be the time taken by the particle to move through a distance S starting from rest and with acceleration a, then,

$$S = \frac{1}{2}at^2$$

$$\Rightarrow \Rightarrow t = \sqrt{\frac{2S}{a}} = \sqrt{\frac{2\times2.50}{166.67}}$$

$$\Rightarrow \Rightarrow t = 0.173\text{s}$$

If P be the average power delivered then,

$$P = \frac{W}{t} = \frac{6.25}{0.173}\text{W}$$

$$\Rightarrow P = 36.1\text{ W}$$

2. Let the body hit the ground at a distance x from the foot of the hill.

As the shape of the curved portion of the hill is arbitrary, the force of reaction of the hill acting on the body keeps on changing in magnitude and direction and therefore, we cannot apply Newton's of motion in this case. However, we have to use the principle of conservation of energy.

Taking the gravitational potential energy at the ground to be zero, the total energy of the body at A, which is entirely potential, is

$$E_A = mgH$$

When the body reaches the point B, it loses potential energy but gains kinetic energy. If v be the velocity of the body at B, then its total energy at B, which is partly kinetic and partly potential, is

$$E_B = mgh + \frac{1}{2}mv^2$$

From the principle of conservation of energy,

$$E_A = E_B$$

$$\Rightarrow \quad \Rightarrow mgH = mgh + \frac{1}{2}mv^2$$

$$\Rightarrow \quad \Rightarrow v = \sqrt{2g(H-h)}$$

This velocity at B is directed horizontally, as the portion of the hill near B is horizontal. Hence, as the body leaves the edge B of the hill, it behaves like a projectile, projected horizontally with an initial velocity v. The downward velocity of the body at this moment is obviously zero.

The body, therefore, falls freely under gravity and reaches the ground after time t.

Hence, for the vertical motion of the body,

$$h = \frac{1}{2}gt^2$$

$$\Rightarrow \quad t = \sqrt{\frac{2h}{g}}$$

3. When the string is horizontal, the bob is at a height of $l = 1.00$ m from the lowest point C of its path. If the potential energy at this lowest position is taken to be zero then, the bob has only potential energy at A, both kinetic as well as potential energy at B and only kinetic energy at the point C.

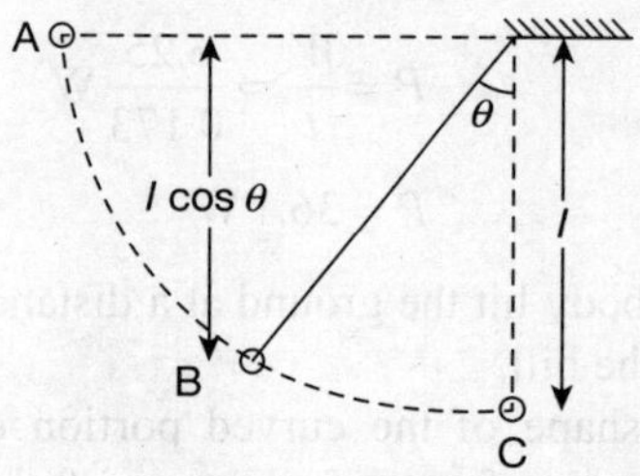

(i) If v_c be the velocity of the bob at C, then total energy at C $= \frac{1}{2}mv_c^2 + 0$

$$\Rightarrow E_c = \frac{1}{2}mv_c^2$$

and total energy at

$$A = 0 + mgl$$

$$\Rightarrow E_A = mgl$$

From the principle of conservation of energy,

$$E_C = E_A$$

$\Rightarrow$ Kinetic energy at C $= mgl = 0.1 \times 9.8 \times 1.00$

Kinetic energy at C = 0.98J

(ii) When the bob falls from A to B, it loses potential energy equal to $mgl \cos\theta$ but gains kinetic energy equal to $\frac{1}{2}mv_B^2$ where v_B is the velocity of the bob at the point B.

From the principle of conservation of energy,

Gain in kinetic energy = Loss in potential energy

$$\Rightarrow \quad \frac{1}{2}mv_B^2 = mgl\cos\theta$$

$$\Rightarrow \quad \frac{1}{2}mv_B^2 = 0.1 \times 9.8 \times 1.00 \times \cos 30°$$

$\Rightarrow$ Kinetic energy at B = 0.848 J

4. Here, $m_1 = m_2 = m$

If $u_1 = v, u_2 = -v$

As the collision is perfectly elastic, the velocities after collision are given by

$$v_1 = \frac{(m_1 - m_2)u_1}{m_1 + m_2} + \frac{2m_2u_2}{m_1 + m_2}$$

$$v_1 = 0 + \frac{2m(-v)}{m+m} = -v$$

$$v_2 = \frac{2m_1u_1}{m_1 + m_2} + \frac{(m_2 - m_1)u_2}{m_1 + m_2}$$

$$v_2 = \frac{2mv}{2m} + 0 = v$$

Hence, after collision, the two balls bounce back with equal speeds.

5. Here, $m_1 = 0.1$ kg $m_2 = ?$

$u_2 = 0$ Let $u_1 = u$;

$$\therefore \qquad v_1 = -\frac{u}{3}$$

As $$v_1 = \frac{(m_1 - m_2)u_1}{m_1 + m_2} + \frac{2m_2u_2}{m_1 + m_2}$$

$$\therefore \qquad -\frac{u}{3} = \frac{(0.1 - m_2)u}{0.1 + m_2}$$

or $$-\frac{1}{3} = \frac{0.1 - m_2}{0.1 + m_2}$$

or $$-0.3 + 3m_2 = 0.1 + m_2$$

$$2m_2 = 0.1 + 0.3 = 0.4$$

$$m_2 = \frac{0.4}{2}\text{ kg.} = 0.2\text{ kg}$$

6. As work done W is equal to area under the F–s curve,

$$W_{AB} = W_{12} + W_{23} + W_{34} + W_{45}$$

= Area under AK + Area under KL + Area under LM – Area enclosed by MB

$$= 10(1) + \left\{10(1) + \frac{5\times 1}{2}\right\} + \frac{15\times 1}{2} - \frac{15\times 1}{2}$$

$$= 10 + 10 + 2.5 + 7.5 - 7.5$$

$$W_{AB} = 22.5\text{ J}$$

7. Here, mass of rocket,
$m_1 = 0.1$ kg;
Mass of fuel, $m_2 = 0.02$ kg.; $t = 3$ s.
Initial velocity $u = 0$
Final velocity $v = 20$ m/s
From $v = u + at$

$$20 = 0 + a\times 3,\ a = \frac{20}{3}\text{ ms}^{-2}$$

Thrust on the rocket, $F = m_1 a$

$$= 0.1\times\frac{20}{3} = \frac{2}{3}\text{ N}$$

If s is the distance moved by the rocket in 3 sec, then from

$$s = ut + \frac{1}{2}at^2$$

$$s = 0 + \frac{1}{2}\times\frac{20}{3}\times 3\times 3 = 30\text{ m.}$$

Energy content of the fuel = Work done on rocket = Thrust × Distance $= \frac{2}{3}\times 30 = 20\text{ J}$.

Energy content per unit mass of the fuel $= \frac{20}{0.02} = 1000$ joule/kg.

8. If a crate is located on the bed of a truck, and the truck accelerated to the east, the static friction force exerted on the crate by the truck acts to the east to give the crate the same acceleration as the truck (assuming the crate doesn't slip). Another example is a car that accelerates because of the friction forces exerted on the car's tires by the road. These forces act in the direction of the car's motion, and the sum of these causes an increase in the car's kinetic energy.

9. (a) Assuming that the person lifts the box with a force of magnitude equal to the weight of the box mg, the work done by the person on the box during the vertical displacement is $W = Fd = (g)(h) = mgh$, because the force in this case is in the same direction as the displacement. For the horizontal displacement, we assume that the acceleration of the box is approximately zero. As a result, the work done by the person on the box during the horizontal displacement of the box is zero because the horizontal force is approximately zero, and the force to support the box's weight in this process is perpendicular to the displacement. Thus, the net work done by the person during the complete process is mgh.

(b) The work done by the gravitational force on the box during the vertical displacement of the box is – mgh, because this force is opposite the displacement. The work done by the gravitational force is zero during the horizontal displacement because this force is perpendicular to the displacement.
Hence, the net work done by the gravitational force for the complete process is $-mgh$. The net work done by all forces on the box is zero, because $+mgh + (-mgh) = 0$.

CHAPTER

4 Gravitation

LEARNING OBJECTIVES

At the end of this chapter, you will be able to:

- Define Newton's law of gravitation
- Define acceleration due to gravity and its variation
- Learn about Satellite, its orbital velocity, time period, and height
- Explain Geostationary Satellite
- Define Kepler's Law and escape velocity

Introduction

The phenomenon of attraction between different bodies in the universe is called gravitation. The force of attraction is called force of gravitation.

Centre of Mass and Centre of Gravity (C.M. and C.G.)

Centre of mass (C.M.) is a point inside or outside a body, at which the whole mass of the body is supposed to be concentrated. An external force applied on the centre of mass will produce the same motion in the body as if the body is a point mass.

Centre of gravity (C.G.) is a point inside or outside a body, at which the whole weight of the body is supposed to be acting.

An external force applied on the centre of gravity, will make the body move in the direction of force just as particle move.

ILLUSTRATION 4.1

Two spheres of mass m are kept at distance l apart, find the centre of mass of the system.

Solution

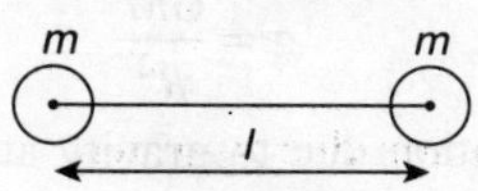

$$\text{C.M.} = \frac{m \times 0 + m \times l}{m + m} = \frac{l}{2}$$

Newton's Universal Law of Gravitation

The law states that everybody in this universe attracts every other body with a force which is directly proportional to the product of their masses and inversely proportional to the square of the distance between their centres.

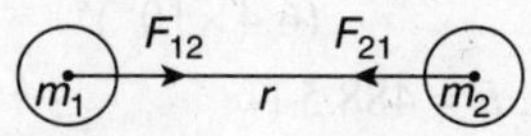

Let two bodies of masses m_1 and m_2 lie with centres at distance r apart. Then force of attraction between them is given by

$$F = \frac{Gm_1m_2}{r^2}$$

where G is constant of proportionality. It is called universal gravitational constant. The force of gravitation is mutual in accordance with Newton's third law of motion. Here action takes place from a distance and so is the reaction.

Universal Gravitational Constant

In the relation,

$$F = \frac{Gm_1m_2}{r^2}$$

If $m_1 = m_2 = 1$, $r = 1$, then $F = G$. Hence, universal gravitational constant may be defined as the force of attraction between two bodies of unit mass each, when kept with their centres a unit distance apart.

Units

S.I.: $G = 6.67 \times 10^{-11}$ Nm^2/kg^2

C.G.S: $G = 6.67 \times 10^{-8}$ dyn cm^2/g^2

ILLUSTRATION 4.2

Suppose you and your friend have mass 50 kg each. Suppose also that both of you are standing such that your centre of gravity are 1 m apart. Calculate the force of gravitation between you and your friend. Calculate also force of gravity acting on you using equation

Force of gravity = $\frac{GMm}{R^2}$ (Take $M = 6 \times 10^{24}$ kg, $R = 6.4 \times 10^6$ m)

Solution

$$m_1 = m_2 = 50 \text{ kg}, r = 1\text{m}$$

$$F = \frac{G m_1 m_2}{r^2}$$

$$\therefore \quad F = \frac{6.67 \times 10^{-11} \times 50 \times 50}{(1)^2}$$

Force of gravitation, $F = 1.67 \times 10^{-7}$ N

Also, $M = 6 \times 10^{24}$ kg, $R = 6.4 \times 10^6$ m

$$F = \frac{GMm}{R^2}$$

$$\therefore \quad F = \frac{6.67 \times 10^{-11} \times 6 \times 10^{24} \times 50}{(6.4 \times 10^6)^2}$$

Force of gravity, $F = 488.5$ N

By direct calculation, $F = mg = 50 \times 9.8 = 490$

$$\therefore \quad F = 490 \text{ N}$$

Acceleration Due to Gravity

The acceleration produced in a body by the force of gravity, is called acceleration due to gravity. It is represented by the symbol g.

Force of gravity between planet having radius R of mass M and body of mass m, when the body lies on the surface of the planet, is given by:

$$F = \frac{G Mm}{R^2}$$

$$\text{Acceleration of body} = \frac{\text{Force on body}}{\text{Mass of body}}$$

$$g = \frac{G Mm / R^2}{m}$$

$$g = \frac{GM}{R^2}$$

For Earth

$$M = 6 \times 10^{24} \text{ kg}, R = 6.4 \times 10^6 \text{ m}$$

$$G = 6.7 \times 10^{-11} \text{ Nm}^2/\text{kg}^2$$

Hence, $g = \frac{6.7 \times 10^{-11} \times 6 \times 10^{24}}{(6.4 \times 10^6)^2} = 9.8$

$$g = 9.8 \text{ ms}^{-2}$$

The body moves towards the earth with this acceleration.

ILLUSTRATION 4.3

Find the acceleration due to gravity on moon's surface. $M = 7.35 \times 10^{22}$ kg, $R = 1.7 \times 10^6$ m, $G = 6.7 \times 10^{-11}$ Nm²/kg²

Solution

$$g = \frac{GM}{R^2}$$

$$= \frac{6.7 \times 10^{-11} \times 7.35 \times 10^{22}}{(1.7 \times 10^6)^2} = 1.7$$

$$g_{moon} = 1.7 \text{ ms}^{-2}$$

Variation of Acceleration due to Gravity

The value of acceleration due to gravity changes with height (i.e., altitude), depth and shape of the earth. The effect of each of the above factors on the value of g is discussed below.

Effect of Altitude

Consider earth to be a sphere of mass M, radius R with centre O. Let g be the value of acceleration due to gravity at a point A on the surface of earth as shown in Fig. 4.1.

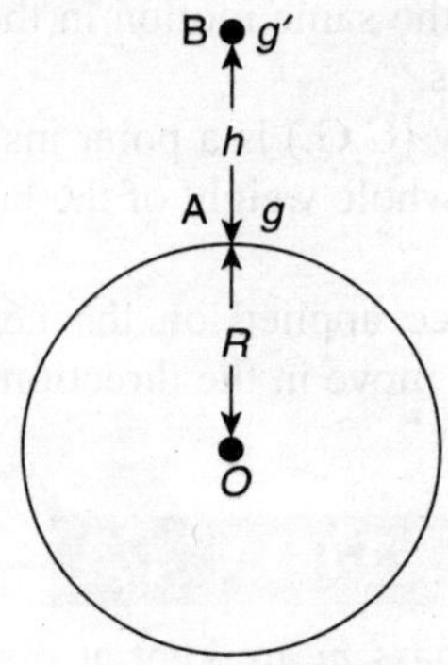

Fig. 4.1

$$\therefore \quad g = \frac{GM}{R^2} \quad \text{...(i)}$$

If g' is the acceleration due to gravity at a point B, at a height h above the surface of earth, then

$$g' = \frac{GM}{(R+h)^2} \quad \text{...(ii)}$$

Dividing (ii) by (i), we get

$$\frac{g'}{g} = \frac{GM}{(R+h)^2} \times \frac{R^2}{GM} = \frac{R^2}{(R+h)^2} \quad \text{...(iii)}$$

ILLUSTRATION 4.4

What is the value of acceleration due to gravity at a height equal to the radius of earth?

(given $h = R = 6400$ Km), from (iii),

$$g' = \frac{gR^2}{(R+R)^2} = \frac{gR^2}{4R^2} = \frac{g}{4}$$

Effect of Depth

Consider earth to be a homogenous sphere of radius R and mass M with centre at O. Let g be the value of acceleration due to gravity at a point A on the surface of earth, as shown in Fig. 4.2.

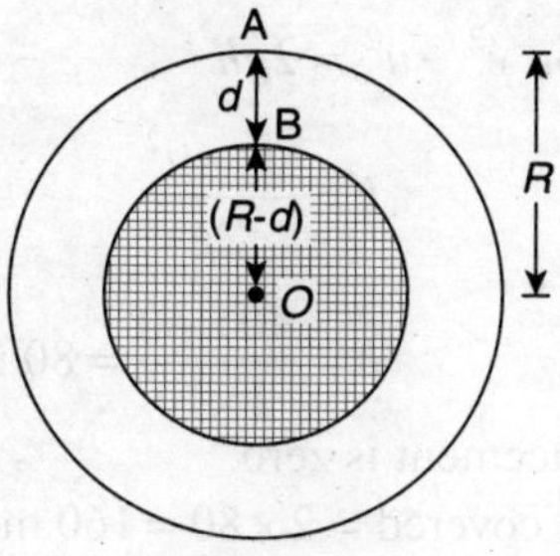

Fig. 4.2

Then,

$$g = \frac{GM}{R^2}$$

If ρ is uniform density of material of the earth, then

$$M = \frac{4}{3}\pi R^3 \rho$$

$$\therefore \quad g = \frac{G \times \frac{4}{3}\pi R^3 \rho}{R^2} = \frac{4}{3}\pi G R \rho \quad \text{...(i)}$$

Let g' be the acceleration due to gravity at the point B at a depth d below the surface of earth. The body at B will experience gravity pull due to the shaded portion of earth whose radius is $(R-d)$ and mass is M'.

$$\therefore \quad g' = \frac{GM'}{(R-d)^2}$$

and

$$M' = \frac{4}{3}\pi (R-d)^3 \rho$$

$$\therefore \quad g' = \frac{G \times \frac{4}{3}\pi (R-d)^3 \rho}{(R-d)^2}$$

$$= \frac{4}{3}\pi G(R-d)\rho \quad \text{...(ii)}$$

Dividing (ii) by (i), we get

$$\frac{g'}{g} = \frac{\frac{4}{3}\pi G(R-d)\rho}{\frac{4}{3}\pi G R \rho}$$

$$= \frac{R-d}{R} = \frac{R}{R} - \frac{d}{R}$$

or

$$g' = g\left(1 - \frac{d}{R}\right) \quad \text{...(iii)}$$

From (iii), we note that the value of acceleration due to gravity decreases with depth.

At the centre of the earth, $d = R$, $g' = g_0$ (say).

From (iii), we get

$$g_0 = g\left(1 - \frac{R}{R}\right) = 0$$

It means the acceleration due to gravity is zero at the centre of earth. Therefore the weight of the body of mass m *at the centre of earth* = m g_0 = *0*, but the mass of the body will not be zero.

Thus the *value of acceleration due to gravity is maximum at the earth's surface and becomes zero at the centre of the earth.*

Effect of Shape of Earth

Earth is not a perfect sphere. It is flattened at the poles and bulges out at the equator. Equatorial radius R_e of the earth is about 21 km greater than the polar radius R_p, (Fig. 4.3).

Now,

$$g = GM / R^2$$

Since G and M are constants

$$\therefore \quad g \propto 1/R^2$$

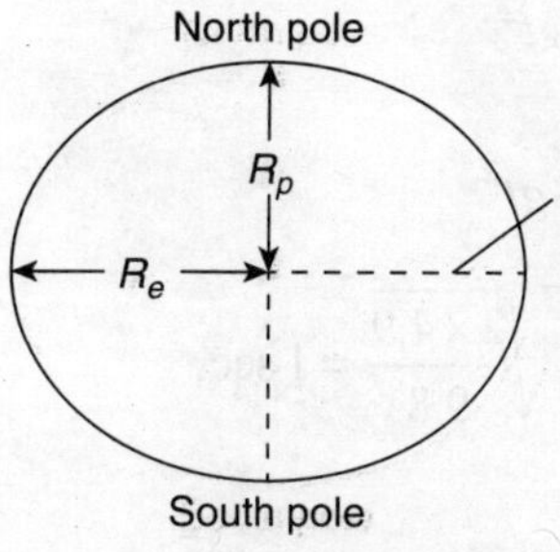

Fig. 4.3

Thus we conclude that the value of g is least at the equator and maximum at the pole. It means that *the value of acceleration due to gravity increases as we go from equator to the pole.*

At sea level, the value of g at pole is greater than its value at equator by 1.80 cm s^{-2}.

Free Fall

When bodies (object) are left free at some height from the earth's surface, they all fall along parallel paths pointing towards the centre of the earth.

Corresponding to the equation for accelerated motion on horizontal surfaces, we can have equation of motion under gravity as:

$$v = u - gt$$

$$s = ut - \frac{1}{2}gt^2$$

$$v^2 = u^2 - 2gs$$

All quantities acting upwards are taken as positive and all quantities acting downwards are taken as negative.

Important Hints for Numericals

(i) For a body dropped, $u = 0$

(ii) Time for upward journey = Time for downward journey

(iii) Velocity of return = Velocity of throw

ILLUSTRATION 4.5

A stone is dropped from the edge of a roof.

(a) How long does it taken to fall 4.9 m?

(b) How fast does it move at the end of that fall?

(c) How fast does it move at the end of 7.9 metres?

(d) What is its acceleration after 1s and after 2s?

Solution

$u = 0$

$g = 9.8\ ms^{-2}$

(a) $s = -h = -4.9$ m (downward, so –ve sign)

$$s = ut - \frac{1}{2}gt^2$$

$$-h = 0 - \frac{1}{2}gt^2$$

$$t = \sqrt{\frac{2h}{g}} = \sqrt{\frac{2 \times 4.9}{9.8}} = 1 \text{ sec.}$$

(b) $t = 1$ sec.

$u = 0$

$v = u - gt = 0 - 9.8 \times 1 = -9.8\ ms^{-1}$

(–ve sign indicates that velocity is downward)

(c) $v^2 = u^2 - 2gs$

$v^2 = 0 - 2 \times 9.8 \times -7.9$

$v = \sqrt{2 \times 9.8 \times 7.9}$

$v = \pm 12.44\ ms^{-1}$

$v = -12.44\ ms^{-1}$ (the body is going down)

(d) The acceleration due to gravity remains same at all time and all places. It is given as $g = 9.8\ ms^{-2}$

ILLUSTRATION 4.6

A stone is thrown vertically with an initial velocity of 40 m/s. Taking $g = 10\ m/s^2$, find the maximum height reached by the stone. What are the net displacement and the total distance covered by the stone?

Solution

Given: $u = 40$ m/s, $g = 10\ m/s^2$, $v = 0$

Using equation $v^2 - u^2 = 2gh$

$$h = \frac{v^2 - u^2}{2g}$$

$$h = \frac{40 \times 40}{2 \times 10} = 80 \text{ m}$$

The net displacement is zero.

Total distance covered = $2 \times 80 = 160$ m

ILLUSTRATION 4.7

A stone is allowed to fall from the top of a tower 100 m high and at the same time another stone is projected vertically upwards from the ground with a velocity of 25 m/s. Calculate when and where the two stones will meet.

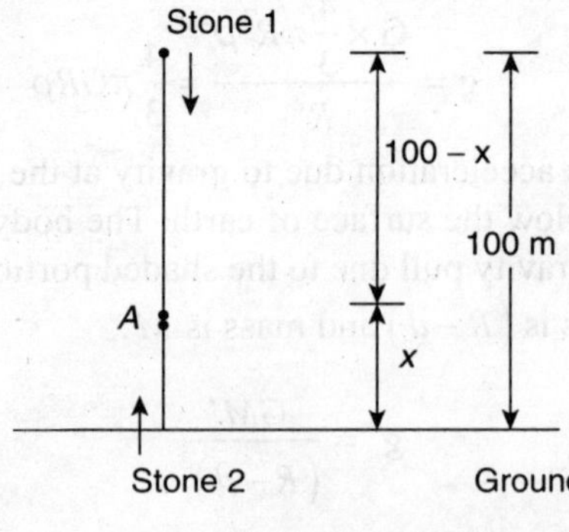

Solution

$h = 100$ m

Let the stones meets after t seconds at a point A, above the ground

For stone 1

$$u = 0$$

$$h = (100 - x) \text{ m}, g = 9.8 \text{ m/s}^2$$

From $$h = ut + \frac{1}{2}gt^2$$

$$(100 - x) = 0 + \frac{1}{2} \times 9.8 \times t^2$$

$$(100 - x) = 4.9t^2 \quad \ldots\text{(i)}$$

For stone 2

$$u = 25 \text{ m/s}, h = x, g = (-9.8 \text{ m/s}^2)$$

From $$h = ut + \frac{1}{2}gt^2$$

We get $$x = 25t + \frac{1}{2}(-9.8)t^2 \quad \ldots\text{(ii)}$$

Add equations (i) and (ii)

$$(100 - x) + x = 4.9t^2 + 25t - 4.9t^2$$

$$100 = 25t$$

$$t = \frac{100}{25} = 4 \text{ s}$$

Substituting the value of t in equation (i)

$$(100 - x) = 4.9 \times 4 \times 4$$

$$100 - x = 78.4$$

$$x = 100 - 78.4 = 21.6 \text{ m}$$

Mass and Weight

Mass: Amount of matter possessed by a body, is called the mass of the body. It is represented by the symbol m. It is a scalar quantity.

A body never has a zero mass.

Weight: The force, with which a body is attracted towards the centre of the earth, is called the weight of the body. It is represented by the symbol W.

It is a vector quantity having direction towards the centre of the earth. A body has zero weight at the centre of the earth.

ILLUSTRATION 4.8

A body has mass m on earth. Find the mass and weight of body on moon.

Solution

Mass on moon = m

Weight on moon = mg_{moon} = 1.7 m

Weightlessness

Weightlessness is defined as the state in which a body loses its weight due to free fall.

Astronauts feel weightlessness inside a spaceship as the spaceship is a freely falling body orbiting round the earth.

Projectiles and Projection

Projectile: Projectile is the name given to a body moving in space by itself without being propelled by any engine. The only force which governs the motion of the projectile is the force of gravity.

Projection: The act of putting a projectile into its trajectory is called projection.

It is of two types:

(i) Horizontal projection

(ii) Inclined projection

Horizontal Projection

A ball is projected horizontally with speed u, from top of a tower of height h. Under the effect of gravity, the ball follows a curved trajectory known as parabola.

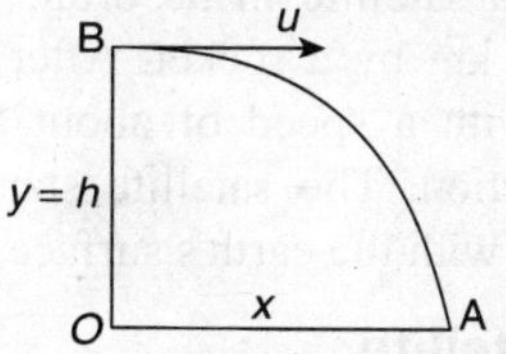

Let the ball hit the ground after time t. The distance OA is called the range of the projectile.

For vertical motion under acceleration due to gravity,

$$y = ut + \frac{1}{2}gt^2$$

$$= 0 + \frac{1}{2}gt^2$$

$$t = \sqrt{\frac{2y}{g}}$$

t is known as the **time of flight** of projectile.

$$\text{OA} = ut = u\sqrt{\frac{2y}{g}}$$

Where OA is range of projectile

Also $$t = \frac{x}{u}$$

So, $$y = \frac{1}{2}g\left(\frac{x}{u}\right)^2$$

$$\therefore \qquad y = \left(\frac{g}{2u^2}\right)x^2$$

This curve represents a parabola.

Satellites

A satellite is a body which is continuously revolving around a bigger body. Satellite may be regarded as a 'secondary body'.

The centripetal force required by a satellite to move in a circular orbit is provided by the gravitational force of attraction between the satellite and the body around which it revolves.

In our solar system, different planets are revolving around the sun. So planets can be said to be the natural satellites of the sun. Moon revolves around the Earth in a nearly circular orbit of radius 3.85×10^5 km and completes one revolution in 27.3 days. So, Moon is a natural satellite of the Earth.

Satellites as Projectiles

Satellites are bodies moving round a planet. For placing a satellite in orbit around a planet, it is to be projected horizontally with very high velocity from a great height.

For placing a satellite in its orbit, first the satellite is taken to 800 km by a rocket. After this the satellite is discharged with a speed of about 8 km/sec. in the horizontal direction. The satellite starts moving in an orbit concentric with the earth's surface.

Motion of Satellite

(i) All satellites today get into orbit by riding on a rocket or by riding in the cargo bay of a space shuttle. Several countries have rocket-launch capabilities. Satellites, weighing several tons, make it safely into orbits on a regular basis.
In most of the launches, a rocket is aimed straight up at first. This gets the rocket through the thickest, part of the atmosphere most quickly and also minimises fuel consumption.

(ii) Once the rocket reaches extremely thin air (at 193 km), the rocket's navigational system fires small rockets. This puts the launch vehicle into a horizontal position. The rockets are once again fired to ensure separation between the launch vehicle and the satellite. Thus, the satellite is put on the path described in the flight plan.

(iii) For understanding the orbital motion of satellites, we must use of Newton's laws of motion and the law of universal gravitation.

For a simple analysis of the motion of a satellite, we make the following two assumptions.

(a) We consider the gravitational force only between the satellite and Earth. The disturbing effect of the gravitational force of the other bodies is ignored.

(b) The centre of mass of the Earth-satellite system is at the centre of the Earth.

Orbital Velocity

Orbital velocity is the velocity which is given to an artificial satellite a few hundred kilometre above the Earth's surface so that it may start revolving round the earth. It is denoted by v_0.

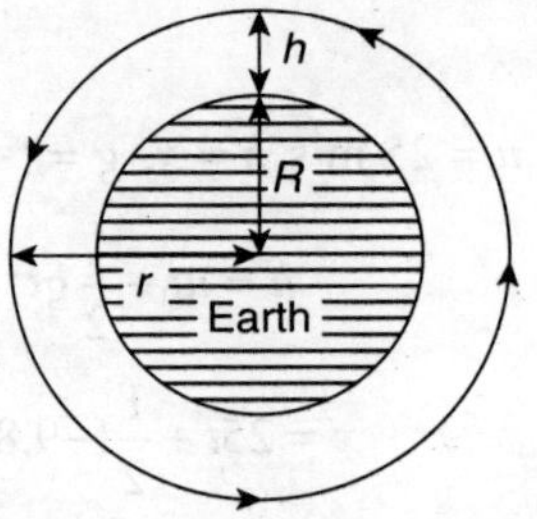

Expression for orbital velocity:

Let m = mass of satellite
r = radius of the circular orbit of satellite
h = height of the satellite above the surface of Earth
R = radius of Earth
v_0 = orbital velocity
M = mass of the Earth

The centripetal force $\frac{mv_0^2}{r}$ required by the satellite to keep moving in a circular orbit is produced by the gravitational force $G\frac{Mm}{r^2}$ between the satellite and the Earth.

Then, $\frac{mv_0^2}{r} = \frac{GMm}{r^2}$ or $v_0^2 = \frac{GM}{r}$

or $$v_0 = \sqrt{\frac{GM}{r}} \qquad \ldots(1)$$

But $r = R + h$

$$\therefore \qquad v_0 = \sqrt{\frac{GM}{R+h}} \qquad \ldots(2)$$

Let g_h be the value of acceleration due to gravity at a height h above the free surface of Earth.

Then, $$mg_h = \frac{GMm}{(R+h)^2}$$

or $$GM = g_h(R+h)^2$$

From equation (2), $$v_0 = \sqrt{\frac{g_h(R+h)^2}{R+h}}$$

or $$v_0 = \sqrt{g_h (R+h)} \qquad \ldots(3)$$

If g be the value of acceleration due to gravity on the surface of Earth, then $g = \dfrac{GM}{R^2}$

Also, $$g_h = \frac{GM}{(R+h)^2}$$

$$\therefore \frac{g_h}{g} = \frac{GM}{(R+h)^2} \times \frac{R^2}{GM} = \frac{R^2}{(R+h)^2} \text{ or } g_h = \frac{gR^2}{(R+h)^2}$$

From equation (3), $$v_0 = \sqrt{\frac{gR^2}{(R+h)^2}(R+h)}$$

or $$v_0 = R\sqrt{\frac{g}{R+h}} \qquad \ldots(4)$$

Equations (1), (2), (3) and (4) give different expressions for the orbital velocity of a satellite.

Period of Revolution or Time Period of Satellite

Period of revolution of a satellite is the time taken by the satellite to complete one revolution round the Earth. *It is denoted by T.*

$$\therefore \quad T = \frac{\text{Circumference of circular orbit}}{\text{Orbital velocity}}$$

or $$T = \frac{2\pi r}{v_0} \qquad \ldots(5)$$

or $$T = \frac{2\pi(R+h)}{v_0} \quad [\because r = R+h] \qquad \ldots(6)$$

or $$T = 2\pi(R+h)\sqrt{\frac{R+h}{GM}} \quad \left[\because v_0 = \sqrt{\frac{GM}{R+h}}\right]$$

or $$T = 2\pi\sqrt{\frac{(R+h)^3}{GM}} \qquad \ldots(7)$$

Also, $$T = 2\pi\sqrt{\frac{(R+h)^2(R+h)}{GM}}$$

or $$T = 2\pi\sqrt{\frac{R+h}{g_h}} \quad \left[\because g_h = \frac{GM}{(R+h)^2}\right] \qquad \ldots(8)$$

But $GM = gR^2$

$$\therefore \text{ From equation (7), } T = 2\pi\sqrt{\frac{(R+h)^3}{gR^2}} \qquad \ldots(9)$$

Equations (5), (6), (7), (8) and (9) give different expressions for the time period of a satellite.

If the satellite is orbiting very close to Earth, then equations (7) and (8) become

$$T = 2\pi\sqrt{\frac{R^3}{GM}} \text{ and } T = 2\pi\sqrt{\frac{R}{g}}$$

Height of Satellite

Squaring equation (9), we get

$$T^2 = 4\pi^2\frac{(R+h)^3}{gR^2} \quad \text{or} \quad (R+h)^3 = \frac{gR^2T^2}{4\pi^2}$$

$$R+h = \left[\frac{gR^2T^2}{4\pi^2}\right]^{1/3} \quad \text{or} \quad h = \left[\frac{gR^2T^2}{4\pi^2}\right]^{1/3} - R$$

Uses of Artificial Satellites

1. To study various phenomena in the outer regions of the Earth's atmosphere.
2. To study various phenomena connected with sea.
3. For communication purposes.
4. For weather forecasting.

Geostationary or Synchronous Satellite

A geostationary satellite is so named because it appears to be stationary to an observer on the Earth. This satellite is also named as synchronous satellite because the angular speed of satellite is synchronised with angular speed of Earth about its axis. When such a satellite is used for communication purposes, it is also known as communication satellite.

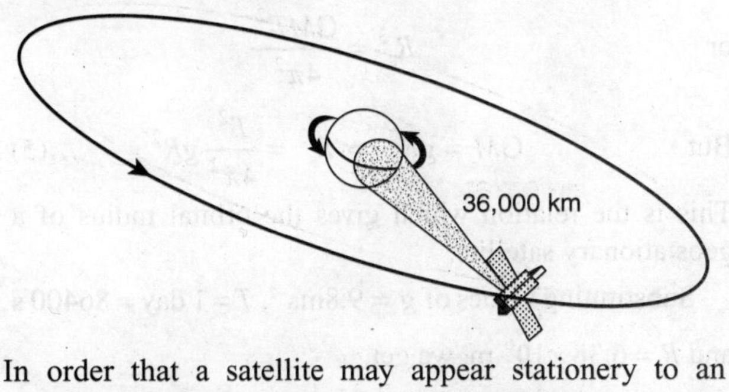

In order that a satellite may appear stationery to an observer on the Earth, the following conditions must be satisfied.

(i) It should be at a height of nearly 36,000 km above the equator.

(ii) It should revolve in an orbit which is concentric and coplanar with the equatorial plane. So, the plane of the orbit of the satellite is normal to the axis of rotation of the Earth.

(iii) The direction of rotation of the satellite should be the same as that of the Earth about its own axis. So, the satellite should orbit around the Earth from west to east as does the Earth. Its orbital velocity is nearly 3.1 km s^{-1}.

(iv) The orbital period of the satellite, i.e., the time taken by the satellite to complete one revolution around the Earth should be the same as that of the Earth about its own axis, i.e., 24 hours.

Height of the geostationary satellite from the Earth's surface

Let m and R_s be the mass and distance of a geostationary satellite from the centre of the Earth respectively. Also, let M and R be the mass and radius of the earth respectively, which is assumed to be a sphere of uniform density.

The orbital speed of the satellite, $v = \dfrac{2\pi R_s}{T}$ ...(1)

where T is its period of rotation (= 1 day).

The centripetal acceleration $= a = \dfrac{v^2}{R_s} = \dfrac{4\pi^2 R_s}{T^2}$ ...(2)

The gravitational force between the Earth and the satellite

$$= F_s = \frac{GMm}{R_s^2} \quad ...(3)$$

Since the centripetal force (ma) is provided by the gravitational force,

$$ma = F_s$$

$$\frac{4\pi^2 m R_s}{T^2} = \frac{GMm}{R_s^2} \quad ...(4)$$

[using equations (2) and (3)]

or $$R_s^3 = \frac{GMT^2}{4\pi^2}$$

But $$GM = gR^2 \Rightarrow R_s^3 = \frac{T^2}{4\pi^2} gR^2 \quad ...(5)$$

This is the relation which gives the orbital radius of a geostationary satellite.

Substituting values of $g = 9.8 \text{ms}^{-2}$, $T = 1 \text{ day} = 86400$ s and $R = 6.38 \times 10^6$ m, we get

$$R_s = \left[\frac{(86400)^2}{4\pi^2} \times 9.8 \times \left(6.38 \times 10^6\right)^2\right]^{1/3}$$

$$= 42237 \times 10^3 \text{ m} = 42,237 \text{ km}$$

Here, R_s is the distance of satellite from the centre of Earth.

∴ Height of the satellite above Earth's surface, $h = R_s - R = 42,237 - 6380 = 35,857$ km

The geostationary satellite should be taken to a height of 35,857 km above the Earth's surface at the equator and then given the required orbital velocity in the horizontal direction.

Planets

Planets are large natural bodies rotating around a star in definite orbits. Kepler, after a life-time of study, worked out three empirical laws which govern the motion of these planets. These are known as *Kepler's laws of planetary motion*. These are:

(i) The Law of Orbits: *All planets move in elliptical orbits with the sun at one focus.*

(ii) The law of Areas: *A line that connects a planet to the sun sweeps out equal areas in equal time, i.e., the areal velocity of the planet is always constant.*

(iii) The Law of Periods: *The square of the period of revolution of any planet is proportional to the cube of the semi-major axis of the orbit, i.e.,* $T^2 \propto r^3$.

Newton showed that these empirical laws followed from his law of gravitation. These laws are found to hold equally well for satellites, either natural or artificial, orbiting around a planet [Earth or any other massive central body].

ILLUSTRATION 4.9

Derive Kepler's third law from Newton's Law of Gravitation.

Solution

$$\frac{GMm}{r^2} = \frac{mv^2}{r}$$

Since velocity is $\dfrac{\text{Distance}}{\text{Time}}$ and the distance here is the circumference of the orbit,

$$\frac{GM}{r} = \left(\frac{2\pi r}{T}\right)^2$$

$$\frac{GM}{r} = \left(\frac{2\pi}{T}\right)^2 r^2$$

$$\frac{GM}{r} = \frac{4\pi^2}{T^2} r^2$$

$$\Rightarrow \quad T^2 = \frac{4\pi^2}{GM} r^3$$

$\Rightarrow T^2 \propto r^3$ ($\dfrac{4\pi^2}{GM}$ is a constant for a given planet)

Escape Velocity

Escape velocity is defined as the minimum velocity with which a body has to be thrown vertically upward so that it may just escape the gravitational pull of the planet.

Note: The derivation of the escape velocity is beyond the scope of your syllabus. This, you would be able to derive in XI class, when you are exposed to Calculus a branch of Mathematics.

The expression for escape velocity is

$$v_e = \frac{\sqrt{2GM}}{R}$$

where M is the mass of the planet and R is the radius

For earth, $M = 6 \times 10^{24}$ kg, $R = 6400$ km.

Universal Gravitational constant $G = 6.67 \times 10^{-11}$ Nm^2/kg^2

Substituting these values in the expression $v = \sqrt{\frac{2GM}{R}}$,

we see find that the escape velocity from the surface of earth = 11.2 km/s.

POINTS TO REMEMBER

1. Newton's law of gravitation states that any two particle attract each other with a force given by
$$F = G.\frac{m_1 m_2}{r^2} \qquad \text{(A)}$$
where m_1 and m_2 are the gravitational masses of the two particles and r is the distance between them. G is a universal constant, called gravitational constant, and its value is experimentally found to be 6.67×10^{-11} N-m^2/kg^2 (or m^3/kg-s^2).
2. The mass defined by eq. (A) is known as gravitation mass whereas mass defined by Newton's second law of motion is known as inertial mass and therefore we use a common word "mass" for both inertial and gravitational masses.
3. Equation (A) holds only for point masses. The extended bodies behave, for an external point, as if their entire mass were concentrated at their centres of gravity.
4. In our daily life, we do not observe this gravitational force between two bodies, given by equation (A), because of the negligibly small value of G.
5. The weight of a body is the force with which the earth attracts it towards its centre. The force of attraction between the earth of mass M and radius R and a particle of mass m at a height h above the surface of the earth is given by
$$F = \frac{GMm}{(R+h)^2} \approx \frac{GMm}{R^2} \text{ as } h << R.$$
If g be the acceleration of the particle then
$$g = \frac{F}{m} = \frac{GM}{R^2} \; (h << R)$$
This acceleration is known as acceleration due to gravity.
6. Since the earth is not a perfect sphere, its radius at the pole and the equator are different. If R_p and R_e be the radii of the earth at the pole and at the equator and g_p and g_e be the corresponding values of acceleration due to gravity at those places, then
$$g_P = \frac{GM}{R_p^2} \text{ and } g_e = \frac{GM}{R_e^2}$$
As, $R_p < R_e, g_P > g_e$
7. If h is not small compared to R, then
$$g = \frac{GM}{(R+h)^2}$$
Hence, the value of acceleration due to gravity at a point varies with h.
8. Following are Kepler's laws of planetary motion:
   (a) All planets move in elliptical orbits with the sun at one of its foci.
   (b) The radius vector from the sun to the planet sweeps out equal area in equal interval to time, or the areal velocity of the radius vector joining the sun to the planet is a constant.
   (c) The square of the time period of the planet round the sun is directly proportional to the cube of the semi-major axis of the ellipse. If a be the semi-major axis of the ellipse, then, $T^2 \propto a^3$.
9. Gravitational force is independent of nature of intervening medium.
10. It is independent of the presence or absence of other bodies.
11. It is independent of nature and size of the bodies, as long as their masses remain the same and the distance between their centres is fixed.
12. The sun and the planets form action and reaction pair, i.e. the gravitational forces between the sun and any given planet is equal in magnitude and opposite in direction and hence they obey **Newton's third law of motion.**

COMMON MISCONCEPTIONS

1. The G and g are not the same. G is the universal gravitational constant whereas g is acceleration due gravity, which varies from place to place.
2. If the mass of an object is m and acceleration due to gravity is g, then weight of the object on the surface is mg but when the object is falling, the weight is zero because weight is basically a reaction force.
3. When a body is falling freely, the acceleration is g. Now the same reaction force is effective on Earth also, but because of large mass of Earth, the acceleration produced on Earth is very small, so the movement of earth is insignificant. The general misconception is the Earth does not move theoretically.
4. Gravitational force is independent of intervening medium. The misconception is that it also dependents upon medium like electrostatic and magnetic force.
5. The value of acceleration due to gravity decreases as we go above the surface of earth or below the surface of earth. The general misconception is that as we go below the surface of Earth, the value of g increases as the distance of the object centre decreases.

PRACTICE QUESTIONS

Very-short Answer Type Questions

1. What will be the mass and weight of a unit mass at the centre of the earth?
2. Name the place on the earth's surface where the weight of a body is maximum and minimum?
3. What would happen if the centrifugal force between moon and earth vanishes away?
4. At which place on the earth, the acceleration due to gravity is zero?
5. If gravitational force acts between all objects, why don't they move towards each other?
6. What happens to the force between two objects if the mass of both objects are doubled?
7. In vacuum, which of the two, a feather and a coin, will reach the ground first on being dropped?
8. What is the relation between mass and weight of a body?
9. How does the force of gravitation between two objects change when the distance between them is reduced to half?
10. What happens to the force between two objects, if the distance between the objects is doubled and tripled?
11. Gravitational force acts on all objects in proportion to their masses. Why then, a heavy object does not fall faster than a light object?
12. What is the acceleration of free fall?
13. What happens to the force between two objects, if the mass of one object is doubled?
14. What do we call the gravitational force between the earth and an object?

Short-answer Type Questions

1. Derive an expression for acceleration due to gravity (g).
2. How are the equations of motion represented for objects under the influence of gravitational force of Earth?
3. How will the weight of a body of mass change if it is taken from equator to the poles? Give reasons.
4. What is the magnitude of the gravitational force between Earth and a 1 kg object on its surface? (Mass of the earth is 6×10^{24} kg and radius of the earth is 6×10^{6} m)
5. The Earth and the Moon are attracted to each other by gravitational force. Does the Earth attract the moon with a force that is grater or smaller or the same as the force with which the moon attracts the earth? Why?
6. If the moon attracts the earth, why does the earth not move towards the moon?
7. What is the importance of universal law of gravitation?
8. Ishan buys a few grams of gold at the poles as per the instruction of one of his friends. He hands over the same when he meets him at the equator. Will the

friends agree with the weight of gold bought? If not, why? [**Hint**: The value of g is greater at the poles than at the equator.]

Descriptive Type Questions

1. State Newton's universal law of gravitation. Derive it mathematically.
2. How is Newton's third law of motion applied to Newton's law of gravitation? Give an example.
3. Derive mathematically that the weight of an object on the moon is one-sixth of its weight on the earth.
4. (a) State:
   (i) Newton's Law of Gravitation
   (ii) Kepler's third law.
   (b) Derive Kepler's third law from Newton's law of gravitation.
5. Derive the expression for acceleration due to gravity in the following cases:
   (a) When we go above the surface of Earth
   (b) Below the surface of Earth.
6. Calculate the following:
   (a) Mass of earth
   (b) Density of earth.

NCERT Corner

1. Gravitational force on the surface of the moon is only $\frac{1}{6}$ as strong as the gravitational force on Earth. What is the weight in Newtons of a 10 kg object on the moon and on Earth?
2. The earth is acted upon by gravitation of the sun even then it does not fall into the sun. Why?
3. Suppose gravity of earth suddenly becomes zero, then in which direction will the moon begin to move if no other celestial body affects it?
4. What is the source of centripetal force that a planet requires to revolve around the sun? On what factors does that force depend?
5. Calculate the average density of earth in terms of g, G and R
6. The weight of any person on the moon is about 1/6th of that on earth. He can lift a mass of 15 kg on the earth. What will be the maximum weight which can be lifted by the same force applied by the person on the moon?
7. Identical packets are dropped from two aeroplanes, one above the equator and the other above the North Pole, both at height h. Assuming all conditions are identical, will the packets take same time to reach the surface of earth. Justify your answer.

Numericals

Level – I

1. Calculate the gravitational force between a 10 kg ball and a 20 kg ball placed at a separation of 5 m.
2. A body weighs 12 kg on a weighing machine on the earth. If it is taken to the moon, find its weight measured by the same weighing machine.
3. A ball is dropped from a cliff. Find:
   (a) its speed at 2 s after it is dropped
   (b) its speed when it has fallen through 78.4 m
   (c) the time taken in falling through 78.4 m.
4. A ball is thrown upwards with a speed of 39.2 m/s. Calculate:
   (a) the maximum height it reaches and
   (b) the time taken in reaching the maximum height.
5. A body thrown upwards takes 4 s to reach the maximum height. Find:
   (a) the initial speed with which it is thrown and
   (b) the maximum height reached.
6. Find the percentage decrease in the weight of the body when taken to a depth of 32 km below the surface of earth. Radius of the earth is 6400 km.
7. A sphere of mass 40 kg is attracted by a second sphere of mass 15 kg, when their centres are 20 cm apart, with a force of 0.1 milligram weight. Calculate the value of gravitational constant.
8. Calculate the gravitational attraction of two cars 5 m apart if the masses of the cars are 1000 kg and 1200 kg. (A very approximate treatment)
9. A ball is thrown up with velocity of 19.6 m/s.
   (i) How long will it take to reach the maximum height?
   (ii) How high will it go?
10. A ball, when thrown vertically upwards, reaches a maximum height of 19.6 m. Find the velocity with which this ball was thrown to reach this height and the time taken by it. (take $g = 9.8\ m/s^2$)
11. A particle weights 120 N on the surface of the earth. At what height above the earth's surface, will its weight be 30 N? Radius of the earth = 6400 km.

12. A stone dropped from top of a building takes 2 s to reach the ground. How high is the building? (Take $g = 9.8$ m/s^2)

13. A ball is dropped from a height of 20 m.
(i) Calculate its final velocity before it hits the ground.
(ii) Calculate the time taken by the ball to reach the ground.

Level – II

1. A sphere of mass 40 kg is attracted by another sphere of mass 15 kg with a force of 1/10 mg wt. Find the value of constant of gravitation if centres of the spheres are 0.2 m apart.

2. The acceleration due to gravity at the Moon's surface is 1.67 m s^{-2}. If the radius of the Moon is 1.74×10^6 m, calculate the mass of the Moon. Use the known value of G.

3. A ball is dropped from a height of 20 m with a velocity u. A second ball is thrown from the same height after one second. If both the balls reach the ground simultaneously, calculate the speed with which the second ball was thrown.

4. Suppose the earth shrinks to decrease its radius to half the present value. What will be the acceleration due to gravity on the surface of the earth?

5. A ball is dropped from the edge of a roof. There is a window of height 2 m. Its top end is 19.6 m below the roof. Calculate the time taken by the ball to cross the window.

6. A ball is dropped from a height of 40 m and simultaneously another ball is thrown upwards from the ground with a speed of 40 m/s. When and where do the balls meet?

7. What will be the acceleration due to gravity on the surface of the moon if its radius were $\left(\frac{1}{4}\right)$th the radius of the earth and its mass is $\left(\frac{1}{80}\right)$th of the mass of the earth?
What will be the escape velocity on the surface of the moon if it is 11.2 km/s on the surface of the earth?

8. The moon goes around the earth in a nearly circular orbit of radius 3.84×10^5 km in 27.3 days. Determine the mass of the earth from the data provided.

9. An artificial satellite is moving in a circular orbit around the earth with a speed equal to half the magnitude of escape velocity from the earth. Determine the height of the satellite above the earth's surface.

10. If the radius of the earth were to shrink by one percent, its mass remaining the same, what would happen to the acceleration due to gravity on the earth's surface?

11. Calculate the acceleration due to gravity at the surface of Mars if its diameter is 6760 km and mass one-tenth that of earth. The diameter of earth is 12742 km and acceleration due to gravity on earth is 9.8 m/s^2.

12. Estimate the mass of the sun, assuming the orbit of the earth round the sun to be a circle. The distance between the sun and earth is 1.49×10^{11} m and $G = 6.66 \times 10^{-11}$ Nm^2/kg^2.

13. If the earth be one half of its present distance from the sun, how many days will there be in one year?

14. The moon revolves around the earth 13 times per year. If the ratio of the distance of the earth from the sun to the distance of the moon from the earth is 392, find the ratio of mass of the sun to the mass of the earth.

Fill in the Blanks

1. The force of gravitation on one body by the other is F. If the mass of each body is doubled, the force will become ______.

2. The force of gravitation between two objects is F when they are kept at some separation on the earth's surface. If the objects are kept at the same separation on the moon's surface, the force of gravitation between them will be ______.

3. If the distance between two objects is doubled, the force of gravitational attraction between them becomes ______ of its initial value.

4. The force of gravitation between two spherical bodies is Gm_1m_2/r^2 where r is the separation between their ______.

5. If the earth shrinks to decrease its radius, the value of g at its surface will ______.

6. The unit of weight is ______.

7. The weight of a body of mass 2 kg is ______ newton.

8. A load that weighs 9.8 kg on earth is hung from a spring balance which is fixed in a satellite. The reading of the balance will be ______.

9. The _____ of an object is constant and does not change from place to place.

10. The value of g is _____ at the poles.

11. 1 kg wt = _____ N.

12. Force acting perpendicular to a surface is called _____.

True or False

1. The force of gravitation by the earth on a ball is Gm_1m_2 / r^2 where r is the distance of the ball from the earth's surface, m_1 is the mass of earth and m_2 is the mass of the ball.
2. Due to gravitational forces, all bodies in the universe attract each other.
3. To calculate the gravitational force on an outside particle, a spherical body may be replaced by a point particle of the same mass placed at its centre.
4. To calculate the gravitational force, 'a square plate' may be replaced by a point particle of the same mass placed at its centre.
5. The force of gravitation between two particles is along the line joining the two particles.
6. The value of G cannot be measured in a laboratory.
7. G and g are two ways of writing the same quantity.
8. The value of G at the moon is about one-sixth of its value at the earth.

COMPETITION VISTA

Match the Columns

1. Match Column A with Column B.

Column A	Column B
(P) Mass per unit volume	(U) Relative density
(Q) Thrust per unit area	(V) Acceleration
(R) Force per unit mass	(W) g
(S) GM/R^2	(X) Pressure
(T) $\frac{\text{Density of object}}{\text{Density of water}}$	(Y) Density

2. Match Column A with Column B.

Column A	Column B
(P) Directly proportional to the product of masses	(T) Newton's law of gravitation
(Q) Inversely proportional to square of the distance between masses	(U) m/s
(R) Universal Gravitational constant	(V) m/s^2
(S) Acceleration due to gravity	(W) Nm^2/kg^2

3. Match Column A with Column B.

Column A	Column B
(P) Universal Gravitational constant	(T) Constant
(Q) Acceleration due to gravity	(U) G
(R) Weight of an object	(V) g
(S) Mass of an object	(W) Varies

4. Match Column A with Column B.

Column A	Column B
(P) Acceleration due to gravity at the centre of the earth	(S) Decreases
(Q) Acceleration due to gravity at height from the surface of the earth	(T) Increases
(R) Acceleration due gravity at depth from the surface of the earth	(U) Zero

5. Match Column A with Column B.

Column A	Column B
(P) Gravitational force	(T) Contact
(Q) Magnetic force	(U) Independent of medium
(R) Electrostatic force	(V) Depends upon medium
(S) Muscular force	(W) Weakest

Assertion–Reason Questions

Read the assertions and reasons carefully to mark the correct option out of the options given below:

(a) If both assertion and reason are true and the reason is the correct explanation of the assertion.
(b) If both assertion and reason are true but reason is not the correct explanation of the assertion.
(c) If assertion is true but reason is false.
(d) If the assertion and reason, both are false.
(e) If assertion is false, but reason is true.

1. **Assertion:** Gravitational force between two particles is negligibly small compared to the electrical force.
   Reason: The electrical force is experienced by charged particles only .

2. **Assertion:** When a clock controlled by a pendulum is taken from the plains to a mountain, it becomes slow.
   Reason: Air is less dense on a mountain, so length of the pendulum increases.

3. **Assertion:** According to Newton's law of gravitation, the acceleration on the moon due to earth should be one-sixth times the acceleration action on a body near the surface of earth.
   Reason: Distance of the moon from the centre of the earth is 60 times the distance of a body placed on the surface of the earth from the centre.

4. **Assertion:** The difference in the value of acceleration due to gravity at the poles and the equator is proportional to the square of angular velocity of Earth.
   Reason: The value of acceleration due to gravity is minimum at the equator and maximum at the poles.

5. **Assertion:** The time period of revolution of a satellite closer to surface of the earth is smaller than that revolving away from surface of Earth.
   Reason: The square of time period of revolution of a satellite is directly proportional to cube of its orbital radius.

6. **Assertion:** A body becomes weightless at the centre of Earth.
   Reason: As the distance from centre of Earth decreases acceleration due to gravity increases.

Comprehension Type Questions

1. If in Newton's law of gravitation one body (say Earth) is taken as 'reference body', the force with which the reference body (usually Earth) attracts any other body towards its centre is called force due to gravity (of reference body) and the phenomenon, 'gravity'. Now if the body is free to move, force of gravity (of reference body) in accordance with Newton's II law (i.e., $F = ma$) will produce an acceleration in it. The acceleration produced in a body by the force of gravity of reference body (usually Earth) is called acceleration due to gravity and is represented by g.
   (a) What is the reference body in this passage?
   (b) What is force due to gravity?
   (c) What is acceleration due to gravity?
   (d) Is the Newton's second law applicable in the case of a freely falling body?

2. Read the following passages carefully and answer the question.
   Newton gave the universal Law of Gravitation that force between two objects is proportional to product of their masses and inversely proportional to square of distance between them. After that, many experiments were performed to know the other factors affecting the force. It was found that the force is independent of the intervening medium between the objects and action–reaction force. Also the force is attractive in nature.
   (a) How will the gravitational force between two objects vary if, instead of in atmosphere, they are kept in vacuum?
   (b) Is Newton's third law applicable on force between the objects?
   (c) What will be effect on gravitation if mass of both objects are doubled and the distance between them is reduced to half?
   (d) In what way is it different from electrostatic and magnetic force?

3. The Danish astronomer Tycho Brahe (1546 –1601) made accurate astronomical measurements over a period of 22 years and provided the basis for the structural model of the Solar System. It is surprising to note that these precise observations, made on the planets and 777 stars, were carried out with nothing more elaborate than a large sextant and compass; the telescope had not yet been invented.

The German astronomer Johannes Kepler, who was Brahe's assistant, acquired Brahe's astronomical data and spent about 16 years trying to deduce a mathematical model for the motions of the planets. After many marathon calculations, he found that Brahe's precise data on the revolution of Mars about the Sun provided the answer. Kepler's analysis first showed that the concept of circular orbits about the Sun in the heliocentric model had to be abandoned. He discovered that the orbits of Mars could be accurately described by a curve called an *ellipse*. He then generalized this analysis to include the motions of all planets. The complete analysis is summarized in three statements, known Kepler's laws.

Newton demonstrated that these laws are consequences of the gravitational force that exists between any two masses. Newton's law of universal gravitation, together with his laws of motion, provides the basis for a full mathematical representation of the motion of planets and satellites.

(a) How do you account for the creditable work of Tycho Brahe?

(b) What correction was made by Kepler in the motion of planet around the sun?

(c) What is the meaning of Heliocentric theory?

(d) What was Newton's proposition?

4. Difference between mass and weight of the body:

Mass (m)	Weight (W)
1. It is a quantity of matter contained in a body.	It is the attractive force exerted by earth on any body.
2. It is a scalar quantity.	It is a vector quantity.
3. Its unit is kilogram.	Its unit is Newton.
4. Its value does not change with g.	Its value changes with g.
5. It is determined by a physical balance.	It is determined by a spring balance.

Answer the questions on the basis of above comparison below mass and weight.

(a) What is the difference between mass and weight on the basis of directional property?

(b) What is the value of weight of an object at the centre of Earth?

(c) Why the weight of an object is zero at the centre of Earth

(d) Can the mass of an object ever be zero?

5. In an imaginary planetary system, the central star has the same mass as our sun, but is much brighter so that only a planet twice the distance between the earth and the sun can support life. According to biological evolution (including aging processes etc.) on that planet similar to ours, what would be the average life span of a 'human' on that planet in terms of its natural life? The average life span of a human on the earth may be taken to be 70 years.

(a) What law will you use to find the solution of above problem?

(b) What is the ratio of one year of earth to 1 year of the planet?

Objective Questions

One Option Correct

1. The trajectory of a projectile is a ______.

(a) circle
(b) ellipse
(c) parabola
(d) straight line

2. Which of the following is a projectile?

(a) Bomb kept inside a flying aeroplane
(b) Ball in the hand of a running bowler
(c) Bullet fired from a gun
(d) Rocket

3. The ratio of the inertial mass to the gravitational mass is ______.

(a) 0.5 (b) 1
(c) 2 (d) 4

4. The mass of a body is measured by a ______.

(a) vernier scale
(b) stop clock
(c) beam balance
(d) spring balance

5. The mass of a body

(a) changes with change of place.
(b) does not change with change of place.
(c) becomes more at night.
(d) becomes more during the day.

6. A ball when thrown up attains a maximum height of 10 m. The velocity of projection of the ball is ______. (Take $g = 9.8\ m/s^2$)

(a) 10 m/s
(b) 14 m/s
(c) 20 m/s
(d) 28 m/s

7. A body is dropped from the top of a building. It takes 4 seconds to reach the ground. The height of the building is approximately _____.
(a) 40 m (b) 60 m
(c) 80 m (d) 100 m

8. Bodies of different masses are falling freely. They have the same _______.
(a) force (b) acceleration
(c) velocity (d) momentum

9. The ratio of the value of g on the surface of moon to that on the earth's surface is _____.
(a) 6 (b) $\sqrt{6}$
(c) $\frac{1}{6}$ (d) $\frac{1}{\sqrt{6}}$

10. The value of acceleration due to gravity near the earth's surface is _____.
(a) 8.9 m/s^2 (b) 9.8 m/s^2
(c) 8.9 m/s (d) 9.8 m/s

11. The value of G in S.I. units is _____.
(a) 6.67×10^{-9} (b) 6.67×10^{-10}
(c) 6.67×10^{-11} (d) 6.67×10^{-12}

12. When the medium between two bodies changes, force of gravitation between them _____.
(a) will increase
(b) will decrease
(c) will change according to the environment
(d) will not change under any condition

13. When an apple falls from a tree,
(a) only the earth attracts the apple.
(b) only apple attracts the earth.
(c) both the earth and the apple attract each other.
(d) none attracts each other.

14. Force of attraction between two bodies does not depend upon
(a) the shape of the masses.
(b) the distance between their centres.
(c) the magnitude of their masses.
(d) the gravitational constant.

15. The value of the universal gravitational constant
(a) changes with change of place.
(b) does not change from place to place.
(c) becomes more at night.
(d) becomes more during the day.

16. The S.I. unit of G is _____.
(a) Nm^2/kg^2 (b) Nm/kg^2
(c) N kg^2/m^2 (d) N kg/m^2

17. The value of acceleration due to gravity
(a) is same everywhere.
(b) becomes more at night.
(c) becomes more during the day.
(d) changes from place to place.

18. Two balls, identical in size and made of iron and wood respectively, are released from the same height in vacuum. The time taken by them to reach the ground is _____.
(a) zero (b) not equal
(c) roughly equal (d) exactly equal

19. A body is thrown vertically upwards. Acceleration of the body
(a) is same everywhere.
(b) increases as it goes up.
(c) decreases as it goes up.
(d) becomes zero only at the highest point.

20. A body is thrown up with a velocity of 20 m/s. The maximum height attained by it is approximately _____.
(a) 80 m (b) 60 m
(c) 40 m (d) 20 m

21. A body is dropped from some point. After 2 seconds, the velocity of the body will be _____. (Take g = 9.8 m/s^2)
(a) 19.6 m/s (b) 10 m/s
(c) 9.8 m/s (d) 5 m/s

22. The weight of a body
(a) changes with change of place.
(b) does not change from place to place.
(c) becomes more at night.
(d) becomes more during the day.

23. The weight of a body is measured by a _____.
(a) vernier scale (b) stop clock
(c) beam balance (d) spring balance

24. The weight of a body at the centre of the earth is _____.
(a) zero (b) same
(c) half (d) double

25. The acceleration due to gravity at a height 1 km above the earth is the same as at a depth d below the surface of earth. Then
(a) $d = 1$ km (b) $d = \frac{3}{2}$ km
(c) $d = 2$ km (d) $d = \frac{1}{2}$ km

26. When an object falls towards earth,
(a) force excreted by the earth is greater.
(b) force excreted by the object is more.
(c) both earth and the object exert equal and opposite force.
(d) the force exerted by the object is first greater and then reduces.

27. The weight of an object in a coal mine, sea level and at the top of a mountain are respectively, W_1, W_2 and W_3. Then,
(a) $W_1 = W_2 = W_3$ (b) $W_1 > W_2 > W_3$
(c) $W_1 < W_2 > W_3$ (d) $W_1 < W_2 < W_3$

28. Four students A, B, C and D find the acceleration due to gravity at the top of Ooty, Nainital, Mount Everest and Shimla. The acceleration due to gravity is the least
(a) at Ooty since it is the highest.
(b) at Mount Everest as it is the highest.
(c) at Nainital as only latitude has the effect and not height of the peak.
(d) at Shimla as it is the coldest.

29. $g =$ ______
(a) $\frac{GR^3}{M}$ (b) GMR
(c) $\frac{GM}{R}$ (d) $\frac{GM}{R^2}$

30. The SI unit of G and g are
(a) Nm^2/kg, m/s^2 (b) m/s^2, Nm^2/kg^2
(c) Nm^2, m/s^2 (d) Nkg^2/m^2, N/m^2

31. All freely falling objects in vacuum will
(a) have the same speed.
(b) have different velocity.
(c) have different acceleration.
(d) have same acceleration.

32. A stone is dropped from a building of height 9.6m. The time taken by it to reach the ground is ______.
(a) 1 s. (b) 2 s.
(c) 4 s. (d) 6 s.

33. The force of gravity on an object thrown vertically upwards
(a) decreases and becomes zero at the highest point.
(b) acts in the opposite direction of motion.
(c) increases with height.
(d) is in the same direction as that of motion.

34. Two planets have same density but different radii. The acceleration due to gravity would be
(a) same on both the planets.
(b) greater on the smaller planet.
(c) greater on the larger planet.
(d) dependent of the distance of planet from the sun.

35. If a planet of given density were made larger, its force of attraction for an object on its surface would increase because of the planet's greater mass but would decrease because of the greater distance from the object to the centre of the planet. Which effect predominates?
(a) Increase in mass.
(b) Increase in radius.
(c) Both affect attraction equally.
(d) None of these.

36. The time period of a satellite in a circular orbit of radius R is T. The radius of the orbit in which time period is $8T$ is ______.
(a) $2R$ (b) $3R$
(c) $4R$ (d) $5R$

37. The acceleration due to gravity g and mean density of the earth ρ are related by which of the following relations (where G is the gravitation constant and R is radius of the earth)?
(a) $\rho = \frac{4\pi gR^2}{3G}$ (b) $\rho = \frac{4\pi gR^3}{3G}$
(c) $\rho = \frac{3g}{4\pi GR}$ (d) $\rho = \frac{3g}{4\pi GR^3}$

38. If both the mass and radius of the earth decrease by 1% the value of
(i) acceleration due to gravity would decrease by nearly 1%.
(ii) acceleration due to gravity would increase by 1%.
(iii) escape velocity from the earth's surface would decrease by 1%.
(iv) the gravitational potential energy of a body on earth's surface will remain unchanged.
(a) (i) and (ii) are correct.
(b) (ii) is correct.
(c) (ii) and (iv) are correct.
(d) None of these is correct.

39. The weight of a body decreases by 51% of the weight on the earth's surface at a height of ______.
(a) $\frac{10}{7}R$ (b) $\frac{7}{10}R$
(c) $\frac{3}{7}R$ (d) $\frac{2}{7}R$

40. A satellite is orbiting round the earth. While in orbit a small part separates from the satellite. The separated part

(a) falls directly to the earth.
(b) moves in a spiral path and reaches the earth after few revolutions about the earth.
(c) continues to move in the same orbit
(d) moves gradually farther from the earth.

41. Two satellites are orbiting around the earth in circular orbits of the same radius. The mass of satellite A is five times greater than the mass of satellite B. Their periods of revolution are in the ratio:
(a) 1 : 1 (b) 1 : 10
(c) 5 : 1 (d) 1 : 5

42. The value of universal gravitational constant G depends upon
(a) the nature of material of two bodies.
(b) the heat content of two bodies.
(c) the acceleration of two bodies.
(d) None of these.

43. The orbital speed of Jupiter is
(a) lesser than orbital speed of Earth.
(b) greater than orbital speed of Earth.
(c) equal to orbital speed of Earth.
(d) It cannot be said.

44. The period of Moon's rotation round the Earth is nearly 29 days. If Moon's mass were two-fold its present value, and all other things remain un-changed, the period of Moon's rotation would be nearly:
(a) $29\sqrt{2}$ days (b) $\frac{29}{\sqrt{2}}$ days
(c) 29×2 days (d) 29 days

45. Mars has about $\frac{1}{10}$ th as much mass as the Earth and half as great a diameter. The acceleration of a falling body on Mars is nearly:
(a) 9.8m s^{-2} (b) 1.96 m s^{-2}
(c) 3.92 m s^{-2} (d) 4.9 m s^{-2}

46. If the radius of the earth were to shrink by one percent, its mass remaining the same, the acceleration due to gravity on the earth's surface would:
(a) decrease (b) remain unchanged
(c) increase (d) be zero

47. When a body is taken from equator to the poles, its weight
(a) remains the same.
(b) increases.
(c) decreases.
(d) increases at North pole and decreases at South pole.

48. A body weighs 200 N on the surface of the earth. How much will it weigh half-way down to the centre of the earth?
(a) 100 N (b) 150 N
(c) 200 N (d) 250 N

49. Two objects of different masses falling freely near the surface of moon would
(a) have same velocities at any instant.
(b) have different acceleration.
(c) experience forces of same magnitude.
(d) undergo a change in their inertia.

50. The value of acceleration due to gravity
(a) is same at the equator and the poles.
(b) is least at the poles.
(c) is least at the equator.
(d) increase from the pole to the equator.

51. In the relation $F = GMm / d^2$, the quantity G
(a) depends on the value of g at the place of observation.
(b) is used only when the earth is one of the two masses.
(c) is greatest at the surface of the earth.
(d) is universal constant of nature.

52. Two particles are placed at some distance. If the mass of each of the two particles is doubled keeping the distance between them unchanged, the value of gravitational force between them will be ______.
(a) 1/4 times (b) 4 times
(c) 1/2 times (d) unchanged

53. A satellite is orbiting around the earth with orbital radius R and time period T. The quantity which remains constant is ______.
(a) $\frac{T^2}{R^3}$ (b) $\frac{T}{R}$
(c) $\frac{T^2}{R^2}$ (d) $\frac{T^2}{R}$

54. An apple falls from a tree because of gravitational attraction between the earth and the magnitude of force exerted by the apple on earth F_1 and the magnitude of force exerted by the earth on the apple be F_2, then
(a) F_1 is very much greater than F_2
(b) F_2 is very much greater than F_1
(c) F_1 is only a little greater than F_2
(d) F_1 and F_2 are equal.

55. Two bodies of masses 2 kg and 8 kg are separated by a distance of 9 m. Then, the point where the resultant gravitational field is zero is at a distance of

(a) 6 m from 8 kg.
(b) 3 m from 8 kg.
(c) 6 m from 2 kg.
(d) 4.5 m from each mass.

56. If m_e is the mass of a body on the surface of the earth and m_m is the mass of the same body on the moon, then ______.

(a) $m_e = 6\ m_m$ (b) $m_e < m_m$
(c) $m_e > m_m$ (d) $m_e = m_m$

57. Where will it be profitable to purchase one kilogram of sugar?

(a) At the poles (b) At equator
(c) At 45° latitude (d) At 40° latitude

58. There is no atmosphere on Moon as

(a) it is closer to Earth.
(b) it revolves round the earth.
(c) it gets light from the Sun.
(d) the gases have less requirement of velocity or energy to escape from its surface.

59. Four planets A, B, C and D made of same material have radius of $\frac{r}{2}$, r, $2r$ and $4r$ respectively. The order of the planets in increasing order of the acceleration due to gravity (on their surface) is ______.

(a) A, B, C, D (b) B, C, D, A
(c) A, C, B, D (d) D, C, B, A

60. Acceleration due to gravity is

(a) dependent on mass of the planet for points outside the planet.
(b) dependent directly on the radius of the planets under comparison and made of the same material.
(c) inversely proportional to the square of radius for all points on the surface of the planet.
(d) All of these.

61. Pick the wrong among the following:

(a) Gravitational force acts along the line joining the masses.
(b) It is independent of the other masses.
(c) It is always attractive in nature
(d) It does not form action–reaction pair.

62. An earth-like planet has a radius of double the earth's radius. The acceleration due to gravity on its surface will be:

(a) g (b) $\frac{g}{2}$
(c) $2g$ (d) g^2

More Than One Option Correct

1. Which of the following are true?

(a) The universal gravitational constant remains same on all planets.
(b) The acceleration due to gravity is maximum at the poles.
(c) The acceleration due to gravity increases as we go above the surface of earth.
(d) The acceleration due to gravity decreases as we go below the surface of earth.

2. The unit of universal gravitational constant is ______.

(a) kg m²/s²
(b) N m²/kg²
(c) m³/kg s²
(d) Unit of acceleration × Unit of square of length/ Unit of square of mass

3. Two objects of different masses fall freely on the surface of the moon. Then

(a) they have same velocity at any instant of time.
(b) they will have same mass during free-fall.
(c) they will have different mass during free-fall.
(d) The weight of both objects will be zero during free-fall.

4. The value of acceleration due to gravity

(a) is least at the poles.
(b) is maximum at the poles.
(c) is least at the equator.
(d) is maximum at the equator.

5. If M_1 is the mass of an object on the surface of Earth and M_2 is the mass on the surface of Moon, then

(a) $M_1 > M_2$
(b) $M_1 = M_2$
(c) weight of the object on Moon will greater than the weight of the object on Earth.
(d) weight of the object on earth will be greater than the weight of the object on moon.

6. An apple is falling on the earth because of the attraction of Earth. As a consequence, the apple also attracts the earth. Then why do we not see the earth moving towards apple?

(a) In the case of Earth, the reaction force is not applicable
(b) The earth also moves towards the apple.
(c) Because of larger mass of the earth, the acceleration of earth towards apple is negligible.
(d) Gravitational force between earth and apple is $F = m_a a_a = M_e a_e$

7. Which of the following phenomenon are based on universal Law of Gravitation?

(a) The tide due to the moon and the sun.
(b) The motion of the moon around the earth.
(c) The motion of planets around the earth.
(d) The force because of which we are bound to Earth.

8. Which of the following is a projectile?
(a) Bomb dropped from a flying aeroplane.
(b) Ball after being delivered by the bowler.
(c) Bullet fired from a gun.
(d) Rocket.

9. Which of the following statements are true about acceleration due to gravity?
(a) g decreases in moving away from the centre if $r > R$.
(b) g decreases in moving away from the centre if $r < R$.
(c) g is zero at the centre of earth.
(d) g decreases if earth stops rotating on its axis.

10. Which of the following are true for a satellite in an orbit?
(a) It is a freely falling body.
(b) Its speed is constant.
(c) Its suffers no acceleration.
(d) It does not require energy for its motion in the orbit.

Scholar Zone

1. An apple of mass 0.25 kg falls from a tree. What is the acceleration of the apple towards Earth? Also calculate the acceleration of the Earth towards the apple. Given: Mass of Earth = 5.983×10^{24} kg, Radius of Earth = 6.378×10^6 metre, $G = 6.67 \times 10^{-11}$ Nm^{2}kg^{-2}.

2. A spherical mass of 20 kg lying on the earth's surface is attracted by another spherical mass of 150 kg with a force equal to the weight of 0.25 mg. The centres of the two masses are 0.30 m apart. Calculate the mass of the Earth. Given: Radius of Earth = 6×10^6 m, g = 9.80 m s^{-2}

3. A man can jump 2.0 m high on Earth. Calculate the approximate height he might be able to jump on a planet whose density is one-quarter that of the Earth and whose radius is one-third of the Earth's radius.

4. If the radius of Earth were increased by a factor of 3, by what factor would its density have to be changed to keep g the same?

5. Two lead spheres whose diameters are 0.20 m and 0.02 m respectively, are placed with their centres 1.00 m apart. Calculate the force of gravitational between two spheres.
Given: Radius of Earth = 6.37×10^6 m, Mean density of Earth = 5530 kg m^{-3}, Specific gravity of lead = 11.5. Take g = 9.8 m s^{-2} and $\pi = 3.14$.
If the lead spheres are replaced by brass spheres of the same radii, would the force of attraction be the same?

6. The astronauts in the satellite orbiting the earth feel weightless. Does the weightlessness depend upon the distance of the satellite from the earth? If so, how? Explain your answer.

7. A satellite of the earth has to move in an equatorial plane close to the surface of the earth, either in the direction of rotation of the earth or against it.

8. Two satellites S_1 and S_2 revolve around a planet in coplanar circular orbit in the same sense. Their periods of revolution are 1 hour and 8 hour respectively. The radius of the orbit of S_1 is 10^4 km. When S_2 is closest to S_1, find
(i) the speed of S_2 relative to S_1
(ii) the angular speed of S_2 actually observed by an astronaut in S_1.

9. Find the distance of a point from the centre of the moon at which the strengths of the earth's and the moon's gravitational fields are equal and opposite and the resultant field is equal to zero. Take the mass of the earth to be η= 81 times greater than that of the moon and the distance between the centres of these planets to be n = 60 times greater than the radius of the earth R.

Answer Keys

Objective Questions

One Option Correct

1. (c)	**2.** (c)	**3.** (b)	**4.** (c)	**5.** (b)	**6.** (b)	**7.** (c)	**8.** (b)	**9.** (c)	**10.** (b)
11. (c)	**12.** (d)	**13.** (d)	**14.** (a)	**15.** (b)	**16.** (a)	**17.** (d)	**18.** (d)	**19.** (c)	**20.** (d)
21. (a)	**22.** (a)	**23.** (d)	**24.** (a)	**25.** (c)	**26.** (c)	**27.** (c)	**28.** (b)	**29.** (d)	**30.** (a)
31. (d)	**32.** (b)	**33.** (b)	**34.** (c)	**35.** (c)	**36.** (c)	**37.** (c)	**38.** (c)	**39.** (c)	**40.** (c)
41. (a)	**42.** (d)	**43.** (a)	**44.** (d)	**45.** (c)	**46.** (c)	**47.** (b)	**48.** (a)	**49.** (a)	**50.** (c)
51. (d)	**52.** (b)	**53.** (a)	**54.** (d)	**55.** (b)	**56.** (d)	**57.** (b)	**58.** (d)	**59.** (a)	**60.** (d)
61. (d)	**62.** (c)								

More Than One Option Correct

1. (a), (b) and (d)	**2.** (b), (c) and (d)	**3.** (a), (b) and (d)	**4.** (b) and (c)	**5.** (b) and (d)
6. (b), (c) and (d)	**7.** (a), (b), (c) and (d)	**8.** (a), (b) and (c)	**9.** (a) and (c)	**10.** (a), (b) and (d)

Hints and Solutions

Very-short Answer Type Questions

1. At the centre of earth, mass remains same, but weight $= mg = m \times 0 = 0$.
2. Weight is maximum at the poles and minimum at the equator.
3. If the centrifugal force between Moon and Earth vanishes away, then the moon moves along a straight-line path with a uniform speed.
4. At the centre of the earth, the acceleration due to gravity is zero.
5. The gravitational force is very weak, hence objects kept on a surface don't move towards each other.
6. The force will become four times more.
7. In vacuum both feather and coin will reach the ground simultaneously because air resistance is absent.
8. The relation between mass and weight of a body is
$$w = mg$$
where, w is the weight of a body and g is acceleration due to gravity.
9. The force of gravitation will become four times, as it is inversely proportional to the square of the distance between the two objects
10. The force will become four times and nine times respectively.
11. The reason is that a falling object experiences acceleration due to gravity which does not depend upon the mass of the falling object.
$$g = \frac{GM}{R^2}$$
12. When an object falls freely under the force of gravity it experiences a uniform accration of free-fall or gravity.
13. Force also gets doubled.
14. Force of gravity

Short-answer Type Questions

1. According to Newton's second law of motion, $F = ma$ which can be written as $F = mg$ where F is force of gravity.

Newton's law of gravitation gives the formula for force of gravitation as
$$F = \frac{GmM}{r^2}$$
where m = mass of object
M = mass of earth
r = distance between the object and earth
G is universal gravitation constant

From equations 1 and 2 we get
$$mg = \frac{GmM}{r^2}$$
$$\therefore \quad g = \frac{GM}{r^2}$$

2. Equation 1 $v = u + at$
Represented as $v = u + gt$ where g is acceleration due to gravity

Equation 2 $$s = ut + \frac{1}{2}at^2$$

Represented as $$b = ut + \frac{1}{2}gt^2$$

where h is the height from the ground

Equation 3 $$v^2 - u^2 = 2as$$

Represented as $$v^2 - u^2 = 2gh$$

3. Since the radius of the earth decreases when we move from the equator to the poles, the value of g increases. Thus, the weight of body will increase at poles.

4. Given: $m = 1$ kg, $M = 6 \times 10^{24}$ kg, $R = 6.4 \times 10^6$ m and $G = 6.67 \times 10^{-11}$ Nm^2/kg^2

Therefore $$F = \frac{GmM}{R^2} = \frac{6.67 \times 10^{-11} \times 1 \times 6 \times 10^{24}}{(6.4 \times 10^6)^2}$$

$$= 9.8 \text{ N}$$

5. The earth attracts the moon with the same force as the earth is attracted by the moon since every action has an equal and opposite reaction.

6. The mass of the earth is greater than the moon and hence the acceleration produced due to the attraction of moon will be much less than that experienced by the moon.

7. The universal law of gravitation applies to everything in this universe. It explains the motion of planets around the sun and the moon around Earth, rainfall on Earth and flow of water etc.

8. As g is greater at poles, the weight of any object will be greater at the poles ($W = mg$).
Hence Ishan's will find the weight of gold as lesser at the equator.

Descriptive Type Questions

1. Any two bodies in the universe attract each other with a force whose magnitude is directly proportional to the product of their masses and inversely proportional to the square of the distance between their centres. The direction of the force is along the line joining the centres of the two bodies.
If masses of two bodies are m_1 and m_2 respectively, separated by a distance d, they attract each other with a force of gravitation F, such that

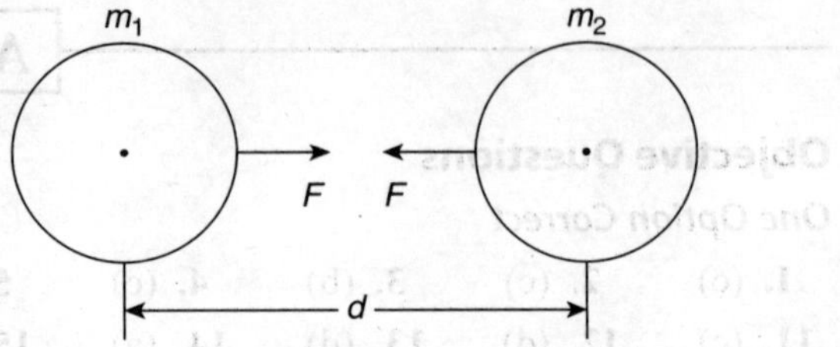

$$F \propto m_1 m_2 \text{ and } F \propto \frac{1}{d^2}$$

or $$F \propto \frac{m_1 m_2}{d^2}$$

or $$F = G\frac{m_1 m_2}{d^2}$$

where G is constant of proportionality, called the universal gravitation constant.
G can be defined as that force of gravitation which is exerted mutually between two bodies each of unit mass and separated from each other by a unit distance.
The SI unit of G is Nm^2 kg^{-2}. The value of G was found out by Henry Cavendish. The accepted value of G = 6.67×10^{-11} Nm^{2}kg^{-2}

2. According to Newton's third law, every action has an equal and opposite reaction. This means if one object exerts some force on another object then the second object will also exert a force on the first object. Newton's law of gravitation states that every object in this universe attracts every other object. Take the common example of an apple falling towards the earth. Earth exerts a force on the apple. The apple exerts equal and opposite force on the earth. But the apple is seen to move towards earth and not the earth towards the apple.
According to Newton's third law

$$F_{\text{earth}} = F_{\text{apple}}$$

$$M_E a_E = m_a a_a$$

Because the mass of apple is very less as compared to the mass of earth, the acceleration produced in it will be much greater than that produced in the earth.

3. Let the mass of an object be m. Let its weight on the moon be W_m. Let the mass of moon be M_m and its radius be R_m. By applying the universal law of gravitation, the weight of the object on the moon will be

$$W_m = G\frac{M_m \times m}{R_m^2}$$

Let the weight of the same object on the earth be W_e. The mass of the earth is M and its radius is R.

$$W_e = G\frac{M \times m}{R^2}$$

$$W_m = G\frac{7.36\times10^{22}\ \text{kg}\times m}{(1.74\times10^6\ \text{m})^2}$$

$$W_m = 2.431\times10^{10}\ G\times m$$

$$W_e = G\frac{5.98\times10^{24}\ \text{kg}\times m}{(6.37\times10^6\ \text{m})^2}$$

$$W_e = 1.474\times10^{11}\ G\times m$$

Diving equation 1 by 2, we get

$$\frac{W_m}{W_e} = \frac{2.431\times10^{10}}{1.474\times10^{11}}$$

or $$\frac{W_m}{W_e} = 0.165 \approx \frac{1}{6}$$

$$\frac{\text{Weight of the object on the moon}}{\text{Weight of the object on the earth}} = \frac{1}{6}$$

Weight of the object on the moon = $(1/6)\times$ Its weight on the earth

4. Solution – Refer theory portion

5. Solution – Refer theory portion

6. Given $R = 6400$ km, $g = 9.8$ m/s

$G = 6.67\times10^{-11}\ \text{Nm}^2/\text{kg}^2$

(a) The acceleration due to gravity on earth surface is given as

$$g_s = \frac{GM_e}{R_e^2}$$

or $$M_e = \frac{g_s R_e^2}{G} = \frac{9.8\times\left(6.37\times10^6\right)^2}{6.67\times10^{-11}}$$

$$= 6\times10^{24}\ \text{kg}$$

(b) If ρ be the density of earth, then

$$M = \frac{4}{3}\pi R^3\times\rho$$

or $$\rho = \frac{3M}{4\pi R^3} = \frac{3\times\left(6\times10^{24}\right)}{4\times3.14\times\left(6.37\times10^6\right)^3}$$

$$= 5.5\times10^3\ \text{kg/m}^3$$

NCERT Corner

1. Weight of the object on the earth

$$= W = m\times g = 10\times9.8 = 98\ \text{N}$$

Weight of the object on the moon

$$= W = \frac{1}{6}\times98 = 16.3\ \text{N}$$

2. The earth does not fall into the sun because the earth remains in its circular orbit due to the gravitational force acting on it.

3. The moon will begin to move in a straight line in the direction in which it was moving at that instant because the circular motion of moon is due to centripetal force provided by the gravitational force of the earth.

4. The motion of the planet around the sun is due to the centripetal force. This centripetal force is provided by the gravitational force between the planet and the sun. This force depends on the mass of the sun and mass of the planet and the distance between the sun and the planet.

5. According to the formula, Density = $\frac{\text{Mass}}{\text{Volume}}$

If radius of earth is R, volume = $\frac{4}{3}\pi R^3$

[∵ shape of earth is spherical]

Mass of earth, $M = \frac{gR^2}{G}$ $\left[\because g = \frac{GM}{R^2}\right]$

$$\therefore \quad \text{Density} = \frac{gR^2}{G} / \frac{4}{3}\pi R^3 = \frac{3g}{4\pi RG}$$

6. Maximum weight which can be lifted = mg

$$= 15\ \text{kg}\times9.8\ \text{m/s}^2 = 147\ \text{N}$$

Weight which can be lifted on the moon

$$= 147\ \text{N}\times\frac{1}{g_{\text{earth}}/6}$$

[∵ At moon, acceleration due to gravity is $\frac{1}{6}$ th to that of the earth]

$$= \frac{147\times6}{9.8} = \frac{147}{1.63} = 90\ \text{kg}$$

7. No, the packets do not take same time to reach the surface of earth.
The earth is not a perfect sphere. As the radius of the earth increases from the poles to the equator, the value of g becomes greater at the poles than at the equator.

Numericals

Level - I

1. $F = \dfrac{Gm_1m_2}{r^2}$

Given $m_1 = 10$ kg, $m_2 = 20$ kg
$r = 5$ m

$$F = \frac{6.67\times10^{-11}\times10\times20}{25}$$
$$= 6.67\times10^{-11}\times8$$
$$= 5.336\times10^{-10}$$
$$= 5.34\times10^{-10}$$

2. It will remain same as mass does not change.

3. (a) $v = u + gt$
$= 0 + 9.8\times20$
$= 19.6$ m/s

(b) $v^2 = u^2 + gh$
$= 0 + 2\times9.8\times78.4$
$= 2\times9.8\times9.8\times8$
$v = 4\times9.8$
$= 39.8$ m/s

(c) $h = ut + \dfrac{1}{2}gt^2$

$$78.4 = 0 + \frac{1}{2}\times9.8\times t^2$$
$$78.4 = 4.9t^2$$
$$\Rightarrow t^2 = 16$$
$$t = 4\text{s}$$

4. (a) $v^2 = u^2 + 2gh$

$$0 = (39.2)^2 - 2\times9.8\times h$$
$$\Rightarrow h = \frac{39.2\times39.2}{2\times9.8}$$
$$= 78.4 \text{ m}$$

(b) $v = u + gt$

$$0 = 39.2 - 9.8\times t$$
$$\Rightarrow t = \frac{39.2}{9.8} = 4 \text{ s}$$

5. (a) $v = u + gt$
$0 = u - 9.8\times4$
$\Rightarrow u = 39.2$ m/s

(b) $v^2 = u^2 + 2gh$
$0 = (39.2)^2 - 2\times9.8\times h$

$$\Rightarrow h = \frac{39.2\times39.2}{9.8}$$
$$= 156.8 \text{ m}$$

6. Here, $d = 32$ km; $R = 6400$ km.
Weight of body at depth d is

$$= mg' = mg\left(1 - \frac{d}{R}\right)$$

% decrease in weight

$$= \frac{mg - mg'}{mg} = \frac{d}{R}\times100$$
$$= \frac{32}{6400}\times100 = 0.5\%.$$

7. Here, $m_1 = 40$ kg ; $m_2 = 15$ kg; $r = 0.20$ m;
F = 0.1 milligram wt. $= 0.1\times10^{-6}$ kg.wt.
$= 0.1\times10^{-6}\times9.8$ N

Now, $G = \dfrac{Fr^2}{m_1m_2}$

$$= \frac{(0.1\times10^{-6}\times9.8)\times(0.20)^2}{40\times15}$$
$$= 6.53\times10^{-11} \text{ Nm}^2 \text{ kg}^{-2}$$

8. $F = \dfrac{6.67\times10^{-11}\times1000\times1200}{25}$ N

$$= 3.2\times10^{-6} \text{ N}$$

9. Initial velocity, $u = 19.6$ m/s; $g = -9.8$ m/s^2

(i) Final velocity, $v = 0$ at maximum height

$$\therefore \quad t = \frac{u}{g} = \frac{19.6}{9.8} = 2 \text{ s}$$

(ii) $u^2 = 2gh \Rightarrow h = \dfrac{u^2}{2g} = \dfrac{(19.6)^2}{2\times9.8} = 19.6$ m

10. Given that $v = 0, h = 19.6, g = 9.8$ m/s^2

$u = ?$

Using $v^2 - u^2 = 2gh$

$$u^2 = v^2 - 2gh$$
$$u^2 = 0 - 2(-9.8 \text{ m/s}^2)\times19.6 \text{ m}$$

$$u^2 = 19.6 \times 19.6$$

$\therefore$ $$u = 19.6 \text{ m/s}$$

Using $$t = \frac{v-u}{g}$$

$$t = \frac{0-19.6}{-9.8} = \frac{-19.6}{-9.8} = 2 \text{ s}$$

11. $g = \dfrac{GM}{R^2}$

$$g' = \frac{GM}{(R+h)^2}$$

$$\frac{g'}{g} = \frac{R^2}{(R+h)^2} \quad \text{...(i)}$$

$$mg = 120$$

$$mg' = 30$$

Hence $$\frac{g'}{g} = \frac{1}{4} \quad \text{...(ii)}$$

From equations (i) and (ii),

$$\frac{1}{4} = \left(\frac{R}{R+h}\right)^2$$

$$\Rightarrow \quad \frac{R}{R+h} = \frac{1}{2}$$

$$\Rightarrow \quad 2R = R+h$$

$$R = h$$

The weight will be 30 N at a height equal to radius of earth = 6400 km.

12. Given that $u = 0, t = 2, g = 10 \text{ m/s}^2$

From the question

$$h = ut + \frac{1}{2}gt^2$$

$$h = 0 \times 2 \text{ s} + \frac{1}{2} \times 10 \text{ m/s}^2 \times 2 \text{ s} \times 2 \text{ s} = 20 \text{ m}$$

13. i. Given that $u = 0, v = ?, g = 10 \text{ m/s}^2,\ h = 20 \text{ m}$

$\therefore v^2 = u^2 + 2gh = 0 + 2 \times 10 \text{ m/s}^2 \times 20 \text{ m} = 400$

$\therefore v = \sqrt{400} = 20 \text{ m/s}$

ii. Given that $u = 0, v = 20 \text{ m/s}, g = 10 \text{ m/s}^2, t = ?$

$$t = \frac{v-u}{g} = \frac{20-0}{10} = 2 \text{ s}$$

Level – II

1. $m_1 = 40 \text{ kg},\ m_2 = 15 \text{ kg},\ r = 0.2 \text{ m}$

$$F = \frac{1}{10} \text{ mg wt} = \frac{1}{10} \times 10^{-3} \text{ g wt}$$

$$= \frac{1}{10} \times 10^{-6} \text{ kg wt} = 10^{-7} \text{ kg wt} = 9.8 \times 10^{-7} \text{ N}$$

[$\because$ 1 kg wt = 9.8 newton]

As $F = G\dfrac{m_1 m_2}{r^2}$

$$G = \frac{Fr^2}{m_1 m_2} = \frac{9.8 \times 10^{-7} \times (0.2)^2}{40 \times 15} \text{Nm}^2\text{kg}^{-2}$$

$$= 6.533 \times 10^{-11} \text{Nm}^2\text{kg}^{-2}$$

2. $g = \dfrac{GM}{R^2}$ or $M = \dfrac{gR^2}{G}$

This relation is true not only for Earth but also for any heavenly body which is assumed to be spherical. Now, g = 1.67 m s^{-2}, $R = 1.74 \times 10^6$ m and $G = 6.67 \times 10^{-11} \text{Nm}^2\text{kg}^{-2}$

$\therefore$ Mass of Moon,

$$M = \frac{1.67 \times (1.74 \times 10^6)^2}{6.67 \times 10^{-11}} \text{kg} = 7.58 \times 10^{22} \text{ kg}$$

$\therefore$ Mass of Moon = 7.58×10^{22} kg

3. For the first ball, initial velocity, $u = 0$

$$h = 20 \text{ m}$$

$$g = 10 \text{ m/s}^2$$

To find v and t

Using equation $v^2 - u^2 = 2gh$

We get $$v^2 = u^2 + 2gh$$

$$= 0 + 2 \times 10 \times 20 = 400$$

$\therefore$ $$v = \sqrt{400} = 20 \text{ m/s}$$

Using $$v = u + gt$$

We get $$t = \frac{v-u}{g} = \frac{20-0}{10} = 2 \text{ s}$$

For the second ball

$t = 2 - 1 = 1$ s as it was thrown 1 s after the first ball was thrown

$$h = 20 \text{ m}, t = 1, g = 10 \text{ m/s}^2$$

$$u = ?$$

Using $$h = ut + \frac{1}{2}gt^2$$

We get $$20 \text{ m} = u \times 1 \text{ s} + \frac{1}{2} \times 10 \text{ m/s}^2 \times 1 \text{ s} \times 1 \text{ s}$$

$$20 = u + 5$$

$$u = 20 - 5 = 15 \text{ m/s}$$

The speed with which the second ball was thrown is 15 m/s.

4. $g = \frac{GM}{R^2}$ ...(i)

$$g' = \frac{GM}{\left(\frac{R}{2}\right)^2} \quad \text{...(ii)}$$

From equations (i) and (ii)

$$\frac{g'}{g} = 4$$

$$g' = 4g$$

The acceleration due to gravity will become 4 times more.

5. Using $h = ut + \frac{1}{2}gt^2$

$$19.6 = \frac{1}{2}gt_1^2 \quad \text{...(i)}$$

$$21.6 = \frac{1}{2}gt_2^2 \quad \text{...(ii)}$$

Finding the value of t_1 and t_2 from equations (i) and (ii)

$$19.6 = \frac{1}{2} \times 9.8 t_1^2$$

$$19.6 = 4.9 t_1^2$$

$$\frac{19.6}{4.9} = t_1^2$$

$$4 = t_1^2$$

$$\Rightarrow \quad t_1 = 2\text{s}$$

Similarly $\quad t_2^2 = \frac{21.6}{4.9}$

$$t_2^2 = 4.41$$

$$t_2 = 2.1$$

Therefore, $\quad t_2 - t_1 = 2.1 - 2 = 0.1$

6. Suppose they meet after a time t.
Then the height lost h_1 by the dropped ball in time t,

$$h_1 = \frac{1}{2}gt^2 \quad \text{...(i)}$$

Similarly, height gained h_2 by the thrown ball in time t,

$$h_2 = 40t - \frac{1}{2}gt^2 \quad \text{...(ii)}$$

From (i) and (ii)

$$h_1 + h_2 = \frac{1}{2}gt^2 + 40t - \frac{1}{2}gt^2$$

$$h_1 + h_2 = 40t$$

$$40 = 40t$$

$$\Rightarrow \quad t = 1\text{s}$$

Now from equation (i)

$$h_1 = \frac{1}{2} \times 9.8 \times (1)^2$$

$$h_1 = 4.9 \text{ m}$$

$$h_2 = 40 - 4.9 = 35.1 \text{ m}$$

So the two balls meet after 1 s at a height of 35.1 m from the ground.

7. The acceleration due to gravity on the surface of a planet of mass M and radius R is given by

$$g = \frac{GM}{R^2}$$

For the earth, $\quad g_e = \frac{GM_e}{R_e^2}$

For the moon, $\quad g_m = \frac{GM_m}{R_m^2}$

Hence, $\quad \frac{g_m}{g_e} = \frac{M_m}{M_e} \times \left(\frac{R_e}{R_m}\right)^2 \quad (1)$

Given $\quad \frac{M_m}{M_e} = \frac{1}{80}$ and $\frac{R_m}{R_e} = \frac{1}{4}$

Putting these values in (1), we get

$$\frac{g_m}{g_e} = \frac{1}{80} \times (4)^2 = \frac{16}{80} = \frac{1}{5}$$

$$\Rightarrow \quad g_m = \frac{1}{5}g_e = \frac{1}{5} \times 9.8$$

$\Rightarrow$ Acceleration due to gravity on the surface of the moon is

$$g_m = 1.96 \text{ m/s}^2$$

The escape velocity from the surface of the earth and from the surface of the moon are respectively

$$v_e = \sqrt{\frac{2GM_e}{R_e}} \text{ and } v_m = \sqrt{\frac{2GM_m}{R_m}}$$

Hence, $\quad \frac{v_m}{v_e} = \sqrt{\frac{M_m}{M_e}\frac{R_e}{R_m}}$

$$= \sqrt{\frac{1}{80} \times 4} = \frac{1}{\sqrt{20}}$$

$$\Rightarrow \quad v_m = \frac{1}{\sqrt{20}}v_e = \frac{1}{\sqrt{20}} \times 11.2$$

$$\Rightarrow \quad v_m = 2.5 \text{ km/s}$$

8. From Kepler's third law,

$$T = \frac{2\pi}{\sqrt{GM}} \cdot a^{3/2}$$

$$\Rightarrow \quad M = \frac{4\pi^2}{GT^2} \cdot a^3 \quad (1)$$

Here, $G = 6.67 \times 10^{-11}\ \text{N-m}^2/\text{kg}^2$

$T = 27.3 \times 24 \times 60 \times 60$ s

$a = 3.84 \times 10^5$ km

$= 3.84 \times 10^8$ m

Putting these values in (1), we get the mass of the earth as

$$M = 6.02 \times 10^{24}\ \text{kg}$$

9. Let M and R be respectively the mass and radius of the earth. Then, the escape velocity from the earth is

$$v_e = \sqrt{\frac{2GM}{R}} \quad (1)$$

Hence the velocity of the satellite is

$$v_s = \frac{1}{2} \cdot v_e = \sqrt{\frac{GM}{2R}} \quad (2)$$

If h be the height of the satellite from the earth's surface, then the velocity of the satellite in the orbit is

$$v_s = \sqrt{\frac{GM}{R+h}} \quad (3)$$

From (2) and (3),

$$\sqrt{\frac{GM}{2R}} = \sqrt{\frac{GM}{R+h}}$$

$$\Rightarrow \quad 2R = R + h$$

$$\Rightarrow \quad h = R = 6400\ \text{km}$$

10. Consider the case of a body of mass m placed on the earth's surface (mass of the earth M and radius R). If g is acceleration due to gravity, then we know that

$$g_s = \frac{GM_e}{R_e^2} \quad (1)$$

Now, when the radius is reduced by 1%, radius becomes 0.99 R. Let acceleration due to gravity be g', then

$$g' = \frac{GM_e}{(0.99R)^2} \quad (2)$$

From equation (1) and (2), we get

$$\frac{g'}{g} = \frac{R^2}{(0.99R)^2} = \frac{1}{(0.99)^2}$$

or $\quad g' = g \times \left(\frac{1}{0.99}\right)^2$

or $\quad g' = 1.02$ g

Thus, the value of g is increased by 2%.

11. We know that $g = \left(\frac{GM}{R^2}\right)$

So $\quad \frac{g_M}{g_E} = \left(\frac{M_M}{M_E}\right)\left(\frac{R_E}{R_M}\right)^2 = \left(\frac{1}{10}\right)\left(\frac{12742}{6760}\right)^2$

$$\Rightarrow \quad \frac{g_M}{g_E} = 0.35 \text{ or } g_M = 9.8 \times 0.35 = 3.48\ \text{m/s}^2$$

12. Here, the revolving speed of earth can be given as

$$v = \sqrt{\frac{GM}{r}} \quad \text{[Orbital speed]}$$

where M is the mass of sun and r is the orbit radius of earth.

We know time period of earth around sun is $T = 365$ days, thus we have

$$T = \frac{2\pi r}{v}$$

or $\quad T = 2\pi r\sqrt{\frac{r}{GM}}$

or $\quad M = \frac{4\pi^2 r^3}{GT^2}$

$$= \frac{4(3.14)^2 \times (1.49 \times 10^{11})^3}{(365 \times 24 \times 3600)^2 \times (6.66 \times 10^{-11})}$$

$$= 1.972 \times 10^{22}\ \text{kg}$$

13. In the previous example we've discussed that time period is given as

$$T = 2\pi r\sqrt{\frac{r}{GM}} = \frac{2\pi}{\sqrt{GM}} r^{3/2} \quad (1)$$

If radius of earth's orbit is r and if radius changes to $r' = \frac{r}{2}$, new time period becomes

$$T' = \frac{2\pi}{\sqrt{GM}} r'^{3/2} \quad (2)$$

From equation (1) and (2) we have

$$\frac{T}{T'} = \left(\frac{r}{r'}\right)^{3/2}$$

or $$T' = T\left(\frac{r'}{r}\right)^{3/2}$$

$$= 365\left(\frac{1}{2}\right)^{3/2} = \frac{365}{2\sqrt{2}} \text{ days}$$

14. The time period T_e of earth around sun of mass M_s is given by

$$T_e^2 = \frac{4\pi^2}{GM_s} \times r_e^3 \quad (1)$$

where r_e is the radius of the earth.

Similarly, time period T_m of moon around earth is given by

$$T_m^2 = \frac{4\pi^2}{GM_s} \times r_m^3 \quad (2)$$

Dividing equation (1) by equation (2), we get

$$\left(\frac{T_e}{T_m}\right)^2 = \left(\frac{M_e}{M_s}\right)\left(\frac{r_e}{r_m}\right)^3$$

or $$\left(\frac{M_s}{M_e}\right) = \left(\frac{T_m}{T_e}\right)^2 \times \left(\frac{r_e}{r_m}\right)^3$$

Substituting the given values, we get

$$\left(\frac{M_s}{M_e}\right) = \left\{\frac{1/3}{1}\right\} \times (392)^3 = 3.56 \times 10^5$$

Fill in the Blanks

1. 4 times
2. same
3. $\frac{1}{4}$ times
4. centre is r
5. increases
6. newton (N)
7. 19.6
8. zero
9. Mass
10. Maximum
11. 9.8 N
12. Normal Reaction

True or False

1. True, where M is mass of the earth m is mass of the ball
2. True
3. True
4. True
5. True
6. False
7. False, G is universal Gravitational constant and g is acceleration due to gravity.
8. False, The value of G is constant for the whole universe.

Match the Columns

1. (P) – (Y), (Q) – (X), (R) – (V), (S) – (W), (T) – (U)
2. (P) – (T), (Q) – (T), (R) – (W), (S) – (V)
3. (P) – (U), (Q) – (V), (R) – (W), (S) – (T)
4. (P) – (U), (Q) – (S), (R) – (S)
5. (P) – (U), (Q) – (V), (R) – (V), (S) – (T)

Assertion-Reason Questions

1. (b)
2. (c)
3. (a)
4. (b)
5. (a)
6. (c)

Comprehension Type Questions

1. (a) Earth
   (b) The force with which reference body (usually earth) attracts any other body towards its centre is called force due to gravity (of the reference body) and the phenomenon is called 'gravity'.
   (c) The acceleration produced in a body by the force of gravity of reference body (usually earth) is called acceleration due to gravity and is represented by g.
   (d) Yes

2. (a) It will not vary as gravitation force is independent of medium between the objects.
   (b) Yes, it is action–reaction pair
   (c) It will remain the same.

$$F_1 = \frac{Gm_1m_2}{R_1^2}$$

Now, $$m_1 = 2m_1$$

$$m_2 = 2m_2$$

$$R_2 = 2R_1$$

$$F_2 = \frac{Gm_1m_2}{(2R_1)^2}$$

$$= \frac{G2m_1 \times 2m_2}{4R_1^2}$$

$$F_2 = \frac{Gm_1m_2}{R_1^2}$$

$$\Rightarrow \quad F_1 = F_2$$

(d) (i) Gravitational force is independent of medium whereas electrostatic and magnetic forces depend upon medium.
(ii) Gravitational force is attractive whereas electrostatic and magnetic force could be both attractive as well as repulsive.

3. (a) Tycho Brahe studied the motion of planets without the telescope since telescopes were not invented at the time.
(b) He proposed that the orbit of planets around the sun is not circular but elliptical.
(c) The planets move around the sun, the sun being at the centre.
(d) Newton demonstrated that these laws are consequences of the gravitational force that exists between any two masses.

4. (a) Mass is a scalar quantity whereas weight is a vector quantity
(b) Zero
(c) Because weight = mg. At the centre of earth g is zero; therefore the weight is zero
(d) No

5. $T_2 = T_1\left(\frac{R_2}{R_1}\right)^{3/2} = T_1\left(\frac{2R_1}{R_1}\right)^{3/2} = 2\sqrt{2}T_1.$

It means $2\sqrt{2}$ years of earth are equal to 1 natural year of the imaginary planet.

$\therefore$ 1 year of earth = $\frac{1}{2\sqrt{2}}$ planet year.

$\therefore$ Age of human = 70 years of earth = $\frac{70}{2\sqrt{2}}$

= 24.75 planets years.

(a) Kepler's third law

(b) $\frac{1}{2\sqrt{2}}$

Objective Questions

One Option Correct

6. (b) Using $h = ut + \frac{1}{2}gt^2$

at maximum height $v = 0$
$h = 10$ m (given)
g is $-$ve $= -9.8$
Therefore $0 = u^2 - 2\times 9.8\times 10$

$$\Rightarrow \quad u^2 = 2\times 9.8\times 10$$

$$u = 2\times 7 = 14 \text{ m/s}$$

7. (c) Using $\quad h = ut + \frac{1}{2}gt^2$

$u = 0$

$t = 4$ s

$g = 10$ s

$$h = 0 + \frac{1}{2}\times 10\times (4)^2$$

$$= 80 \text{ m}$$

20. (d) $v^2 = u^2 + 2gh$

$$0 = (20)^2 - 2\times 10\times h$$

$$\Rightarrow 20h = 20\times 20$$

$$h = 20$$

21. (a) $v = u + gt$
$= 0 + 9.8\times 2$
$= 19.6$ m/s

27. (c) The weight at sea level is W_2. As we go above or below the surface of earth, the value of acceleration due to gravity decreases. Hence, $W_1 < W_2 > W_3$.

28. (b) g decreases with height. So choice (b) is correct and choice (a) is incorrect.
Latitude change does not bring a large variation between these places. So choice (c) is incorrect. Temperature does not affect g. So choice (d) is incorrect.

36. (c) $T^2 \times R^3$

$$\left(\frac{T_2}{T_1}\right)^2 = \left(\frac{R_2}{R_1}\right)^3$$

$$(8)^2 = \left(\frac{R_2}{R}\right)^3$$

$$\Rightarrow \quad R_2 = 4R$$

37. (c) $\rho=\frac{M}{\frac{4\pi R^3}{3}}$ ...(i)

$g=\frac{GM}{R_2}$ ...(ii)

$\rho=\frac{3g}{4\pi GR}$ ...(iii)

39. (c) Let g' be acceleration due to gravity at height h, then

$$g'=\frac{GM}{(R+h)^2}$$

$$\frac{W'}{W}=\frac{mg'}{mg}$$

$$=\frac{g'}{g}=\frac{49}{100}$$

$$=\frac{R^2}{(R+h)^2}=\frac{49}{100}$$

$$\frac{R}{R+h}=\frac{7}{10}$$

$$\Rightarrow \quad h=\frac{3}{7}R$$

41. (a) Time period is independent of mass.

43. (a) $V_0=\sqrt{\frac{GM}{R_0}}$ where M is mass of the sun. G is constant. Therefore, $V_0 \propto \sqrt{\frac{1}{R_0}}$. Since the orbital radius of Jupiter is greater than that of Earth, its orbital velocity is lesser.

45. (c) $g'=\frac{GM'}{R'}=\frac{G\times\frac{1}{10}M}{\left(\frac{R}{2}\right)^2}$

$$=\frac{GM}{R^2}\times\frac{2}{5}$$

$$=g\times\frac{2}{5}$$

$$=9.8\times\frac{2}{5}$$

$$=3.92 \text{ m/s}^2$$

48. (a) $g'=g\left(1-\frac{d}{R}\right)$

$$g'=g\left(1-\frac{R}{2R}\right)$$

$$g'=g\times\frac{1}{2}$$

$$mg'=\frac{mg}{2}$$

$$mg'=\frac{200}{2}=100\text{N}$$

49. (a) Two objects of different masses falling freely near the surface of the moon would have same velocities at any instant because they have the same acceleration due to gravity.

50. (c) Acceleration due to gravity is least at the equator and maximum at the poles due to rotation of the earth.

i.e., It is given by $g'=g-\omega^2 R\sin\theta$

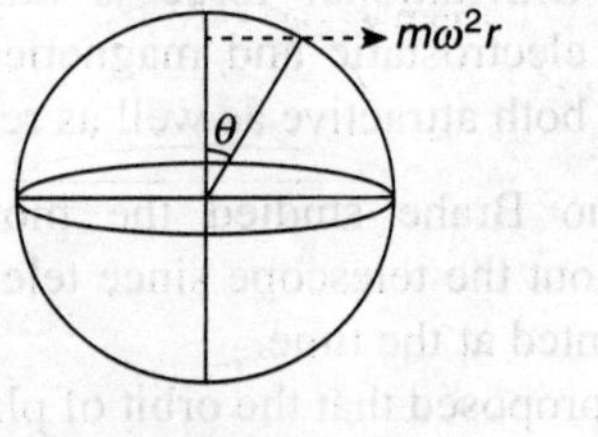

At poles $\theta=0$, so g maximum

At equator $\theta=90°$, so g minimum

51. (d) The quantity G is universal constant of nature. It is applied to all the bodies present in universe. Accepted value of G is 6.67×10^{-11} N-m^2/kg^2

52. (b) We know that the gravitational force

$$F=G\frac{Mm}{r^2}$$

where, F = force between two masses

M = first mass

m = second mass

G = gravitational constant

r = distance between two masses

According to the question,

F' = new force when mass is doubled

If mass of each particle is doubled, i.e.,

$$M'=2M'=2m$$

These values are put in equation (i), and we get

$$F=G\frac{(2M)(2m)}{r^2}\Rightarrow F'=4G\frac{Mm}{r^2}$$

From equation (i), $F'=4F$

53. (a) According to Kepler's law of planetary motion, $T^2\propto R^3$

or, $$\frac{T^2}{R^3}=\text{Constant}$$

54. (d) According to Newton's universal law of gravitation, force exerted by one body on another body is equal in magnitude and given by $\left(F = \frac{Gm_1m_2}{r^2}\right)$ and opposite in direction

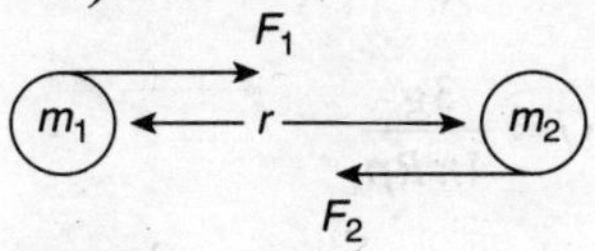

55. (b) Let x be the distance of point P from the 2 kg mass where the resultant gravitational field is zero.

$$\frac{G2}{x^2} = \frac{G8}{(9-x)^2} \Rightarrow \frac{(9-x)^2}{x^2} = 4$$

$$\Rightarrow \quad \frac{9-x}{x} = 2 \quad \Rightarrow x = 3 \text{ m}$$

56. (d) Mass of a body is constant and does not the change from place to place.

$$\therefore \quad m_e = m_m$$

57. (b) Since, the value of g is minimum at the equator, while purchasing 1 kg of sugar at equator using spring balance we will get more mass as compared to that at any other place.

58. (d) Any gas with less speed or energy to escape from gravitational force cannot stay in the near vicinity of celestial bodies to form the atmosphere.

59. (a) $g_A : g_B : g_C : g_D = \frac{r}{2} : r : 2r : 4r = 1 : 2 : 4 : 8$

Hence, $g_A < g_B < g_C < g_D$. So the choice (a) is correct and the rest are incorrect

60. (d) $g = \frac{GM}{r^2}$ for points on the surface and outside with r as the distance from the centre of the planet of mass M. So choice (a) is correct.

$g = G\frac{4}{3}\pi\frac{R^3.\rho}{R^2} \propto R$ for comparing planets made of the same material. So choice (b) is correct.

$g = \frac{GM}{R^2}$ for points on the surface. So choice (c) is correct.

61. (d) For a pair of masses, the gravitational force satisfies the options (a), (b) and (c). Also the force forms action–reaction pair. So choice (d) is wrong.

62. (d) Since the planet is earth-like, density should be taken to be same as of earth

So, $$g = \frac{G\frac{4}{3}\pi R^3\rho}{R^2} = G\frac{4}{3}\pi R\rho \propto R$$

Making a g double that on earth. So choice (c) is correct and the rest are incorrect.

More Than One Option Correct

4. Using $g = \frac{GM}{R^2}$ as $R_p < R_e$, Hence $g_p > g_e$

5. (b) Mass (the matter contained) remains same at all places

(d) Weight = mg.

Since the acceleration due to gravity on earth is 6 times that of moon, the weight of the object on earth will be greater than the weight of the same object on moon.

Scholar Zone

1. Mass of apple, $m = 0.25$ kg

Mass of Earth, $M = 5.983 \times 10^{24}$ kg

Radius of Earth, $R = 6.378 \times 10^6$ metre

Gravitational constant, $G = 6.67 \times 10^{-11}\,\text{Nm}^2\text{kg}^{-2}$

Let F be the gravitational force of attraction between the apple and Earth.

Then, $F = \frac{GmM}{R^2}$

Let a be the acceleration of apple towards Earth.

$$a = \frac{F}{m} = \frac{GmM}{mR^2} = \frac{GM}{R^2}$$

$$\therefore a = \frac{6.67 \times 10^{-11} \times 5.983 \times 10^{24}}{\left(6.378 \times 10^6\right)^2}\text{ms}^{-2} = 9.810\text{ ms}^{-2}$$

Let a' be the acceleration of Earth towards the apple.

$$a' = \frac{F}{m} = \frac{Gm}{R^2} = \frac{6.67 \times 10^{-11} \times 0.25}{\left(6.378 \times 10^6\right)^2}\text{ms}^{-2}$$

$$= 4.099 \times 10^{-25}\text{ ms}^{-2}$$

2. $m_1 = 20$ kg, $m_2 = 150$ kg, $r = 0.30$ m

$F = 0.25$ mg wt $= 0.25 \times 10^{-3}$ g wt $= 0.25 \times 10^{-6}$ kg wt

$= 0.25 \times 9.80 \times 10^{-6}$ N $\quad$ ($\because$ 1 kg wt = 9.80 N)

$$G = \frac{Fr^2}{m_1m_2}$$

or $$G = \frac{0.25 \times 9.80 \times 10^{-6} \times 0.30 \times 0.30}{20 \times 150}\text{Nm}^2\text{kg}^{-2}$$

$= 75 \times 9.80 \times 10^{-13}\ \text{Nm}^2\text{kg}^{-2}$

Now, $g = \dfrac{GM}{R^2}$

$$\therefore M = \frac{g \times R^2}{G} = \frac{9.80 \times (6 \times 10^6)^2}{75 \times 9.80 \times 10^{-13}}\ \text{kg} = 4.8 \times 10^{24}\ \text{kg}$$

3. We know that, $g = \dfrac{4\pi}{3} GR\rho$ (1)

where R and ρ are the radius and the mean density respectively of the earth.

If R' and ρ' are the corresponding values for the planet, the acceleration due to gravity on the planet is given by

$$g' = \frac{4\pi}{3} GR'\rho'$$

Dividing equation (1) by equation (2), we get

$$\frac{g}{g'} = \frac{R}{R'} = \frac{\rho}{\rho'} = 3 \times 4 = 12 \left[\because R' = \frac{R}{3} \text{ and } \rho' = \frac{\rho}{4}\right]$$

If the mass of the man is m and he can jump to a height h on Earth and h' on the planet, we have

$$mgh = mg'h'$$

[It is assumed that the man puts in the same energy in jumping high from the surface of the planet as when jumping from the earth's surface.]

$$\therefore h' = \frac{g \times h}{g'} = 12 \times 2\ \text{m} = 24\ \text{m}$$

∴ The man can jump 24 m high on the planet.

4. $g = \dfrac{GM}{R^2}$

Let ρ be the density of Earth.

$$\rho = \frac{M}{\text{Volume of Earth}} = \frac{M}{\frac{4}{3}\pi R^3} \text{ or } M = \frac{4}{3}\pi R^3 \rho$$

$$\therefore g = \frac{G}{R^2} \times \frac{4}{3}\pi R^3 \rho \text{ or } g = \frac{4}{3}\pi G \rho R$$

$\dfrac{4}{3}, \pi, G$ are constants.

For no change in value of g, $R \propto \dfrac{1}{\rho}$

Thus, if R is made $3R$, ρ must become $\rho/3$.

5. $F = G\dfrac{m_1 m_2}{r^2}$ (1)

Also, $g = \dfrac{GM}{R^2} = \dfrac{G}{R^2}\left(\dfrac{4}{3}\pi R^3 \rho\right) = \dfrac{4}{3}\pi G R \rho$

or $G = \dfrac{3g}{4\pi R \rho}$

∴ Equation (1) becomes $F = \dfrac{3g m_1 m_2}{4\pi R \rho r^2}$

Now, $m_1 = \dfrac{4}{3} \times 3.14 \times 0.10^3 \times 11.5 \times 10^3\ \text{kg}$

and $m_2 = \dfrac{4}{3} \times 3.14 \times 0.01^3 \times 11.5 \times 10^3\ \text{kg}$

Substituting values,

$$F = \frac{3 \times 9.8 \times \frac{4}{3} \times 3.14 \times 0.10^3 \times 11.5 \times 10^3 \times \frac{4}{3} \times 3.14 \times 0.01^3 \times 11.5 \times 10^3}{4 \times 3.14 \times 6.37 \times 10^6 \times 5530 \times 1^2}$$

$= 1.54 \times 10^{-10}\ \text{N}$

If lead spheres be replaced by brass spheres, their masses will change due to change in density and so the value of the force of gravitational attraction will also change.

6. A satellite moves around the earth in a circular orbit under the action of gravity. The acceleration of the satellite is GM/R^2 towards the centre of the earth, where M is the mass of the earth and R is the radius of the orbit of the satellite.

Consider an astronaut of mass m inside the satellite moving round the earth. The force acting on him are:

(i) gravitational pull of the earth,

(ii) normal force of reaction N by the satellite.

Under the action of those two forces, the astronaut moves in a circular path with acceleration GM/R^2. Hence, by Newton's law

$$\frac{GMm}{R^2} - N = m\left(\frac{GM}{R^2}\right)$$

$$\Rightarrow \quad N = 0$$

Thus the satellite does not exert any force on the astronaut and hence the astronaut in the satellite is in a state of weightlessness. This state of weightlessness does not depend upon the distance of the satellite from the earth.

7. The orbital velocity of a satellite is given by $\sqrt{\frac{GM}{R}}$ relative to the earth. Hence the absolute velocity of the satellite when it is moving in the same direction at that of the earth will be

$$V_1 = \frac{2\pi R}{T} + \sqrt{\frac{GM}{R}} = \sqrt{\frac{GM}{R}}\left[1 + \frac{2\pi R}{T}\sqrt{\frac{R}{GM}}\right]$$

Similarly the absolute velocity of the satellite when it moves in a direction opposite to that of the earth is

$$V_2 = \frac{GM}{R} - \frac{2\pi R}{T} = \frac{GM}{R}\left[1 - \frac{2\pi R}{T}\sqrt{\frac{R}{GM}}\right]$$

8. Let r_1 and r_2 be the radii of the orbit of the satellite S_1 and S_2 and T_1 and T_2 be their respective time periods of revolution.
Then from Kepler's law,

$$T_1^2 \propto r_1^3 \text{ and } T_2^2 \propto r_2^3$$

$$\Rightarrow \quad \left(\frac{T_1}{T_2}\right)^2 = \left(\frac{r_1}{r_2}\right)^3$$

$$\Rightarrow \quad r_2^3 = r_1^3 \cdot \left(\frac{T_2}{T_1}\right)^2 \qquad (1)$$

Here $r_1 = 10^4$ km, $T_2 = 8$ hours
$T_1 = 1$ hour
Putting these values in (1), we get

$$r_2^3 = \left(10^4 \text{ km}\right)^3 \left(\frac{8 \text{ hours}}{1 \text{ hour}}\right)^2 = \left(10^4 \times 4\right)^3 \text{ km}$$

$$\Rightarrow \quad sr_2 = 4 \times 10^4 \text{ km} \qquad (2)$$

(i) If v_1 and v_2 be the orbital speed of satellite S_1 and S_2 then

$$v_1 = \frac{2\pi r_1}{T_1} \text{ and } v_2 = \frac{2\pi r_2}{T_2}$$

Hence speed of S_2 relative to S_1 is

$$v = |v_2 - v_1| = \left|2\pi\left[\frac{r_2}{T_2} - \frac{r_1}{T_1}\right]\right|$$

$$\Rightarrow \quad v = \left|2\pi\left[\frac{4 \times 10^4}{8} - \frac{10^4}{1}\right]\right| \text{ km/hour}$$

$$\Rightarrow \quad v = 3.14 \times 10^4 \text{ km/hour}$$

9. Let, mass of the moon = M_m.
Hence, mass of the earth = $M_e = \eta M_m$.

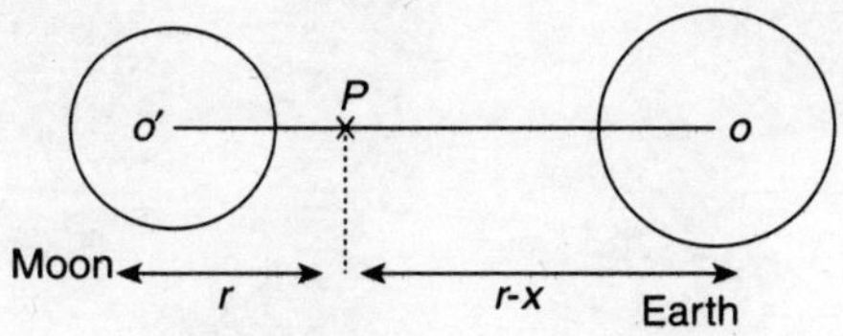

Radius of the earth = R.
Hence, distance between the centres of the moon and the earth = $r = nR$
Let the gravitational field be zero at the point P (fig) which is at a distance of x from the centre of the moon. Then, at the point P, the field due to the moon and the earth must be equal and opposite,

$$\frac{GM_m}{x^2} = \frac{GM}{(r-x)^2}$$

$$\Rightarrow \quad \frac{M_m}{x^2} = \frac{M_e}{(nR-x)^2}$$

$$\Rightarrow \quad \frac{M_e}{M_m} = \left(\frac{nR-x}{x}\right)^2 = \left(\frac{nR}{x} - 1\right)^2$$

$$\Rightarrow \quad \eta = \left(\frac{nR}{x} - 1\right)^2$$

$$\Rightarrow \quad \frac{nR}{x} - 1 = \sqrt{\eta}$$

$$\Rightarrow \quad x = \frac{nR}{\sqrt{\eta} + 1} = \frac{60 \times 6400}{\sqrt{81} + 1} \text{ km}$$

$$\Rightarrow \quad x = 3.84 \times 10^4 \text{ km}$$

CHAPTER

5 Sound

LEARNING OBJECTIVES

At the end of this chapter, you will be able to:

- Define a wave, identify its kind and understand its characteristics
- Describe the characteristics of sound wave and its propagation
- Describe reflection of sound wave and its practical application
- Explain Sonar
- Learn the formula for velocity of sound and list the factors on which it depends
- Define Doppler effect

Introduction

Sound is a form of energy which usually produces the sensation of hearing in our ears. It is produced by vibrating objects. The sound of human voice is also produced due to vibrations in the vocal cords. A vibration is a rapid back-and-forth motion of a particle about its mean position. Vibration is known as oscillation. The matter through which sound is transmitted from the point of source to the listener is called a medium. The medium can be solid, liquid or gas. The disturbance set by sound in the medium travels because of the vibration of the particles of the medium. However, the particles of the medium do not travel from one place to another.

Wave

- The movement of a disturbance through a medium due to the repeated periodic motion of the particles of the medium about their mean positions is known as wave.
- Sound waves are characterized by vibration of particles in the medium and are called mechanical waves.
- Sound, being a mechanical wave, needs material medium like air, water, steel etc. for its propagation. It cannot travel through vacuum.

The medium for wave propagation should have the property of

(i) Elasticity
(ii) Inertia

Essential Characteristics of Wave Motion

(i) Particles of the medium vibrate simple harmonically about their mean position.
(ii) Particles do not leave their mean position and the medium is not transported along with the wave.
(iii) With the wave, energy and momentum are transferred to the neighbouring particles of the medium.
(iv) There is a continuous phase difference amongst successive particles of the medium.

Classification of Waves

1. On the basis of necessity of material medium

(i) Mechanical waves or elastic waves like sound waves, waves on the surface of water, waves on strings. All these waves require a material medium for propagation.
(ii) Electromagnetic waves such as light waves and radio waves which require no medium for propagation.

2. On the basis of vibration of particles

(i) *Longitudinal waves* – in which particles vibrate in the direction of propagation of waves.
(ii) *Transverse waves* – in which particles vibrate in a direction perpendicular to the direction of propagation of waves.

3. On the basis of energy propagation.

(i) *Progressive waves* – along which energy is propagated.
(ii) *Stationary waves* in which energy is confined in a particular region.

Longitudinal Waves

- Sound waves are longitudinal. In these waves, the individual particles of the medium move in a direction parallel to the direction of motion of the disturbances.
- The particles of the medium do not move from one place to another, but they simply oscillate back and forth about their position of rest. This is how sound wave propagates.

Characteristics of Longitudinal Waves

- When a longitudinal wave passes through a medium, the medium is divided into the regions – **compression** and **rarefaction** as shown in Fig. 5.1.

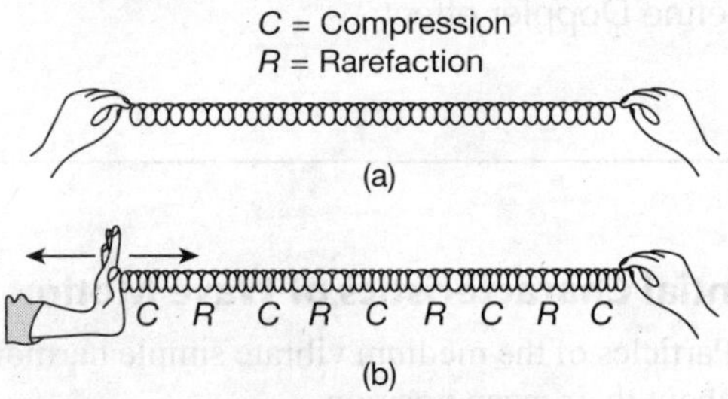

Longitudinal wave in a slinky

Figure 5.1

- The part of a medium where the particles of the medium are very close to each other is known as compression, denoted by C.
- The part of a medium where the particles of the medium are far apart from each other is known as rarefaction, denoted by R.

Transverse Waves

- In transverse waves the individual particles of the medium move about their mean positions in a direction perpendicular to the direction of wave propagation.
- Transverse waves are produced on water surfaces and in the strings of musical instruments.

Characteristics of Transverse Waves

- When a transverse wave travels through medium elevations and depressions are formed, as shown in Fig. 5.2.
- The point of elevation of the medium when its distance from the mean position is maximum is known as **crest** (C).
- The point of depression of the medium when its distance from the mean position is maximum is known as **trough** (T).

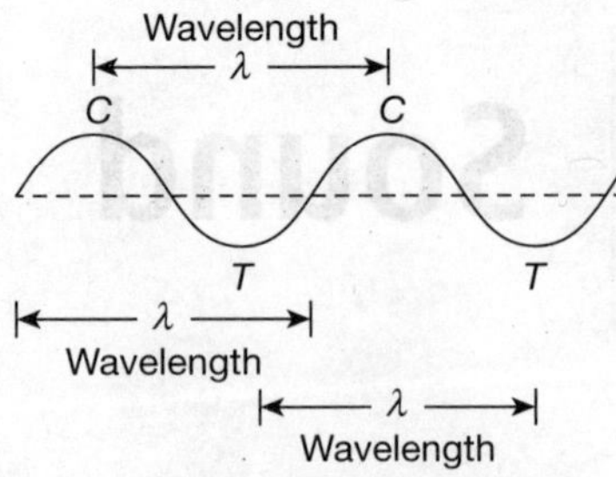

Transverse wave

Figure 5.2

Difference Between Longitudinal and Transverse Waves

Longitudinal waves	Transverse waves
1. Particles of the medium vibrate about their mean position in the direction of propagation of wave.	1. Particles of the medium vibrate about their mean position in a direction perpendicular to the direction of propagation of wave
2. Such waves travel in the form of compressions and rarefactions.	2. Such waves travel in the form of crests and troughs.
3. The medium must possess volume elasticity.	3. The medium must possess elasticity of shape.
4. Density of the medium changes.	4. There is no change in density of the medium.
5. These waves can propagate in all types of media.	5. These waves can travel only through solids and on the surface of liquids.

Propagation of Sound through Air

- Sound propagates through a medium with a finite speed. Its speed is much less than that of light. That is why the sound of thunder is heard only a little after the flash of light is seen during a thunderstorm.
- When a vibrating object moves forward, first it pushes the air in front of it and compresses the air creating a region of high pressure. This region is called area of compression.
- When the same vibrating object suddenly moves backward, it causes a temporary low-pressure region which is called area of rarefaction.

- The region of high pressure is the compression region and the region of low pressure is the rarefaction region. Pressure is related to the number of particles of a medium in a given volume.
- When density of the particles in the medium is more, it gives more pressure and vice-versa. Therefore, the propagation of sound can be visualised as propagation of alternatively fluctuating levels of density or pressure.
- The density and pressure variations respectively as a sound wave propagates in a medium are given in Fig. 5.3.

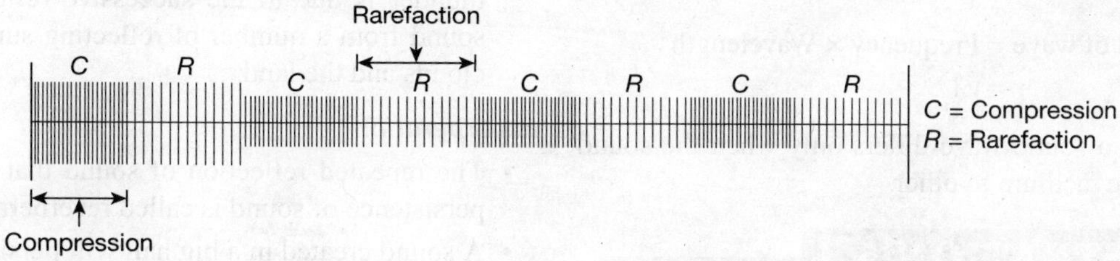

Figure 5.3 Density and pressure variations in a medium as a sound wave travels through it

Characteristics of a Sound Wave

A sound wave can be characterized by
(i) Amplitude **(ii)** Wavelength **(iii)** Frequency
(iv) Pitch **(v)** Loudness **(vi)** Intensity

Amplitude

The maximum displacement of a vibrating particle from its mean position is called amplitude. It is denoted by the letter A. Its SI unit is metre (m).

The loudness or softness of a sound is determined basically by its amplitude. The amplitude of the sound wave depends upon the force with which an object is made to vibrate.

Wavelength

The distance between two consecutive compressions, or two consecutive rarefactions, is the wavelength of that wave. It is represented by lambda (λ) and its SI unit is metre (m).

The distance between a compression and an adjacent rarefaction is equal to half the wavelength.

Frequency

The number of oscillations made by a vibrating body or particles of a medium in 1 second is known as frequency of a wave. It is represented by ν (Greek letter, nu) and its SI unit is Hertz (Hz).

Time Period

Time taken by a vibrating body or a vibrating particle to complete one oscillation is time period. It is denoted by T. Its SI unit is second (s).

Relation Between Frequency and Time Period

Frequency and time period are reciprocals of each other.

$$\text{Frequency} = \frac{1}{\text{Time period}}$$

or,

$$\nu = \frac{1}{T}$$

Pitch

- Pitch is the characteristic of a sound that depends on the frequency received by a human ear.
- A sound wave of high frequency has high pitch and a sound wave of low frequency has a low pitch.

Loudness

Loudness of a sound depends on the amplitude of the vibrating body producing the sound. Loudness depends on the sensitivity of one's ear. A loud sound to a person may be a feeble sound for another person who is hard of hearing.

Intensity

- Intensity of a sound is defined as the sound energy transferred per unit time through a unit area placed perpendicular to the direction of propagation of sound.
- Thus,

$$\text{Intensity of sound} = \frac{\text{Sound energy}}{\text{Time} \times \text{Area}}.$$

- Its SI unit is Joule $s^{-1}m^{-2}$ or watt m^{-2}.

Relation Between Wave Speed, Frequency and Wavelength

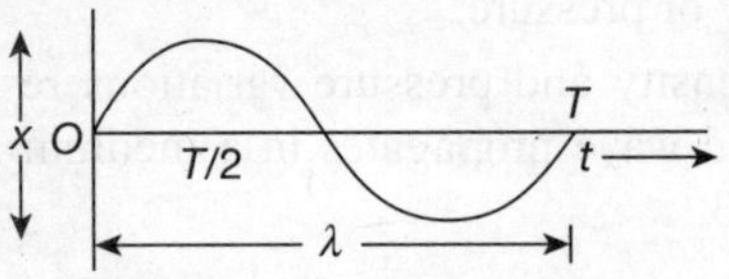

Speed of wave = Frequency × Wavelength

$$v = \nu\lambda$$

The speed of a sound wave differs only when the sound goes form one medium to other.

ILLUSTRATION 5.1

An observer standing at sea coast observes 54 waves reaching the coast per minute. If the wavelength of the waves is 10 m, find the velocity.

Solution

As 54 waves reach the shore per minute,

$$f = \frac{54}{60} = \frac{9}{10}\text{ Hz}$$

And as wavelength of waves is 10 m

$$v = f\lambda = \frac{9}{10} \times 10 = 9 \text{ m/s}$$

Reflection of Sound

- Sound bounces off a solid or a liquid like a rubber ball bounces off a wall. This is called **reflection of sound.** Sound is reflected like light and the same laws of reflection are applicable.
- The directions in which the sound is incident and reflected make equal angles with the normal to the reflecting surface. The incident ray, the normal and the reflected ray, all three lie in the same plane.
- For reflection of sound waves, a polished or rough obstacle of large size is needed.

Echo

- It is the second distinct sound that is heard by the observer after it is reflected from a rigid surface.
- This is a common phenomenon in mountains and valleys where the sound emitted from the source strikes the mountains and returns to the source after a certain interval of time. This similar sound which is heard is called an **echo**.
- The sensation of sound persists in our brain for 0.1 second.
- To hear distinct echoes the minimum distance of the obstacle from the source of sound must be 17 .2 m so that the reflected sound reaches the ear after 0.1 second.
- Echoes may be heard more than once due to successive or multiple reflections. The rolling of thunder is due to the successive reflections of the sound from a number of reflecting surfaces such as clouds and the land.

Reverberation

- The repeated reflection of sound that results in the persistence of sound is called reverberation.
- A sound created in a big hall will persist by repeated reflection from the walls until it is reduced to a value where it is no longer audible.

Practical Applications of the Multiple Reflection of Sound

- Megaphones, loudspeakers, horns and musical instruments such as trumpets, shehnai, etc. are all designed to send sound in a particular direction without spreading in all directions.
- One end of megaphone tube is narrow and its other end is wide.
- When a person speaks into the narrow end, the sound waves produced are prevented from spreading by successive reflections. So sound is heard over a longer distance.
- The ceilings of concert halls and cinema halls are curved so that sound after reflection reaches all the corners of the hall. Sometimes a curved soundboard may be placed behind the stage. Sound, after reflecting from the soundboard, spreads evenly across the width of the hall.
- Stethoscope used by doctors also works on the principle of multiple reflection of sound.

Audible Range of Hearing

- The audible range of sound for humans extends from about 20 Hz to 20000 Hz.
- Sounds of frequencies below 20 Hz are called **infrasonic sounds or infrasound** and those above 20000 Hz are called **ultrasonic sound or ultrasound.**
- Ultrasonic sounds are produced by dolphins, bats and porpoises.

Applications of Ultrasound

- Ultrasound is reflected just like ordinary sound waves and produces echoes. Due to their high frequencies, the ultrasound waves have a high penetrating power.

- Ultrasound is used to clean parts located in hard-to-reach places, for example, spiral tube, odd shaped parts, etc. Objects to be cleaned are placed in a cleaning solution and ultrasonic waves are sent into the solutions. Due to the high frequency, the particles of dust, grease and dirt get detached and drop out.
- Ultrasounds can be used to detect cracks and flaws in metal blocks. Ultrasonic waves are allowed to pass through the metal block and detectors are used to detect the transmitted waves. If there is even a small defect, the ultrasound gets reflected back indicating the presence of the flaw or defect.
- Ultrasonic waves are made to reflect from various parts of the heart and form the image of the heart. This technique is called **echocardiography**.
- Ultrasound scanner is an instrument which uses ultrasonic waves for getting images of internal organs of the human body. In this technique the ultrasonic waves travel through the tissues of the body and get reflected from a region where there is a change of tissue density. These waves are then converted into electrical signals that are used to generate images of the organ. These images are then displayed on a monitor or printed on a film. This technique is called ultrasonography. It is also used for examination of the foetus during pregnancy to detect congenital defects and growth abnormalities.
- Ultrasound may be employed to break small stones formed in the kidneys into fine grains. These grains later get flushed out with urine.

Sonar

- The acronym SONAR stands for Sound Navigation and Ranging.
- Sonar is a device that uses ultrasonic waves to measure the distance, direction and speed of underwater objects.
- It consists of a transmitter and a detector and is installed in a boat or a ship.
- The transmitter produces and transmits ultrasonic waves.
- These waves travel through water and after striking the object on the seabed, get reflected back and are sensed by the detector.
- The detector converts the ultrasonic waves into electrical signals, which are appropriately interpreted.
- The distance of the object that reflects the ultrasonic waves can be calculated by using the formula $2d = v \times t$, where d is the distance, v is speed of sound in water and t is time interval between transmission nand reception of ultrasound.

Shock Waves

- In 1948 first supersonic jet was invented. Bullets and aircrafts also travel sometimes at supersonic speed.
- When a sound-producing source moves with speed higher than sound, it produces **shock waves** in air.
- The shock waves carry a large amount of energy.
- The air pressure variation associated with this type of shock waves produces a very sharp and loud sound called **sonic boom**.
- The shock waves can shatter glass and even damage buildings.

Human Ear

- We are able to hear with the help of an extremely sensitive device called the ear.
- It allows us to convert pressure variations in air with audible frequencies into electric signals that travel to the brain via the auditory nerve.
- The brain interprets them as sound.

Velocity of Sound

In a given medium, the velocity of sound depends on the coefficient of elasticity (E or Y or K) and density (ρ) of the medium.

In solids, $v = \sqrt{\dfrac{Y}{\rho}}$

In liquids; $v = \sqrt{\dfrac{E}{\rho}} = v = \sqrt{\dfrac{K}{\rho}}$

In gases; $v = \sqrt{\dfrac{K_a}{\rho}} = \sqrt{\dfrac{\gamma P}{\rho}}$

where γ is the ratio of two specific heat of a gas,

P = Pressure

T = Temperature

R = Universal gas constant.

Effect of temperature on velocity of sound in air

$$v = \sqrt{\frac{\gamma P}{\rho}} = \sqrt{\frac{\gamma RT}{V\rho}} = \sqrt{\frac{\gamma RT}{M}}$$

i.e., $v \propto \sqrt{T}$ or $\dfrac{v_2}{v_1} = \sqrt{\dfrac{T_2}{T_1}}$

When temperature is increases by 1°C, the velocity of sound in air increases by 0.61 m/s. This is called temperature coefficient of velocity of sound in air.

Effect of density is $v \propto \dfrac{1}{\sqrt{\rho}}$ or $\dfrac{v_2}{v_1} = \sqrt{\dfrac{\rho_1}{\rho_2}}$

Presence of moisture in air decreases the density of air. Therefore, velocity of sound in humid air increases.

At a given temperature, the speed of sound in a medium does not depend upon pressure as (P/ρ) is constant.

ILLUSTRATION 5.2

Speed of sound in air is 332 m/s at NTP. What will be the speed of sound in hydrogen at NTP if the density of hydrogen NTP is (1/16) that of air? [Assume $\rho_{air}/\rho_H = 1$.]

Solution

In terms of density and pressure,

$$\frac{\upsilon_H}{\upsilon_{air}} = \sqrt{\frac{P_H}{\rho_H} \times \frac{\rho_{air}}{P_{air}}} = \sqrt{\frac{\rho_{air}}{\rho_H}} \quad [\text{as } P_{air} = P_H]$$

or $$\upsilon_H = \upsilon_{air} \times \sqrt{\frac{\rho_{air}}{\rho_H}} = 332 \times \sqrt{\frac{16}{1}} = 1328\,\text{m/s}$$

Doppler Effect

When a source of sound and listener of sound separated by a large distance are at rest relative to the medium (usually air), the listener receives the same number of waves per second as that emitted by the source. However, if there is motion of source and/or listener relative to the medium along the line of sight, the number of waves received per second is usually different from that emitted and the frequency or pitch of the source appear to be different. This apparent change in frequency or pitch due to motion of source and listener relative to the medium along the line of sight is called Doppler Effect.

Let us consider the case when both source and listener are moving along the same line with the listener ahead, with speeds v_S and v_L respectively, relative to the medium air, at rest, as shown in Fig.5.4.

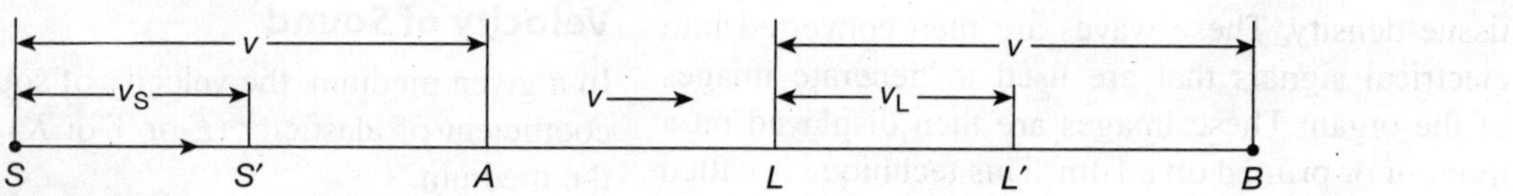

Figure 5.4 Sound source and listener moving along a straight line, with the listener ahead

In one second the first wave emitted by the source as S travels a distance $v \times 1 = v$ and reaches A. Meanwhile, the source S moves a distance v_S and reaches S'. emitting f th wave. Hence if $v_S < v$, the f waves emitted by the source in one sec are contained in a distance $S'A = v - v_S$.

Now the first wave received by the listener at L, covers a distance v in one sec and reaches B. Meanwhile, the listener covers a distance v_L and reaches L′ where he/she receives the wave emitted in the previous second. Hence, if $v_L < v$ the waves received by the listener in one second are contained in a distance $L'B = v - v_L$.

But as waves contained in a distance $v - v_S = f$, the number of waves contained in a distance $v - v_L$ will be

$$= f \times \frac{(v - v_L)}{(v - v_S)}$$

which is the apparent frequency f_A as perceived by the listener.

$$f_A = f\left[\frac{v - v_L}{v - v_S}\right] \quad (1)$$

In this formula, the direction of sound from source to observer is taken to be positive, so speeds v_L and v_S are positive if these are in the direction of sound (from S and L) and negative, if opposite. This will become clearer from the following particular cases:

Source Alone in Motion

In this situation $v_L = 0$ and source is moving towards the listener (i.e., approaching it), v_S is in the assumed direction of v, so $v_S \to +v_S$

$$\therefore \quad f_A = f\left[\frac{v - 0}{v - (+v_S)}\right] = f\left[\frac{v}{v - v_S}\right] \quad \frac{\to v_S}{S \to v\,L}$$

i.e., apparent frequency or pitch will increase as $f_A > f$.

However, if the source moves away from the listener (i.e., recedes from it) v_S will be opposite to v, so $v_S \to -v_S$.

$$\therefore \quad f_A = f\frac{v - 0}{v - (-v_S)} = f\left[\frac{v}{v + v_S}\right] \quad \frac{\leftarrow v_S}{S \to v\,L}$$

i.e., apparent frequency or pitch will decrease as $f_A < f$.

Listener Alone in Motion

In this situation, $v_S = 0$ and the listener is moving away from the source (i.e., receding from it). v_L is in the assumed direction of v, so $v_L \to +v_L$.

$$\therefore \quad f_A = f\left[\frac{v - (+v_L)}{v - 0}\right] = f\left[\frac{v - v_L}{v}\right] \quad \frac{v_L \to}{S \to v\,L}$$

i.e., apparent frequency or pitch will decrease as $f_A < f$.

However, if the listener is moving towards the source (i.e., approaching it) v_L will be opposite to v, so $v_L \to -v_L$.

$$\therefore \quad f_A = f\left[\frac{v-(-v_L)}{v-0}\right] = f\left[\frac{v+v_L}{v}\right] \quad \frac{v_L \leftarrow}{S \to v\ L}$$

i.e., apparent frequency or pitch will increase as $f_A > f$.

Source and Listener, Both in Motion

If the source and listener both move towards each other, i.e., approach each other, v_S will be in the assumed direction of v while v_L opposite to v so that $v_S \to v_S$ and $v_L \to -v_L$

$$\therefore \quad f_A = f\left[\frac{v-(-v_L)}{v-(+v_S)}\right] = f\left[\frac{v+v_L}{v-v_S}\right] \quad \frac{\to v_S\ v_L \leftarrow}{S \to v\ L}$$

which shows maximum increase in frequency or pitch.

However, if both the source and listener move away from each other, i.e., recede from each other, v_L will be in the assumed direction of v while v_S is opposite to v so that $v_L \to +v_L$ and $v_S \to -v_S$

$$\therefore \quad f_A = f\left[\frac{v-(+v_L)}{v-(-v_S)}\right] = f\left[\frac{v-v_L}{v+v_S}\right] \quad \frac{\leftarrow v_S\ v_L \to}{S \to v\ L}$$

which shows maximum decrease in frequency or pitch.

All cases of Doppler Effect can be written in a single formula:

$$f_A = f\left[\frac{v \pm v_L}{v \mp v_S}\right]$$

where v_S, v_L and v are the speeds of source, listener and sound respectively, relative to the medium (and if the medium is at rest, relative to earth). The upper sign on v_S (or v_L) is used when the source (listener) moves towards the listener (source) while the lower sign is used when it moves away.

ILLUSTRATION 5.3

A siren emitting a sound of frequency 2000 Hz moves away from you towards a cliff at a speed of 8 m/s.

(a) What is the frequency of the sound you hear coming directly from the siren?

(b) What is the frequency of sound you hear reflected off the cliff? Speed of sound in air is 330 m/s.

Solution

(a) The frequency of sound heard directly.

$$f_1 = f_0\left(\frac{v}{v+v_s}\right)$$

$$V_s = 8 \text{ m/s}$$

$$\therefore \quad f_1 = \left(\frac{330}{330+8}\right) \times 2000$$

$$f_1 = \frac{330}{338} \times 2000$$

(b) The frequency of the reflected sound is given by

$$f_2 = f_0\left(\frac{v}{v-v_s}\right)$$

$$\therefore \quad f_2 = \left(\frac{330}{330-8}\right) \times 2000$$

$$f_2 = \frac{330}{322} \times 2000 = 2050 \text{ Hz}$$

POINTS TO REMEMBER

1. Sound produced in air is not heard by a diver inside water, because majority of sound energy is reflected from the surface of water.
2. On moon, there is no atmosphere. So no sound can be heard.
3. The sound waves are not dispersed in air.
4. Mechanical waves are of two types:
   (i) Transverse (ii) Longitudinal
5. The weakest audible sound has an intensity $= 10^{-12}$ Wm^{-2}, with loudness equal to 1 decibel.
6. Loudness is the sensation received by the ear due to intensity of sound.
7. Pitch is the sensation received by the ear due to frequency.
8. Quality (or Timbre) is the sensation received by the ear due to 'waveform'.
9. Wave is a means of transferring energy, in which there is no bulk motion of matter. There are two types of waves:
   (i) Mechanical wave
   (ii) Electromagnetic wave
10. Mechanical wave can be produced and propagated in a material medium which possesses the properties of both elasticity and inertia.

11. Electromagnetic waves do not require any material medium for their production or propagation.
12. Relation between frequency, velocity and wavelength: $\nu = v/\lambda$
13. Velocity of longitudinal waves in solids, liquids and gases

$$v = \sqrt{\frac{E}{\rho}}$$

where E is elasticity of the medium and ρ is the density of the undisturbed medium. In the case of solids, E represents Young's modulus of elasticity. In the case of liquids and gases, E represents the bulk modulus of elasticity.
14. Velocity of sound in a gas

$$v = \sqrt{\frac{\gamma P}{\rho}}$$

where γ is the ratio of two specific heats of gas.
15. Factors affecting the change in velocity of sound in gases:
    (a) Pressure – Independent
    (b) Temperature – $v_t \propto \sqrt{T}$

$$\frac{v_t}{v_0} = (1 + \frac{t}{273})^{1/2}$$

where t is in °C

Also $v_t = v_0 + 0.61t$ m/s

    (c) Density – $v \propto \frac{1}{\sqrt{\rho}}$
    (d) Humidity – Sound travels faster in moist air than dry air.
16. Doppler Effect
    The apparent change in frequency of sound when the source of sound, the observer and the medium are in relative motion is called Doppler Effect.
17. The formula for apparent frequency ν observed by the observer and the medium moving in the same direction is

$$\nu' = \left(\frac{v + v_m - v_0}{v + v_m - v_s}\right)\nu$$

COMMON MISCONCEPTIONS

1. We know wave motion is characterized by transfer of a disturbance through space without accompanying transfer of matter. The transfer of disturbance also represents a transfer of energy. But we should not conclude that the energy transfer in all situations is without transfer of matter. In mechanics, we have seen energy transfer is accompanied by matter transfer.
2. The three characteristics of sound, namely, Loudness, Frequency and Quality are sometimes confusing. Remember the difference: Loudness is the sensation received by the ear due to intensity of sound; Pitch is the sensation received by the ear due to frequency; Quality is the sensation received by the ear due to wave form.
3. A common misconception about Doppler Effect is that it depends upon the distance between the source and observer. Although the intensity of sound will vary as the distance changes, the apparent frequency will not. The frequency depends only on speed.

PRACTICE QUESTIONS

Very-short Answer Type Questions

1. Name the physical quantity described by the maximum displacement of a particle from its mean position.
2. A mobile ringing inside a vacuum chamber cannot be heard outside. Why?
3. A vibrating object has a frequency of 250 Hz. Calculate its time period.
4. What do you mean by a wave?
5. What is the difference between a mechanical and an electromagnetic wave?
6. How is ultrasonic sound different from infrasonic sound?
7. How long does sound persist in our brain?
8. In which form are the sound waves transferred to our brain?
9. Which type of waves propagate in the form of compressions and rarefactions?

10. How will the sound produced by a guitar differ from that produced by a drum?

Short-answer Type Questions

1. Mention two characteristics of wave motion.
2. State the difference between transverse and longitudinal waves. Give examples.
3. Represent transverse and longitudinal waves graphically.
4. Derive an expression for the velocity waves graphically.
5. Differentiate between the following:
   (i) Echo and reverberation
   (ii) Loudness and pitch
6. Mention two uses of ultrasounds.
7. On what principle does a megaphone and a stethoscope work?
8. Give reason for the following:
   (i) The ceiling of concert halls are curved.
   (ii) The reverberation time is longer for an empty hall than for a crowded hall.
9. What are the conditions required for an echo to be heard?
10. Mention the factors on which the speed of sound depends. Which characteristics of sound does not change with medium?

Descriptive Type Questions

1. What is sound? How does it propagate through a medium?
2. Discuss the characteristics of sound briefly.
3. What is ultrasound? Mention the applications of ultrasound in different fields of life.
4. Describe reflection of sound. Mention the practical applications of reflection of sound.
5. Show that sound requires a medium to travel through. It can pass through solids, liquids and gases.

NCERT Corner

1. The given graph shows the displacement versus time relation for a disturbance travelling with a velocity of 1500 m s^{-1}. Calculate the wavelength of the disturbance.

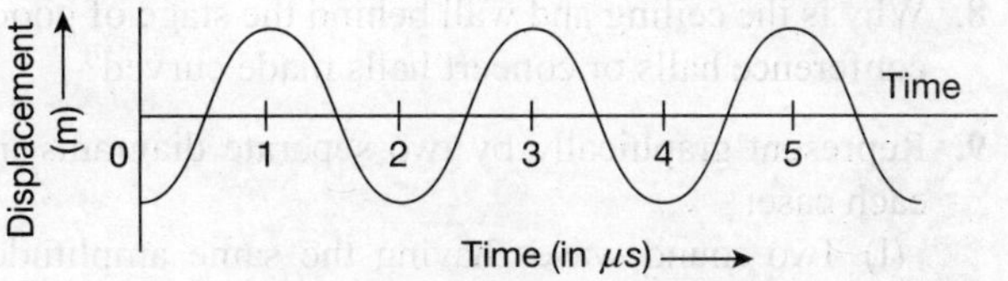

2. Which of the following two graphs (a) and (b), representing the human voice, is likely to be the male voice? Give reason for your answer.

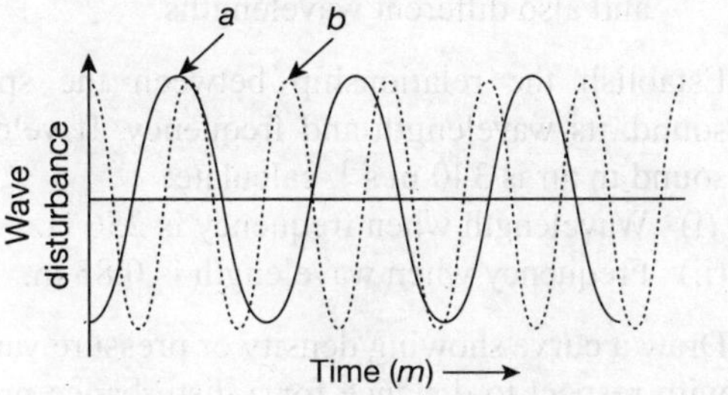

3. A girl is sitting in the middle of a park of dimension 12 m × 12 m. On the left side, there is a building adjoining the park and on right side, there is a road adjoining the park. A sound is produced on the road by a cracker. Is it possible for the girl to hear the echo of this sound? Explain your answer.
4. Why do we hear the sound produced by the humming bees while the sound of vibrations of a pendulum is not heard?
5. If any explosion takes place at the bottom of a lake, what type of shock waves will take place in water?
6. Sound produced by a thunderstorm is heard 10 s after the lightning is seen. Calculate the approximate distance of the thunder cloud. (Given, speed of sound = 340 ms^{-1}.)
7. For the ear to pick up the loudest ticking sound, find the angle x in the figure.

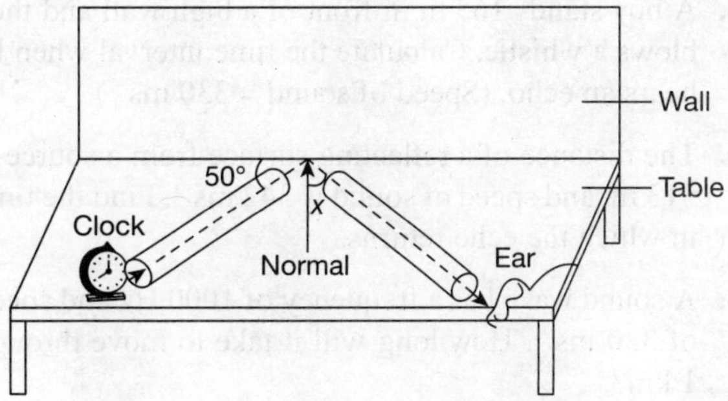

8. Why is the ceiling and wall behind the stage of good conference halls or concert halls made curved?

9. Represent graphically by two separate diagrams in each case:
   (i) Two sound waves having the same amplitude but different frequencies.
   (ii) Two sound waves having the same frequency but different amplitudes.
   (iii) Two sound waves having different amplitudes and also different wavelengths.

10. Establish the relationship between the speed of sound, its wavelength and frequency. If velocity of sound in air is 340 m s^{-1}, calculate:
   (i) Wavelength when frequency is 256 Hz.
   (ii) Frequency when wavelength is 0.85 m.

11. Draw a curve showing density or pressure variations with respect to distance for a disturbance produced by sound. Mark the position of compression and rarefaction on this curve. Also define wavelengths and time period using this curve.

Numericals

Level – I

1. A reflected wave of a SONAR takes 2.2 s to return to the detector. How far is the reflecting surface from the machine? (Take speed of sound = 1531 s)

2. If you hear a thunder 3 s after the lightning is seen, how far are you from the cloud?
Take speed of light = 3×10^8 m/s, Speed of sound = 330 m/s
Why is lightning seen a few seconds before the thunder is heard during a thunderstorm?

3. A sound wave travelling at a speed of 330 m/s has a wavelength of 0.6 m. What will be its frequency? Will it be audible?

4. A boy stands 165 m in front of a high wall and then blows a whistle. Calculate the time interval when he hears an echo. (Speed of sound = 330 ms^{-1})

5. The distance of a reflecting surface from a source is 513 m, and speed of sound is 342 ms^{-1}. Find the time in which the echo returns.

6. A sound wave has a frequency of 1000 Hz and speed of 330 ms^{-1}. How long will it take to move through 1 km?

7. The ultrasonic waves take 4 seconds to travel from the ship to the bottom of the sea and back to the ship (in the form of an echo). What is the depth of the sea? (Speed of sound in water = 1500 m/s)

Level – II

1. In a slinky, 20 vibrations are produced in 10s. Determine the frequency and time period of the vibrations. How can a longitudinal wave be produced using a slinky?

2. A ball is dropped into a pond from a height of 44.1 m. The splash of sound is heard 3.13 s after the ball is dropped. Determined the velocity of sound in air. (Take $g = 9.8$ m/s^2)

3. A car approaches a hill with constant speed. When it is at a distance of 0.96 km from the hill, it blows a horn whose echo is heard by the driver 6 seconds later. If the speed of sound in air is 300 m/s, calculate the speed of the car.

4. A whistle of frequency 540 Hz is moving in a horizontal circle of radius 60 cm at an angular speed of 15 rad/s. What is the highest frequency heard by a listener a long distance away, at rest with respect to the centre of the circle?

5. Compression wave pulses are sent to the bottom of sea from a ship and the echo is heard after 2 s. If bulk modulus of elasticity of water is 2×10^9 N/m^2 and mean temperature is 4° C, what is the depth of the sea?

6. A man is travelling in a train moving away from a station with a speed 30 m/s. Calculate the apparent frequency heard by him of a whistle which is blown at the station with a frequency 1000 c/s.

Fill in the Blanks

1. Mechanical longitudinal waves are propagated through _______.

2. Non-mechanical transverse waves can travel through _______.

3. The velocity of sound in vacuum is _______.

4. Sound waves are propagated _______ through gases.

5. The amplitude of vibration of a particle in air during normal talk is about _______.

6. A point source emits sound equally in all directions in a non-absorbing medium. Two points P and Q are at a distance of 9 metres and 25 metres respectively from the source. The ratio of the amplitudes of the waves between P and Q is _______.

7. A normal ear cannot listen to waves of frequency exceeding _______ cycles/second, which are called _______ waves.

8. In normal talk, the pressure change in air is about ______.

9. Sound waves of frequency 600 Hz fall normally on a perfectly reflecting wall. The velocity of sound is 330 m/sec. The shortest distance from the wall at which the air particles have the maximum amplitude of vibration is ____ metres.

10. Sound travels faster in _______ than in liquids.

11. Frequency above 20000 Hz is _______.

12. Stethoscope works on the _______ of sound.

13. ______ waves require a medium to travel.

14. _______ is used to find the depth of sea.

15. The audible frequency range of the sound for human beings is__________.

16. Jet planes which move with speeds greater than the speed of sound are called__________.

17. If the motion of an object repeats itself at regular intervals of time, it is called________ motion.

18. The velocity of sound is_________ when the density of a gas is quadrupled, with the pressure remaining constant.

19. In a stationary wave, the phase difference between the particles in a given loop is_________.

True or False

1. The distance between a node and adjoining antinode is $\frac{\lambda}{2}$.

2. When a tuning fork vibrates, waves produced in the stem are transverse.

3. A stationary wave transmits energy.

4. There is an apparent change in frequency of the sound heard when the listener and the sounding object move in directions at right angles to each other.

5. The apparent frequency is n, when a source of sound approaches a stationary listener with a velocity x. It is n_2 when the listener approaches the stationary source with equal speed x. then $n_1 > n_2$.

6. Velocity of sound in solids is affected by the solid density.

7. During resonance, the body undergoing forced vibrations vibrates with a larger amplitude.

8. The velocity of sound in a gas is directly proportional to the square root of the temperature of the gas taken in degree Celsius.

9. Light waves are transverse in nature.

COMPETITION VISTA

Match the Columns

1. Match Column A with Column B.

Column A	Column B
(P) Short wavelength	(U) 17.4 m
(Q) SONAR	(V) Wavelength/time
(R) Pitch	(W) Below 20 Hz
(S) Echo	(X) Frequency
(T) Velocity of sound	(Y) Find the depth of sea

2. Match Column A with Column B.

Column A	Column B
(P) Pulse	(T) Pressure waves
(Q) Other name for mechanical waves	(U) Independent of amplitude of vibration
(R) Dolphins	(V) Disturbance for a short duration
(S) Velocity of sound in air	(W) Ultra sonic sounds

(*Continued*)

3. Match Column A with Column B.

Column A	Column B
(P) Supersonic speeds	(T) SONAR
(Q) Very high amplitude	(U) Phase change of π radians
(R) Reflection of sound wave in an open tube	(V) Mach number > 1
(S) Tracking of fish in an ocean	(W) Resonance

4. Match Column A with Column B.

Column A	Column B
(P) Sound cannot propagate through	(T) Decibel
(Q) Unit of frequency	(U) Noise buffers
(R) Unit of loudness	(V) Vacuum
(S) Trees act as	(W) Hertz

5. Match Column A with Column B.

Column A	Column B
(P) Regular vibration produces	(T) Solids
(Q) Speed of sound is maximum in	(U) Noise
(R) Humans cannot hear	(V) Music
(S) Irregular vibration produces	(W) Infrasonic sound

6. Match the following and select the correct answer from the codes given below.

Column A	Column B
(P) Pitch	(S) Wave form
(Q) Quality	(T) frequency
(R) Loudness	(U) intensity

Assertion–Reason Questions

In the following questions, a statement of Assertion (A) is followed by a statement of Reason (R). Mark the correct choice as:

(a) Both A and R are true and R is the correct explanation of A.
(b) Both A and R are true but R is NOT the correct explanation of A.
(c) A is true but R is false.
(d) A is false and R is true.

1. **Assertion (A):** Two persons on the surface of the moon cannot talk to each other.
Reason (R): There is no atmosphere on the moon.

2. **Assertion (A):** Sound passes through air in the form of longitudinal waves.
Reason (R): Longitudinal waves are easier to propagate.

3. **Assertion (A):** Transverse waves travel through air in an organ pipe.
Reason (R): Air possesses only volume elasticity.

4. **Assertion (A):** Compression and rarefaction involve changes in density and pressure.
Reason (R): When particles are compressed, density of the medium increases and when they are rarefied, density of the medium decreases.

5. **Assertion (A):** Sound travels faster on a rainy day than on a dry day.
Reason (R): Moisture increases the pressure.

6. **Assertion (A):** There is no effect of change in pressure on velocity of sound in air, provided temperature remains constant.
Reason (R): Change in pressure also causes change in volume.

7. **Assertion (A):** Under given conditions of pressure and temperature, sound travels faster in a monoatomic gas than in the diatomic gas.
Reason (R): Opposition to travel is more in diatomic gas than in monoatomic gas.

8. **Assertion (A):** An observer standing on a seashore finds that 54 waves are reaching the shore per minute. If the wavelength of waves is 10 m, then their speed will be 9 m/s.
Reason (R): This follows from $v = n\lambda$.

9. **Assertion (A):** In underwater communications, sound waves with low frequency are better than those with high frequency.
Reason (R): When pressure amplitude of sound waves exceeds the atmospheric and hydrostatic pressure, it sets up eddies and turbulent motion in water.

Comprehension Type Questions

1. The number of oscillations or vibrations made by a vibrating body in one second is known as the frequency of the wave. The S.I. unit of frequency is hertz (Hz) which is named after the scientist Heinrich Hertz.
The time taken to complete one vibration by the vibrating body is known as time period. Time period and frequency are related as

$$\text{Time period} = \frac{1}{\text{Frequency}}.$$

(I) The frequency of a source of sound is 20 Hz. Find its time period.

(a) 0.05 s (b) 0.5 s
(c) 0.2 s (d) 0.02 s

(II) A vibrating body vibrates 200 times in 5 second. Calculate its frequency.
(a) 30 Hz (b) 40 Hz
(c) 500 Hz (d) 1000 Hz

(III) A boy heard a sound of frequency 100 Hz at a distance of 500 m from the source of sound. What is the time period of the oscillating particles of the medium?
(a) 0.1 s (b) 0.01 s
(c) 0.2 s (d) 0.02 s

2. The human voice can produce sounds with a frequency between 60 Hz and 1300 Hz. It is interesting to note that a normal human ear can hear sound of frequency between 20 Hz to 20,000 Hz. The sound of frequency greater than 20000 Hz is called ultrasonic and the sound of frequency below 20 Hz is known as infrasonic.

(I) A human can produce sound of frequency of ______.
(a) 20 Hz (b) 200 Hz
(c) 15 kHz (d) 20 kHz

(II) Frequency of ultrasonic sound is ______.
(a) below 20 Hz
(b) above 20 Hz
(c) below 20000 Hz
(d) above 20000 Hz

(III) Frequency of infrasonic sound is ______.
(a) below 20 Hz
(b) above 20 Hz
(c) below 20000 Hz
(d) above 2000 Hz

3. When a sound wave travels through a medium such as air, water or steel, it will set the particles of the medium into vibration as it passes through medium. For this, the medium must possess inertia and elasticity. The velocity of sound is the characteristic of the medium in which it propagates.

The velocity of sound is given by $v = \sqrt{\frac{E}{\rho}}$, where E is the elasticity of the medium and ρ the density. Velocity of sound is maximum in solid and minimum in gases. The velocity of sound in air at normal temperature and pressure is approximately 332 m/s.

(I) Which of the following is incorrect regarding propagation of sound wave in a medium?
(a) The medium must have inertia.
(b) The medium must have elasticity.
(c) The medium should have both inertia and elasticity.
(d) In vacuum sound wave propagates at a very low speed.

(II) The velocity of sound in a medium is
(a) directly proportional to the square of density.
(b) inversely proportion to density.
(c) inversely proportional to the square root of density.
(a) independent of density

(III) The velocity of sound is
(a) least in solids
(b) greatest in liquids
(c) least in gases
(d) least in liquids

(IV) The velocity of sound in air at 40° C is
(a) less than 332 m/s.
(b) more than 332 m/s.
(c) cannot be predicted.
(d) exactly 332 m/s.

Objective Questions

One Option Correct

1. Wave is a disturbance in which ______ is transferred from one point to other.
(a) Particle (b) Energy
(c) Velocity (d) All of these.

2. Sound if a form of
(a) longitudinal wave only.
(b) transverse wave only.
(c) longitudinal and transverse waves.
(d) longitudinal waves, most of the time.

3. Sound wave travels in the form of
(a) crests and troughs.
(b) crest only.
(c) compression and rarefaction.
(d) rarefaction only.

4. Speed of sound can be determined by the expression:
(a) $v = \lambda v$ (b) $v = \frac{\lambda}{2}$
(c) $v = \frac{v}{\lambda}$ (d) λv^2

5. The sound of frequency 15 Hz is ______.
(a) audible
(b) infrasonic
(c) ultrasonic
(d) Both (a) and (b).

6. The density of a medium through which sound travels is minimum in a region which is
(a) at rarefaction.
(b) between rarefaction and compression.
(c) at compression.
(d) after compression.

7. Frequency is measured in ______.
(a) metre (b) m/s
(d) Hz (d) s

8. The distance between two consecutive crests will be ______.
(a) $\frac{\lambda}{2}$ (b) λ
(c) $\frac{3\lambda}{2}$ (d) 2λ

9. The frequency of a sound wave corresponding to 17 m is ______.
(a) 20 Hz (b) 200 Hz
(c) 2000 Hz (d) 20000 Hz

10. A wave 0.2 m long of frequency 500 Hz travels at a speed of 100 m/s. This wave will take ______ to travel 300 m.
(a) 70 s (b) 60 s
(c) 12 s (d) 3 s

11. The angle of incidence of a sound wave is 30. It will reflect back at an angle of
(a) 0° (b) 30°
(c) 60° (d) 90°

12. Which of the following is the correct setup to prove the first law of reflection of sound?
(a) 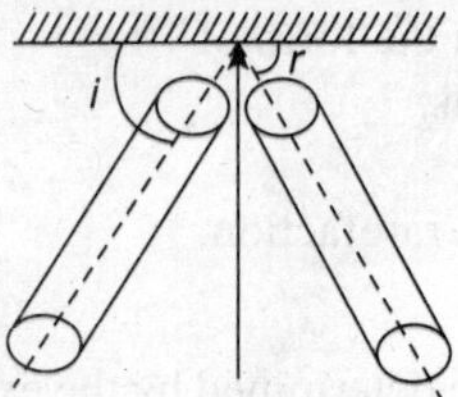

(b) 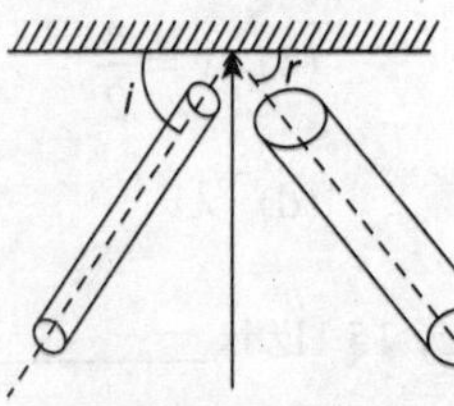

(c) 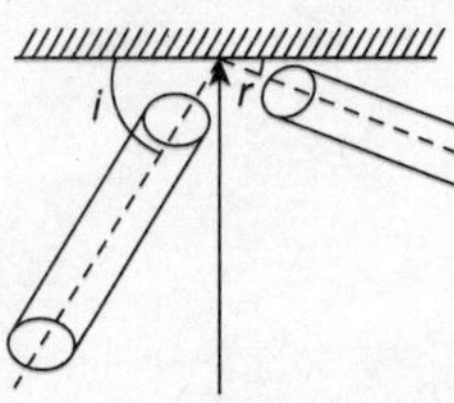

(d) 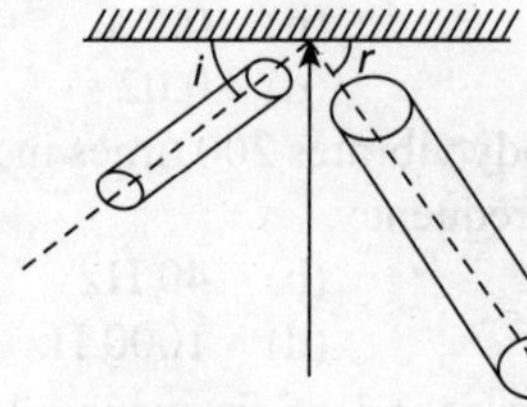

13. The sound of the source used to prove the laws of reflection of sound must be ______.
(a) loud and clear (b) soft but audible
(c) of high pitch (d) of high amplitude

14. If a slinky, used to determine the wave velocity, is stretched 2 m long then the distance travelled by one wave will be ______.
(a) 2 m (b) 3 m
(c) 4 m (d) 1 m

15. An echo returns in 3 s. If the speed of a sound is 342 m/s, then the distance of the reflecting surface from the source is ______.
(a) 1026 m (b) 1000 m
(c) 513 m (d) 342 m

16. When a sound wave goes from one medium to another, the quantity that remains unchanged is ______.
(a) frequency (b) amplitude
(c) wavelength (d) speed

17. Water waves are ______.
(a) longitudinal
(b) transverse
(c) both longitudinal and transverse
(d) neither longitudinal nor transverse

18. In a gas, the velocity of sound varies directly as
(a) square root of isothermal elasticity.
(b) square of isothermal elasticity.
(c) square root of adiabatic elasticity.
(d) adiabatic elasticity.

19. The temperature at which speed of sound in air becomes double its value at 0 °C is ______.
(a) 546 °C (b) 819 °C
(c) 273 °C (d) 1092 °C

20. The ratio of the speed of a body to the speed of sound is called ______.
(a) Mach number (b) Doppler ratio
(c) Sonic index (d) Refractive index

21. Human ear cannot hear those mechanical waves whose frequency lies in the frequency range
(a) less than 1000 Hz but greater than 10000 Hz.
(b) between 1000 Hz and 5000 Hz.

(c) between 500 Hz and 20000 Hz.
(d) less than 16 Hz and more than 20000 Hz.

22. The velocity of sound in air at NTP is 330 m/s. What will be its value when temp. is doubled and pressure is halved?
(a) 330 m/s (b) 165 m/s
(c) $330\sqrt{2}$ m/s (d) $330/\sqrt{2}$ m/s

23. In ordinary talk, the amplitude of vibration is approximately ______.
(a) 10^{-12}m (b) 10^{-11} m
(c) 10^{-8} m (d) 10^{-7} m

24. From a point source, if amplitude of waves at a distance r is A, its amplitude at a distance $2r$ will be ______.
(a) A (b) $2A$
(c) $A/2$ (d) $A/4$

25. The distance between two consecutive crests in a wave train produced in a string is 5 cm. If 2 complete waves pass through any point per second, the velocity of the wave is ______.
(a) 10 cm/sec (b) 2.5 cm/sec
(c) 5 cm/sec (d) 15 cm/sec

26. Under identical conditions of pressure and density, the speed of sound is highest in a
(a) monoatomic gas. (b) diatomic gas.
(c) triatomic gas. (d) polyatomic gas.

27. The wavelength of ultrasonics in air is of the order of ______.
(a) 10^2 m (b) 10^1 m
(c) 10^{-2} m (d) 10^0 m

28. A plain wave of sound travelling in air is incident upon a plain water surface. The angle of incidence is 30°. If the velocity of sound in water is 1400 m/s, and the velocity of sound in air is 330 m/s, there will be ______.
(a) Reflection only.
(b) Refraction only.
(c) both reflection and refraction.
(d) neither reflection nor refraction.

29. The speed of sound in a perfectly rigid rod is ______.
(a) infinite (b) zero
(c) 332 m/s (d) 3×10^8 m/s

30. A radio station broadcasts at 760 kHz. What is wavelength of the station?
(a) 395 m (b) 790 m
(c) 760 m (d) 197.5 m

31. What is the amplitude of wave shown in the figure?

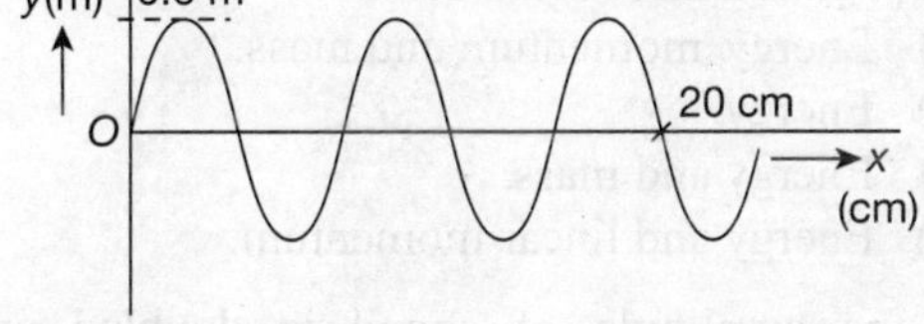

(a) 0.6 m (b) 0.3 m
(c) 20 cm (d) 4 cm

32. What is the wavelength of the wave shown in the figure.

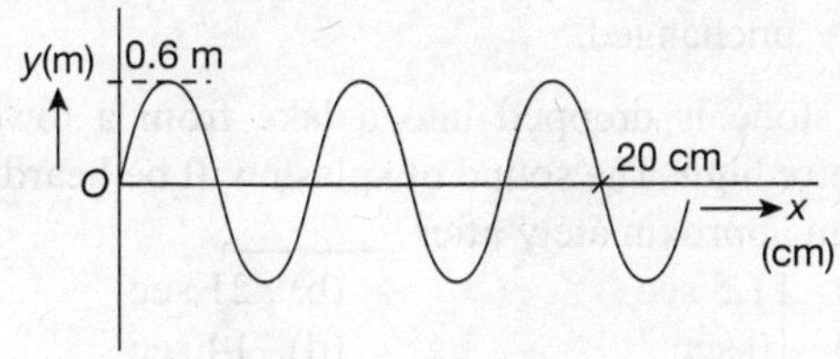

(a) 20 cm (b) 10 cm
(c) 0.08 m (d) 0.6 m

33. What is the frequency of the wave shown in the figure if its speed is 300 m/s?

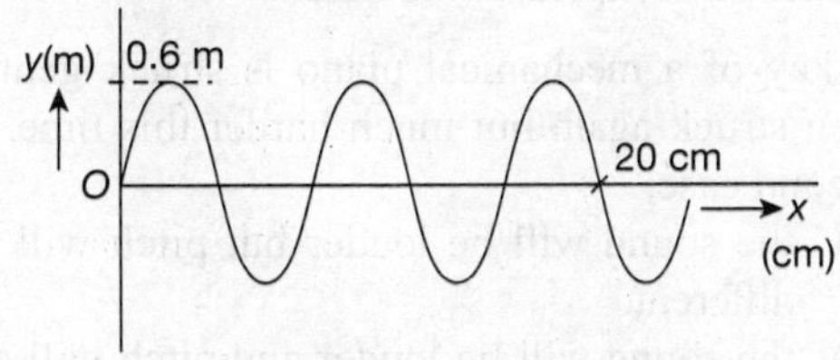

(a) 3750 Hz (b) 375 Hz
(c) 37.5 Hz (d) 300 Hz

34. An underwater sonar source operating at a frequency of 60 kHz directs its beam towards the surface. If velocity of sound in air is 330 m/s, wavelength and frequency of the waves in air are ______.
(a) 5.5 m, 60 kHz (b) 330 m, 60 kHz
(c) 5.5 m, 30 kHz (d) 5.5 m, 80 kHz

35. If the pressure amplitude in a sound wave is tripled, then by what factor is the intensity of sound wave increased?
(a) 3 (b) 6
(c) 9 (d) $\sqrt{3}$

36. The distance between two points differing in phase by 60° on a wave having a wave velocity 360 m/s and frequency 500 Hz is ______.
(a) 0.72 metre (b) 0.12 metre
(c) 0.18 metre (d) 0.36 metre

37. With the propagation of a longitudinal wave through a material medium, the quantities transmitted in the propagation direction are:
(a) Energy, momentum and mass.
(b) Energy.
(c) Energy and mass.
(d) Energy and linear momentum.

38. If the amplitude of sound is doubled and the frequency reduced to one fourth, the intensity of sound at the san1e point will be
(a) increasing by a factor of 2.
(b) decreasing by a factor of 2.
(c) decreasing by a factor of 4.
(d) unchanged.

39. A stone is dropped into a lake from a tower 500 metre high. The sound of splash will be heard by the man approximately after ________.
(a) 11.5 sec (b) 21 sec
(c) 10 sec (d) 14 sec

40. Note is a sound
(a) of mixture of several frequencies.
(b) of mixture of two frequencies only.
(c) of a single frequency.
(d) always unpleasant to listen.

41. A key of a mechanical piano is struck gently and then struck again but much harder this time. In the second case,
(a) the sound will be louder but pitch will not be different.
(b) the sound will be louder and pitch will also be higher.
(c) the sound will be louder but pitch will be lower.
(d) both loudness and pitch will remain unaffected.

42. In SONAR, we use ______.
(a) ultrasonic waves
(b) infrasonic waves
(c) radio waves
(d) audible sound waves

43. Sound travels in air if
(a) particles of the medium travel from one place to another.
(b) there is no moisture in the atmosphere.
(c) disturbance moves.
(d) both particles as well as disturbance travel from one place to another.

44. When we change feeble sound to loud sound, we increase its ______.
(a) frequency (b) amplitude
(c) velocity (d) wavelength

45. In the curve (figure) half the wavelength is

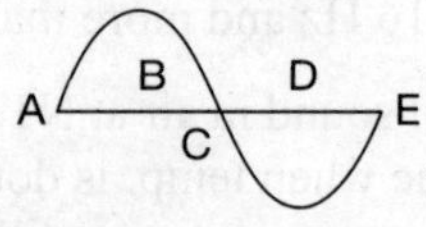

(a) AB (b) BD
(c) DE (d) AE

46. Earthquake produces which kind of sound before the main shock wave begins?
(a) Ultrasound (b) Infrasound
(c) Audible sound (d) None of the above.

47. Infrasound can be heard by ________.
(a) dog (b) bat
(c) rhinoceros (d) human beings

48. Before playing the orchestra in a musical concert, a sitarist tries to adjust the tension and plucks the string suitably. By doing so, he is adjusting
(a) intensity of sound only.
(b) amplitude of sound only.
(c) frequency of the sitar string with the frequency of the other musical instruments.
(d) the loudness of sound.

More Than One Option Correct

1. Which of the following statements are correct about intensity of sound?
(a) It depends only on amplitude of the wave.
(b) It depends both on amplitude and frequency of the wave.
(c) Its practical unit is decibel.
(d) Its practical unit is phonon.

2. Transverse mechanical waves can travel in:
(a) Iron rod (b) Hydrogen gas
(c) Water (d) Stretched strings

3. The velocity of sound is affected by change in ________.
(a) temperature (b) medium
(c) pressure (d) wavelength

4. Which of the following waves are mechanical?
(a) Torsional waves (b) Microwaves
(c) Seismic waves (d) Ripples

5. Which of the following phenomena can take place with sound (in air)?
(a) Interference (b) Diffraction
(c) Polarisation (d) Scattering

6. At displacement nodes:
(a) Displacement is minimum
(b) Pressure is maximum

(c) Sound is maximum
(d) Particle speed is minimum

7. If a satellite moving in an elliptical orbit relative to the earth transmits radio-signals towards earth, the frequency received by a tracking station on the earth
(a) can be higher.
(b) can be lower.
(c) can be equal.
(d) cannot remain constant.

Scholar Zone

1. Sound waves ($v = 343$ m/s) from a musical performance are encoded onto radio waves for transmission from the radio studio to your home receiver. A radio wave travelling at $c = 3.00\times10^8$ m/s with a wavelength of 3.00 m carries a sound wave that has a wavelength of 3.00 m in air. Which has the higher frequency, the radio wave or the sound wave?

2. You are on a cruise and the person next to you at the deck railing falls overboard. During the entire time of fall, the person screams with a constant frequency. How does the apparent frequency of the sound of the scream change with time?

3. You are driving towards a cliff and you honk your horn. The frequency of the echo you hear will be Doppler shifted
(a) up in frequency, equivalent to a moving source.
(b) up in frequency, equivalent to a moving observer.
(c) down in frequency, equivalent to a moving source.
(d) down in frequency, equivalent to a moving observer.
(e) down in frequency, equivalent to a moving source and a moving observer.

4. Distinguish between sound and radio waves of the same frequency (say 15 kHz).

5. Explain why
(a) transverse mechanical waves cannot be propagated in liquids and gases.
(b) waves on strings are always transverse.

6. The frequency of sound produced by a bell is 500 Hz. The velocity of the source relative to still air is 60 m/s. An observer moves at 30 m/s along the same line as the source. Calculate the frequency of sound wave measured by the observer. Consider all the possible cases (speed of sound $v = 340$ m/s)

Answer Keys

Objective Questions

One Option Correct

1. (b)	2. (a)	3. (c)	4. (a)	5. (b)	6. (a)	7. (b)	8. (b)	9. (a)	10. (d)
11. (b)	12. (a)	13. (b)	14. (a)	15. (c)	16. (a)	17. (c)	18. (c)	19. (b)	20. (a)
21. (d)	22. (c)	23. (c)	24. (c)	25. (a)	26. (a)	27. (c)	28. (a)	29. (a)	30. (a)
31. (a)	32. (c)	33. (a)	34. (a)	35. (c)	36. (b)	37. (b)	38. (c)	39. (a)	40. (c)
41. (a)	42. (a)	43. (c)	44. (b)	45. (b)	46. (b)	47. (c)	48. (c)		

More Than One Option Correct

1. (b) and (c)
2. (a) and (d)
3. (a) and (b)
4. (a), (c) and (d)
5. (a) and (b)
6. (a), (b), (c) and (d)
7. (a), (c), (c) and (d)

Hints and Solutions

Very-short Answer Type Questions

1. Amplitude
2. A mobile ringing inside a vacuum chamber cannot be heard outside because sound requires a medium to travel.
3. $T = \frac{1}{f} = \frac{1}{250} = 0.004$ s
4. A disturbance in which energy is carried/transferred from one particle to other is called wave.
5. Mechanical wave needs a medium to propagate whereas an electromagnetic wave does not need a medium to travel.
6. Ultrasonic sound has frequency greater than 20 kHz whereas infrasonic sound has frequency less than 20 Hz. (1 kHz = 1000 Hz).
7. Sensation of sound persists for 0.1 s in our brain.
8. Sound waves are transferred to our brain in the form of electric signals.
9. Longitudinal waves.
10. The sound produced by a guitar will have a greater frequency (pitch) than that produced by a drum.

Short Answer Type Questions

1. (i) In a wave motion, energy is transferred from one point to another, not matter.
   (ii) Wave travels at the same speed in all directions.
2.

S. No.	Transverse waves	Longitudinal waves
i.	They do not require a medium to travel.	They require a medium to travel.
ii.	Particles vibrate up and down, i.e., perpendicular to the direction of propagation of the wave e.g. light.	Particles vibrate to and from, i.e., parallel to the direction of propagation of wave e.g. sound.

3.

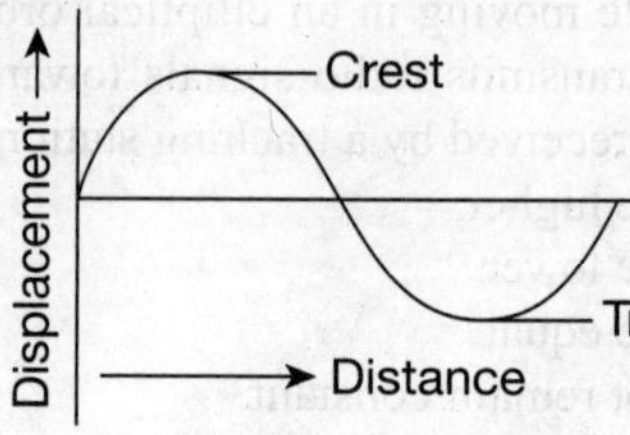

Transverse wave

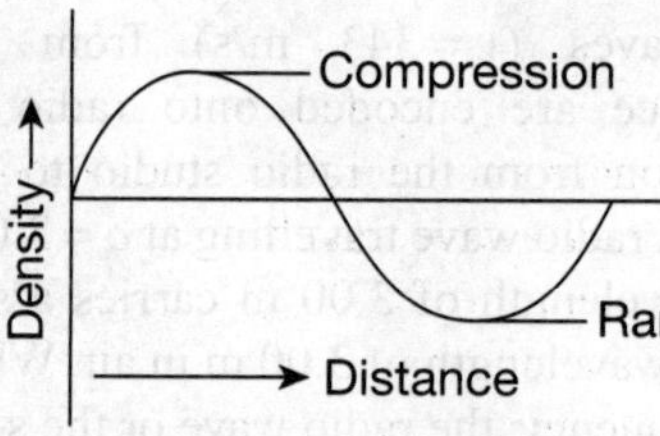

Longitudinal wave

4. Velocity is distance travelled per unit time. Distance travelled by a wave is its wavelength (λ).

$$\therefore \quad v = \frac{\lambda}{T}$$

$$\frac{1}{T} = \nu \text{ (frequency)}$$

$$\therefore \quad v = \lambda\nu$$

Wave velocity = Wavelength × Frequency

5. (i)

Echo	Reverberation
Distinct sound heard by an observer after it gets reflected by a surface.	Repeated reflection of sound that results in persistence of sound.

(ii)

Loudness	Pitch
Loudness in determined by the amplitude of the vibrating particle.	Pitch is determined by the frequency of the wave

6. (i) Ultrasounds are used to scan body parts.
   (ii) Ultrasounds are used to clean parts located in hard-to-reach places, e.g. spiral tube, electronic components etc.

7. Both these instruments work on the principle of multiple reflection of sound. The funnel shaped body of these devices reflects the sound multiple times and increases its amplitude/loudness.
8. (i) Curved ceiling of concert halls help in ensuring that the sound reaches every corner of the hall after reflection.
   (ii) In a crowded hall, the audience absorbs reverberation and hence the reverberation time decreases.
9. (i) The reflected sound must not reach the car before $\frac{1}{10}$ s.
   (ii) The minimum distance between the source of sound and the reflecting surface must be 17.4 m.
10. The speed of sound depends upon the nature and temperature of the medium (material). The frequency of sound does not change with the medium.

Descriptive Type Questions

1. Sound is a disturbance in which energy is transferred from one point to another in the form of vibrations.
   Propagation of sound through medium:
   - Sound propagates through a medium at a finite speed. Its speed is much less than that of light. That is why the sound of thunder is heard only a little later, after a flash of light is seen during a thunderstorm.
   - When a vibrating object moves forward, first it pushes the air in front of it and compresses the air creating a region of high pressure. This region is called area of compression.
   - When the same vibrating object suddenly moves backward, it causes a temporary low-pressure region, which is called area of rarefaction.
   - The area of compression is a region of high pressure and area of rarefaction is a region of low pressure. Pressure is related to the number of particles of a medium in a given volume.
   - More density of the particles in the medium gives more pressure and vice-versa. Therefore, the propagation of sound can be visualised as propagation of alternatively fluctuating levels of density or pressure.
   - The density and pressure variations respectively are given in the figure below, as sound wave propagates in the medium.

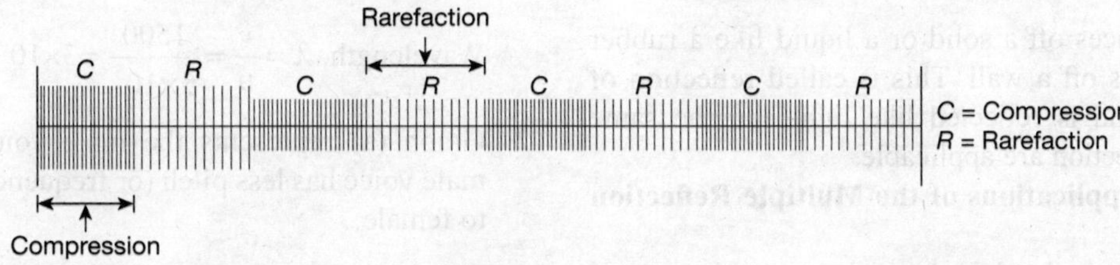

2. The following are the characteristics of sound.
   - **Amplitude:** The magnitude of the maximum displacement of particles of a medium from their mean position. It is denoted by A and measured in metre (SI unit).
   - **Time period:** The time taken for one complete oscillation or vibration is called time period. It is denoted by T and measured in seconds (s).
   - **Frequency:** The number of complete oscillations per unit time is called frequency. It is denoted as ν and its SI unit is Hertz (Hz).
   - **Wavelength:** The distance between two consecutive crests/compressions or one crest/compression and one trough/rarefaction is the wavelength of a wave. It is denoted as λ and measured in metres (SI unit).
   - **Velocity:** Wave velocity is calculated by dividing wavelength with time.

   $$v = \frac{\lambda}{T} = \lambda\nu \left(\because \frac{1}{T} = \nu \right)$$

   It is measured in m/s.
3. The audible range of sound for humans extends from about 20 Hz to 20000 Hz. Sounds of frequencies below 20 Hz are called infrasonic sounds or infrasound and those above 20000 Hz are called ultrasonic sound or ultrasound.

   Applications of Ultrasound
   - Ultrasound is reflected just like ordinary sound waves and produces echoes. Due to their high frequencies, the ultrasound waves have high penetrating power.
   - Ultrasound is used to clean parts located in hard-to-reach places, for example, spiral tube, odd shaped parts, etc. Objects to be cleaned are placed in a cleaning solution and ultrasonic waves are

sent into the solutions. Due to the high frequency, the particles of dust, grease and dirt get detached and drop out.

- Ultrasounds can be used to detect cracks and flaws in metal blocks. Ultrasonic waves are allowed to pass through the metal block and detectors are used to detect the transmitted waves. If there is even a small defect, the ultrasound gets reflected back indicating the presence of the flaw or defect.
- Ultrasonic waves are made to reflect from various parts of the heart and form the image of the heart. This technique is called echocardiography.
- Ultrasound scanner is an instrument which uses ultrasonic waves for getting images of internal organs of the human body. In this technique, the ultrasonic waves travel through the tissues of the body and get reflected from a region where there is a change of tissue density. These waves are then converted into electrical signals that are used to generate images of the organ. These images are then displayed on a monitor or printed on a film. This technique is called ultrasonography. It is also used for examination of the foetus during pregnancy to detect congenital defects and growth abnormalities.
- Ultrasound may be employed to break small stones formed in the kidneys into fine grains. These grains later get flushed out with urine.

4. Sound bounces off a solid or a liquid like a rubber ball bounces off a wall. This is called reflection of sound. Sound is reflected like light and the same laws of reflection are applicable.

Practical Applications of the Multiple Reflection of Sound

- Megaphones, loudspeakers, horns and musical instruments such as trumpets, shehnai, etc. are all designed to send sound in a particular direction without spreading in all directions.
- One end of megaphone tube is narrow and its other end is wide.
- When a person speaks into the narrow end, the sound waves produced are prevented from spreading by successive reflections. So sound is heard over a longer distance.
- The ceilings of concert halls and cinema halls are curved so that sound, after reflection, reaches all the corners of the hall. Sometimes a curved soundboard may be placed behind the stage. Sound, after reflecting from the soundboard, spreads evenly across the width of the hall.
- Stethoscope used by doctors also works on the principle of multiple reflection of sound.

5. **Experiment to demonstrate the need for sound to have a medium to travel**

Fix an electric bell inside a jar connected to a vacuum pump. Close the jar and connect the bell to a source of electrical energy. As you ring the bell and slowly remove the air with the vacuum pump, you could hear sound of reducing intensity and the sound will die down when all the air is pumped out. This demonstrates the requirement for a medium to carry sound.

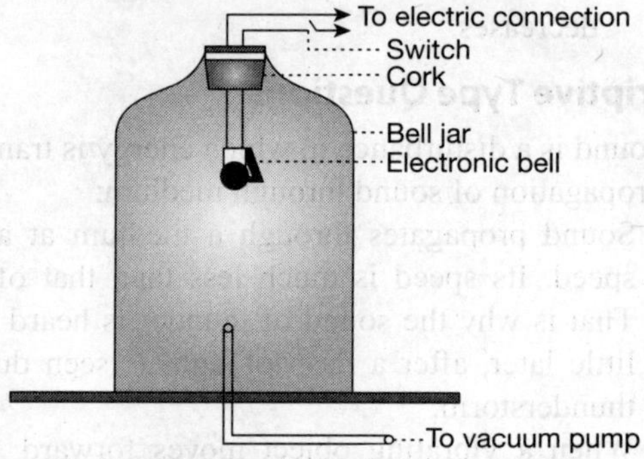

Bell jar experiment showing sound cannot travel in vacuum

NCERT Corner

1. From the graph

Time period, $T = 2 \times 10^{-6}$ s.

Frequency, $\upsilon = 1/T = 5\times10^5$ Hz.

Wavelength, $\lambda = \dfrac{v}{\upsilon} = \dfrac{1500}{5\times10^5} = 3\times10^{-3}$ m.

2. Graph (a) represents the male voice. Usually the male voice has less pitch (or frequency) as compared to female.

3. Echo can be heard only if the time gap between the original sound and reflected sound received by the listener is around 0.1 s.

The minimum distance to be travelled by the reflected sound wave for the echo to be distinctly heard

= Velocity of sound × Time interval

$\cong 344\times0.1$

$\cong 34.4$ m

But in this case the distance travelled by the sound reflected from the building and then reaching to the girl will be (6 + 6) = 12 m, which is much smaller than the required distance. Therefore, no echo can be heard.

4. Humming bees produce sound in the audible range by vibrating their wings. In case of pendulum the frequency is below 20 Hz, which does not come in the audible range.

5. Longitudinal waves.

6. $S = 340\,\text{m s}^{-1} \times 10\,\text{s} = 3400$ m or 3.4 km.

7. $\angle i = \angle r$, so $x = 90° - \angle r = 90° - 50° = 40°$.

8. Ceiling and walls are made curved so that the sound, after reflection, reaches the target audience.

9.

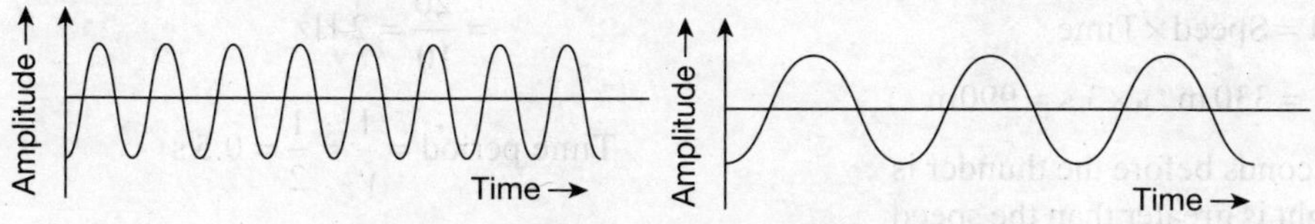

Same amplitude but different frequency

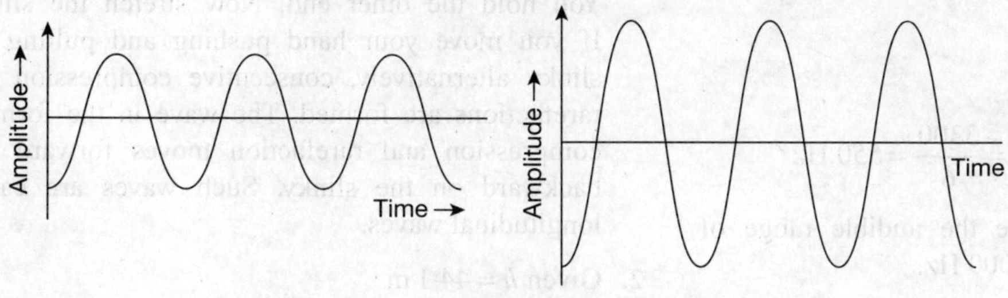

Same frequency but different amplitude

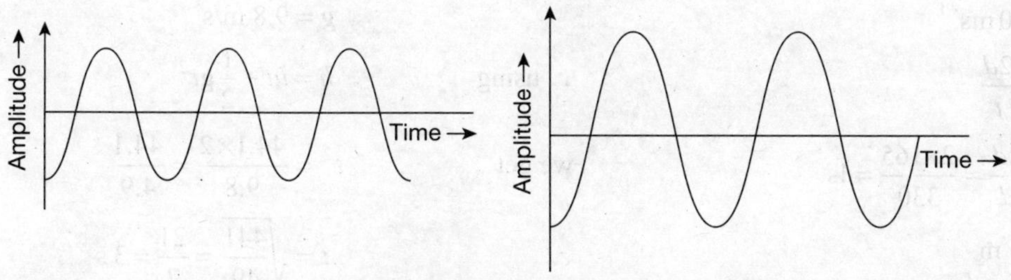

Different amplitudes and different wavelengths

10. **Derivation of formula** $v = \nu\lambda$

Let ν, λ and v be respectively the frequency, wavelength and velocity of a wave.

Frequency ν, is the number of vibrations per second.

In one oscillation the wave travels a distance λ.

Thus, the distance travelled in 1 second = $\nu\lambda$

Therefore, $v = \nu\lambda$.

(i) $340 = 256\lambda$

$\lambda = 1.33$ m.

(ii) $340 = \nu(0.85)$

$\nu = 400$ Hz

11.

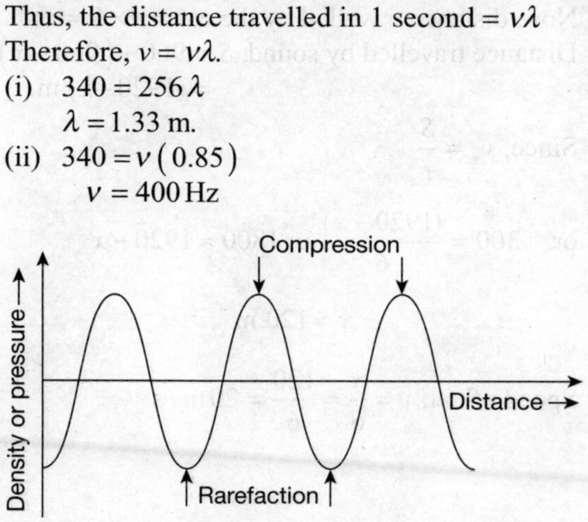

Wavelength is the distance between two consecutive compressions or two consecutive rarefactions. Time period is the time taken to travel to travel the distance between any two consecutive compressions or rarefactions from a fixed point.

Numericals

Level – I

1. Given Speed = 1531 m/s

Time = 2.2 s

Distance travelled = 2 d

Using $\quad v = \dfrac{d}{T}$

we get $\quad v = \dfrac{2d}{T}$

$$1531\,\text{m/s} = \frac{2d}{2.2\,\text{s}}$$

$$d = \frac{1531\,\text{m/s} \times 2.2\,\text{s}}{2} = 1531 \times 1.1 = 1684.1\,\text{m}$$

2. Speed of sound = 330 m/s
Time = 3 s
Presuming that light takes no time to reach us

$$\text{Speed of sound} = \frac{\text{Distance}}{\text{Time}}$$

$\therefore$ Distance of cloud = Speed × Time

$$= 330\,\text{m/s} \times 3\,\text{s} = 990\,\text{m}$$

Lightning is seen a few seconds before the thunder is heard because speed of light is greater than the speed of sound.

3. Given $\lambda = 0.6$ m

$v = 330\,\text{m/s}$

$$\therefore \quad \nu = \frac{v}{\lambda} = \frac{330\ \text{m/s}}{0.6\ \text{m}} = \frac{3300}{6} = 550\ \text{Hz}$$

Yes, it is audible because the audible range of frequencies is 20 Hz to 20,000 Hz.

4. Distance between the wall and the boy, i.e., $d = 165$ m
Speed of sound, $v = 330\,\text{ms}^{-1}$

We have, velocity, $v = \dfrac{2d}{t}$

$$\therefore \quad t = \frac{2d}{d} = \frac{2 \times 165}{330} = 1\text{s}$$

5. Here, distance, $S = 513$ m
Speed of sound = 342 ms^{-1}

Time taken by sound to travel $= \dfrac{513\,\text{m}}{342\,\text{ms}^{-1}} = 1.5$ s

Time taken by echo to return = 1.5×2 = 3 s

6. Distance travelled = 1000 m
Speed of sound = 330 ms^{-1}
Time taken by the wave to move through 1 km

$$= \frac{\text{Distance}}{\text{Speed}} = \frac{1000\,\text{m}}{330\,\text{ms}^{-1}} = 3.03 = 3\ \text{s}$$

7. Since wave takes 4 seconds to travel from the ship to the bottom of the sea and back to the ship, the time taken by the wave to travel from the ship to the bottom of the sea,

$$t = 2 \text{ seconds}$$

Here, $v = 1500$ m/s
Therefore, $d = v \times t = 1500\,\text{m} \times 2$

$$d = 3000 \text{ m}$$

Level - II

1. Number of vibrations = 20
Time for 20 vibrations = 10 s

$$\text{Frequency } (\nu) = \frac{\text{Number of vibrations}}{\text{Time}} = \frac{20}{10} = 2\ \text{Hz}$$

$$\text{Time period} = \frac{1}{\nu} = \frac{1}{2} = 0.5\ \text{s}$$

Take a slinky. Ask your friend to hold one end. You hold the other end. Now stretch the slinky. If you move your hand pushing and pulling the slinky alternatively, consecutive compression and rarefactions are formed. The wave in the form of compression and rarefaction moves forward and backward on the slinky. Such waves are called longitudinal waves.

2. Given $h = 44.1$ m

$u = 0$
$t = ?$
$g = 9.8\ \text{m/s}^2$

$\therefore$ using $h = ut + \dfrac{1}{2}gt^2$

we get $t^2 = \dfrac{44.1 \times 2}{9.8} = \dfrac{44.1}{4.9}$

$$\therefore \quad t = \sqrt{\frac{441}{49}} = \frac{21}{7} = 3\ \text{s}$$

$\therefore$ Time taken by sound to reach the height

$$= 3.13 - 3 = 0.13\ \text{s}$$

$$\therefore \quad v = \frac{44.1}{0.13} = 339.2\ \text{m/s}$$

3. Here, $t = 6$s and speed of sound $v_s = 300$ m/s
Now, distance travelled by the car $= v \times 6 = x$
Distance travelled by sound, $S = 960 + (960 - x)$
$= (1920 - x)$ m.

Since, $v_s = \dfrac{S}{t}$

or $300 = \dfrac{(1920 - x)}{6}$ or $1800 = 1920 - x$

$$x = 120\ \text{m}$$

Speed of car, $v = \dfrac{x}{6} = \dfrac{120}{6} = 20\,\text{m/s}$

4. Frequency $\nu = 540$ Hz
Source velocity, $v_s = 12 \times 0.6 = 9$ m/s
For highest frequency, the source should approach towards observer.

$$\nu' = \nu\left(\frac{300}{300-9}\right)$$

$$= \frac{540 \times 300}{291} \simeq 555\,\text{Hz}$$

5. Let the velocity of sound be v, then

depth $h = \frac{v}{t} = v$

$t = \frac{2}{2} = 1\text{s}$

depth $h = \frac{v}{1} = v$

$$v = \sqrt{\frac{E}{\rho}} = \sqrt{\frac{2\times10^9}{1000}}$$

$$= 10^3\sqrt{2} = 1.414 \times 10^3 = 1414\,\text{m/s}$$

(At 4° C, the density of water is maximum, i.e., 10^3 kg/m³)

6. $$\nu' = \nu\left(\frac{330-30}{330}\right)$$

$$= 1000 \times \frac{300}{330}$$

$$= \frac{1000}{11} = 909\,\text{c/s}$$

Fill in the Blanks

1. solids, liquids and gases.
2. vacuum, solids liquids and gases
3. zero
4. adiabatically
5. 10^{-9} m
6. 25 : 9
7. 20000; ultrasonic
8. 0.2 N m^{-2}
9. $\frac{1}{8}$ metre
10. solids
11. ultrasound
12. reflection
13. Mechanical
14. SONAR
15. 20 Hz to 20 kHz
16. supersonic jets
17. periodic
18. halved
19. zero

True or False

1. False. The distance between a node and adjoining antinode is $\lambda / 4$. The statement is false.
2. False. Vibrations of prongs are transverse but the vibrations of the stem of the fork are longitudinal. The statement is false.
3. False. No, in a stationary wave, there is no transmission of energy. The statement is false.
4. False. When the listener and the sounding object move in directions at 90° to each other, there is no apparent change in frequency. The statement is false.
5. True. When a source of sound approaches a stationary listener with speed x, then $n_1 = \frac{V \times n}{V - x}$

   When the listener approaches the stationary source with equal speed x, then

$$n_2 = \frac{(V+x)n}{V}$$

   Clearly, $n_1 > n_2$.
   The statement is true.
6. True
7. True
8. False
9. True

Match the Columns

1. (P) – (W), (Q) – (Y), (R) – (X), (S) – (U), (T) – (V).
2. (P) – (V), (Q) – (T), (R) – (W), (S) – (U).
3. (P) – (V), (Q) – (W), (R) – (U), (S) – (T).
4. (P) – (V), (Q) – (W), (R) – (T), (S) – (U).
5. (P) – (V), (Q) – (T), (R) – (W), (S) – (U).
6. (P) – (T), (Q) – (S), (R) – (U).

Assertion-Reason Questions

1. (a) Sound waves require a material medium for propagation. So the reason is true and a correct explanation of the assertion.

2. (c) Air possesses volume elasticity, which is the requirement for propagation of longitudinal waves.

3. (d) Air does not have elasticity of shape. So transverse waves cannot be formed in air.

4. (a) Both the assertion and reason are true and the reason is correct explanation of the assertion.

5. (c) Assertion is true, but the reason is false. In fact, presence of moisture in air decreases the density of air.

6. (b) Because of change in pressure, density of air also changes. If T is constant, (P/ρ) becomes constant. $\therefore v = \sqrt{\gamma P/\rho}$ remains constant.
The reason is true, but not a correct explanation of the assertion.

7. (c) The correct formula for velocity of sound in a gas is

$$v = \sqrt{\frac{\gamma P}{\rho}}$$

For monoatomic gas, $\gamma = 1.67$;
For diatomic gas, $\gamma = 1.40$.
$\therefore v$ is larger in case of monoatomic gas compared to its value in diatomic gas.

8. (a) As $v = n\lambda \therefore v = \left(\frac{54}{60}\right) \times 10 = 9$ m/s

9. (d) In underwater communications, sound waves with higher frequency are better, because they carry more energy and can travel longer distances. The assertion is false. The reason too is false.

Comprehension Type Questions

1. (I) (a) (II) (b) (III) (b)

2. (I) (b) (II) (d) (III) (a)

3. (I) (d) (II) (c) (III) (c) (IV) (b)

Objective Questions

One Option Correct

9. (a) $f = \frac{v}{\pi} = \frac{340}{17}$
$= 20$ Hz

10. (d) $v = f\pi$
$= 500 \times 0.2$
$= 100$ m/s
$= S = v \times t$
$300 = 100 \times t$
$\Rightarrow t = 3$

11. (b) Angle of incidence = Angle of reflection

15. (c) Let the distance be d, then
Echo is received after travelling distance $2d$
$2d$ = Velocity × Time

$$d = \frac{342 \times 3}{2} = 513\,\text{m}$$

16. (a) Wave frequency is a function of the source of waves. Therefore, it remains unchanged.

17. (c) Water waves are both longitudinal and transverse.

18. (c) As $v = \sqrt{\frac{\gamma P}{\rho}} = \sqrt{\frac{K_a}{\rho}}$

19. (b) $\frac{v_t}{v_0} = \sqrt{\frac{T}{T_0}} = \sqrt{\frac{273+t}{273+0}} = 2$
Solve to get, $t = 819$°C

20. (a) $\frac{\text{Speed of body}}{\text{Speed of sound}}$ = Mach number

21. (d) The audible frequency range (of human ear) lies between 20 Hz and 20000 Hz.

22. (c) As there is no effect of change in pressure on velocity of sound in air, and velocity $\propto \sqrt{T}$, when temperature is doubled, velocity becomes $330\sqrt{2}$ m/s.

23. (c) In normal talk, the amplitude of vibration is about 10^{-8}m.

24. (c) Intensity $= \frac{\text{Energy / sec}}{\text{Area}} = \frac{\text{Power}}{\text{Area}}$. From a point source, energy spreads over the surface of a sphere of radius r.

$$\therefore \text{Intensity} = \frac{P}{A} = \frac{P}{4\pi r^2} \propto \frac{1}{r^2}$$

But Intensity = (Amplitude)^2

$$\therefore (\text{Amplitude})^2 \propto \frac{1}{r^2} \text{ or Amplitude} \propto \frac{1}{r}$$

At distance 2 r, amplitude becomes $A/2$.

25. (a) Here, $\lambda = 5$ cm,

$$v = \frac{s}{t} = \frac{2\lambda}{1} = \frac{2 \times 5}{1} = 10\,\text{cm / s}$$

26. (a) $v = \sqrt{\frac{\gamma P}{\rho}}$. The speed (v) will be highest for the gas which γ is highest, which is a monoatomic gas.

27. (c) For ultrasonic waves, frequency $n > 20000$ c/s

$$\lambda = \frac{v}{n} < \frac{330}{20000} \approx 10^{-2}\text{ m}$$

28. (a) In this case, sound tends to go from air (lower velocity, v_a) to water (higher velocity v_w). For such a surface, critical angle C is given by

$$\sin C = \frac{1}{\mu} = \frac{v_a}{v_w} = \frac{330}{1400} \therefore C = 14°$$

As angle of incidence = $30° > C$, sound wave undergoes total reflection only.

29. (a) For a perfectly rigid rod, $\eta = \infty$

$\therefore v = \sqrt{\eta / \rho} \rightarrow$ infinite.

30. (a) $\lambda = \frac{v}{n} = \frac{3\times10^8}{760\times10^3} \approx 395$ m

31. (a) Amplitude of wave = 0.6 m

32. (c) From figure, $\frac{5}{2}\lambda = 20$ cm,

$$\lambda = 8\text{cm} = 0.08\text{ m}$$

33. (a) Frequency $n = \frac{v}{\lambda} = \frac{300}{0.08} = 3750$ Hz.

34. (a) Frequency remains the same i.e., 60 Hz,

$$\lambda_a = \frac{v_a}{v} = \frac{300}{60\times10^3}$$

$$= 5.5\times10^{-3}\text{ m} = 5.5\text{ mm}$$

35. (c) As intensity $\propto$ (amplitude)^2, the intensity of sound becomes 9 times.

36. (b) $\lambda = \frac{v}{n} = \frac{360}{500} = \frac{18}{25}$ m

$$\Delta x = \frac{\lambda}{2\pi}\Delta\phi = \frac{18}{25(2\pi)} \times \frac{\pi}{3} = 0.12\text{ m}$$

37. (b) With the propagation of a longitudinal wave, energy alone is propagated.

38. (c) intensity $\propto$ (amplitude)^2 and also intensity $\propto$ (frequency)^2. Therefore, intensity becomes $\frac{2^2}{4^2} = \frac{1}{4}$th.

39. (a) Time taken by stone

$$= \sqrt{\frac{2s}{g}} = \sqrt{\frac{2\times500}{10}} = 10\text{s}$$

Time taken by sound $= \frac{500}{332} \approx 1.5$ s

$\therefore$ Total time $= 10 + 1.5 = 11.5$ s.

More Than One Option Correct

1. The intensity of sound depends on both amplitude and the frequency of wave. The practical unit of intensity (i.e., loudness) is decibel.
2. Hydrogen gas and water have volume elasticity only and no elasticity of shape. Therefore, transverse mechanical waves cannot travel through them. Iron rod and stretched strings are the only alternatives.
3. Change of medium and change of temperature do affect the velocity of sound. Change in wavelength does not. Further, there is no effect of change in pressure on velocity of sound, provided temperature remains constant.
4. Among the options, only microwaves are electromagnetic waves. All others are mechanical waves.
5. In air, sound cannot undergo scattering. Polarization is not possible as sound waves are longitudinal.
6. At displacement nodes, displacement is minimum, particle speed is minimum, pressure is maximum and intensity of sound is maximum. All alternatives are correct.
7. Because of relative motion between the earth and the satellite, the frequency of radio signals received by a tracking station on earth cannot remain constant (Doppler effect). Depending on the relative directions of motion, frequency can be higher or lower or even equal.

Scholar Zone

1. The radio wave has higher frequency. Although both waves have the same wavelength, the speed of the radio wave is much higher than that of the sound wave.
2. The apparent frequency would be Doppler-shifted downward in frequency because the screaming person is moving away from you. Because the speed of the source increases due to gravitational force, the apparent frequency decreases with time.

3. (c) The sound that leaves your horn in the forward direction is Doppler-shifted to a higher frequency because it is coming from a moving source. As the sound reflects back and comes toward you, you are a moving observer, so a second Doppler shift occurs to an even higher frequency.

4. The following are the points that distinguish a sound wave from a radio wave.
   (i) Sound waves cannot travel through free space while radio waves can.
   (ii) The speed of sound in air is 332 m/s while that of radio waves 3×10^8 m/s.
   (iii) With rise in temperature, the velocity of sound increases while that of radio waves does not change.
   (iv) Sound waves can be detected by the human ear if its frequency lies between 20 Hz – 20 kHz while radio waves of any frequency cannot be heard by humans.
   (v) Sound waves can never be polarised (excluding the transverse mode in solids) while radio waves can always be polarised.

5. (a) To transmit a transverse mechanical wave, the medium must be elastic so as to provide a restoring force when acted on by shearing stress. But liquids and gases flow when acted on by shearing stress, i.e., they cannot sustain shear stress to provide restoring force. Hence, liquids and solids cannot transmit transverse mechanical waves.
   (b) Longitudinal waves are pressure waves, i.e., they are transmitted as compression and rarefaction in a medium. Now as the string is non-stretchable, so it can neither be compressed nor stretched, i.e., compression and rarefaction cannot be produced in strings. This, in turn, implies that longitudinal waves cannot be propagated along a string [So the waves in a string are always transverse, that too when the string is under tension. If tension in the string is zero, transverse mechanical waves will also not propagate, as then

$$v = \sqrt{(T/m)} = 0.$$

6. 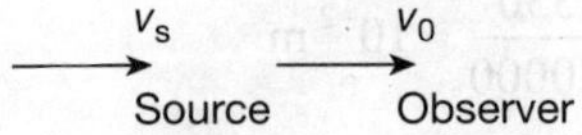

Case (i): The source and the observer move to the right. Since the observer is moving away from the source, there will be an apparent decrease in frequency. The source is approaching the observer; so there will be apparent increase in frequency. Hence the apparent frequency

$$f' = \left(\frac{v - v_0}{v - v_s}\right) f = \left(\frac{340-30}{340-60}\right) \times 500 \text{ Hz} = 553 \text{ Hz}$$

Case (ii): When the source moves to the right and the observer to the left, they approach each other and both contribute to an apparent increase in the frequency. Hence,

$$f' = \left(\frac{v + v_0}{v - v_s}\right) f = \frac{340+30}{340-60} \times 500 \text{ Hz} = 661 \text{ Hz}$$

Case (iii): When the source moves to the left and the observer to the right so that they recede from each other; consequently, both contribute to an apparent decrease in frequency. Hence,

$$f' = \left(\frac{v - v_0}{v + v_s}\right) f = \frac{340-30}{340+60} \times 500 \text{ Hz} = 387 \text{ Hz}$$

You should repeat the example assuming that the observer is to the left of the source. What conclusions can be derived from this example? Do you get a different set of four apparent frequencies?

CHAPTER 6 Electricity and Magnetism

LEARNING OBJECTIVES

At the end of this chapter, you will be able to:

- Define charge, current, potential difference
- Learn about conductors, insulators and semiconductors
- Understand Ohm's Law, specific resistance and combination of resistances
- Describe Cell and Internal resistance
- Learn about electrical measuring instruments
- Understand the magnetic effect of current and Oersted demonstration
- Define magnetism and its properties
- Describe Earth's magnetism and elements of Earth's magnetic field

Introduction

Electricity and Magnetism are among the most fascinating and useful concepts in physics. Electricity is the presence and motion of charged particles, while magnetism is caused by the interplay between moving charges.

Electric Charges

In nature two types of electric charges exist. Positive charge exists on protons and negative charge, on electrons.

A body whose atoms have lost some electrons is said to have a positive charge and a body whose atoms have gained some electrons is said to have a negative charge.

In case of spherical charged conductors, their charge stays on the surface.

Unit: S.I. unit of charge is coulomb, denoted as C.

Mutual Interaction: Like (similar) charges repel each other.

Unlike (opposite) charges attract each other.

Conservation of Charge: Charges can neither be created nor destroyed. They are merely transferred from one body to another.

Electric Force

The force between two point-charges q_1 and q_2 kept at distance r from each other is given as

$$F = \frac{kq_1q_2}{r^2} \text{ (} k \text{ is constant of proportionality)}$$

where $k = 9 \times 10^9 \text{ Nm}^2/\text{C}^2$

This expression is called Coulomb's law. The force F acts along the line joining charges q_1 and q_2.

Potential and Potential Difference

Electric Potential

Electric potential at a point inside the field of a field charge is defined as the work done in moving a unit positive test charge from infinity to that point.

It is represented by the symbol V. Its SI unit is volt (V). It is a scalar quantity.

$$\text{Potential } (V) = \frac{\text{Work done } (W)}{\text{Charge moved } (q)}$$

or $$W = qV$$

Potential Difference

Potential difference between two points inside the field of a field charge is defined as the work done in moving a unit positive test charge from one point to other.

A Voltmeter is used to measure potential difference between the ends of a resistance. For this purpose, it is connected in parallel with the resistance.

Electric Current

The time rate of flow of electric charge, i.e., electric charge flowing per second, is called electric current. Its symbol is I.

$$I = \frac{q}{t}$$

Unit: The unit of electric current is ampere (A). Current through a point is one ampere, if one coulomb charge flows through that point in one second. The conventional current flow is in the direction opposite to the flow of electrons.

An electric current flows between two points only when there is an electric potential difference between the two points. Also,

Current ∝ Potential difference

An electric cell or a battery is used to maintain a constant potential difference between two points. Due to the cell, a continuous current flows between two points.

Important Points about Current

(i) Current is a fundamental quality. In SI units, it is measured in A (ampere). It is a scalar quantity.

(ii) The conventional direction of current is the direction of flow of positive charge.

(iii) For a given conductor, current does not change with change in cross-sectional area.

(iv) Electric current passing through a conductor produces heating, magnetic and chemical effects.

ILLUSTRATION 6.1

How much work is done in moving a charge of 2 coulombs from a point at 118 volts to a point at 128 volts.

Solution

Charge moved, $q = 2$ C

Potential of first point, $V_1 = 118$ V

Potential of second point, $V_2 = 128$ V

Work done, $W = ?$

From relation,

Work done = Charge moved × Potential difference

i.e., $W = q(V_2 - V_1)$

Putting values, we get $W = 2 \times (128 - 118)$

$= 2 \times 10$

Work done, $W = 20$ J

ILLUSTRATION 6.2

The filament of an electric lamp draws a current of 0.4 ampere, which lights for 3 hours. Calculate the amount of charge.

Solution

Current flowing, $I = 0.4$ A

Time taken, $t = 2$ hrs. $= 3 \times 3600 = 10800$ s

Charge flow, $q = ?$

From the relation, $I = q/t$

We have, $q = It$

Putting values, we get, $q = 0.4 \times 10800$

Charge flow, $q = 4320$ coulomb

Symbols for Electrical Components

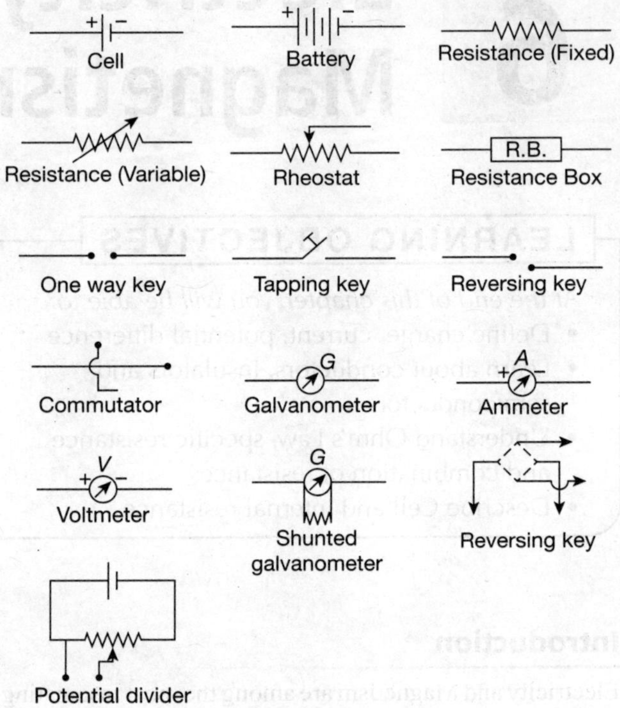

Conductors, Insulators and Semiconductors

Conductors Conductors are those materials which allow an easy flow of electric current through them, when a source of potential difference (electric cell or battery is connected across their ends.

These materials have free electrons inside them, which move, when a potential difference is applied across their ends. This constitutes electric current.

Silver, copper and aluminium, graphite, aqueous solution of salts are good (electrical) conductors.

Insulators Insulators are those materials which do not allow any flow of electric current through them when a source of potential difference is connected across their ends.

These materials do not have free electrons inside them. Since there are no free electrons to move, no electric current flows.

Rubber, wood, glass, porcelain are the best insulators.

Semiconductors Semiconductors are those materials which allow a very small current to flow when a source of large potential difference is connected across their ends.

These materials have only a few free electrons but more electrons become free under the influence of the source of a large potential difference.

Then, some current flows.

Germanium and silicon are some of the common semiconductors. These are now used in large scale in manufacture of transistors and computers.

Ohm's Law

Ohm's law states that the electric current flowing through a conductor is directly proportional to the potential difference (voltage) V across its ends (temperature remaining constant).

i.e., Potential difference $\propto$ Current

$$V \propto I$$

or

$$V = RI$$

where R is a constant of proportionality. It is called the resistance of the conductor. The unit of resistance, volt per ampere, is given a special name ohm and a symbol Ω (omega). A plot of current flowing through a circuit vs. the applied potential difference is a straight line with a positive slope (Fig. 6.1).

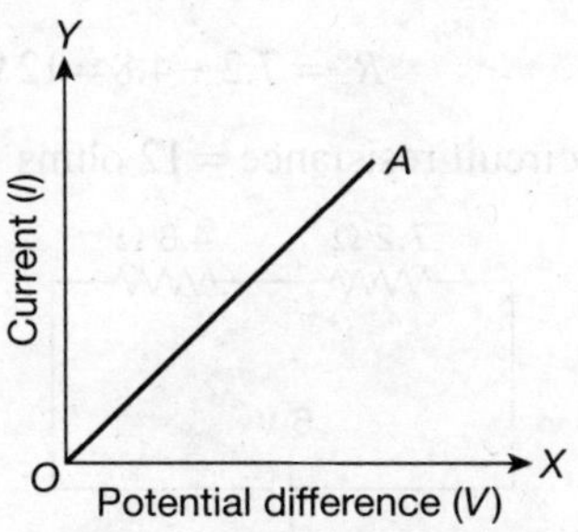

Figure 6.1 Potential difference–Current graph

Specific Resistance or Resistivity

The resistance of a conductor depends upon the following:

1. Length (l) of the conductor, $R \propto l$
2. Area of cross-section (A) of the conductor, $R \propto \frac{1}{A}$

Thus $R = \frac{\rho l}{A}$

where ρ is constant of proportionality. It is called the specific resistance or the resistivity of the material of the conductor.

The resistivity of the material of a conductor is defined as the resistance of the conductor per unit length and per unit area of cross-section.

Unit

S.I. unit of resistivity is ohm-m.

Effect of Temperature

With rise in temperature, the resistance of a resistor increases, according to the formula.

$$R_t = R_0 (1 + at)$$

where R_0 = Resistance 0 °C

R_t = Resistance at temperature t °C

a = temperature coefficient of resistance

ILLUSTRATION 6.3

Potential difference between two points of wire carrying 2 ampere current is 0.1 volt. Calculate the resistance of the wire.

Solution

Current, $I = 2$ A

Potential difference, $V = 0.1$ V

Resistance $R = ?$

From relation (Ohm's law), $R = \frac{V}{I}$

Putting the values, we get $R = \frac{0.01}{2} = 0.05$ ohm

The wire resistance is 0.05 Ω

Combination of Resistances

There are primarily two combinations of resistances:

1. Series combination
2. Parallel combination

Series Combination

When the same current flows through conductors joined together, then they form a series combination.

The equivalent resistance (R_{eq}) is given as

$$R_{eq} = R_1 + R_2 + R_3 + ... + R_n$$

Parallel Combination

When the same potential difference develops across resistors joined together, then they form a parallel combination.

The equivalent resistance (R_{eq}) is given as

$$\frac{1}{R_{eq}} = \frac{1}{R_1} + \frac{1}{R_2} + \frac{1}{R_3} + \ldots + \frac{1}{R_n}$$

ILLUSTRATION 6.4

resistance of 6 ohms is connected in series with another resistance of 4 ohms. A potential difference of 20 volts is applied across the combination. Calculate the current through the circuit and potential difference across the 6 ohm resistance.

Solution

We use proper symbols for electrical components

Resistances are shown connected in series, and a 20 V battery is connected across its positive and negative terminals. Direction of current flow is also shown from positive terminal of the battery towards its negative terminal.

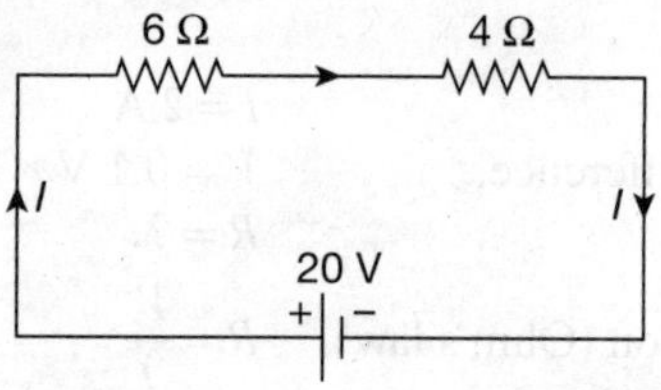

Here, First resistance $R_1 = 6\Omega$

Second resistance, $R_2 = 4\Omega$

Potential difference $V = 20$ V

Current in circuit, $I = ?$

Potential difference across 6Ω $V_1 = ?$

From series formula $R = R_1 + R_2$

Putting values, circuit resistance $= 6\,\Omega + 4\Omega = 10\,\Omega$

From Ohm's law, $R_s = \frac{V}{I}$

We have, $I = \frac{V}{R_s}$

Putting values, we get $I = \frac{20}{10} = 2$

Circuit current, $I = 2$ ampere or (2A)

Again, from relation, $R = \frac{V}{I}$

We have, $R_1 = \frac{V_1}{I}$

or $V_1 = IR_1$

Putting values, we get $V_1 = 2 \times 6 = 12$ volts.

Potential difference across 6Ω resistance = 12 V

ILLUSTRATION 6.5

In the circuit diagram given, find

(i) Total resistance of the circuit

(ii) Total current flowing in the circuit

(iii) Potential difference across R_1

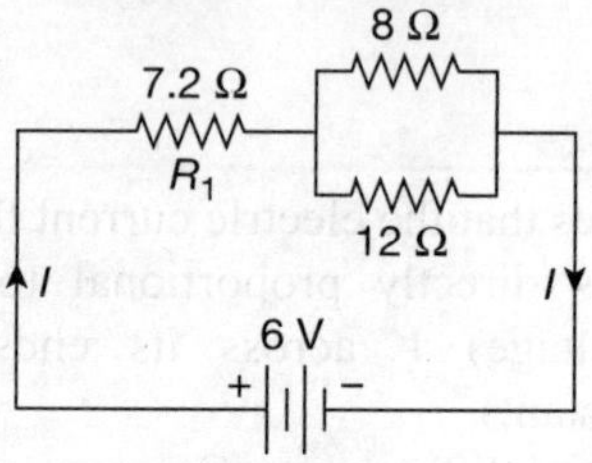

Solution

(i) For total resistance

8Ω and 12Ω are connected in parallel.

Their equivalent resistance,

$$R_p = \frac{8 \times 12}{8 + 12} = \frac{96}{20} = 4.8\Omega$$

Now, this equivalent resistance comes in series with 7.2 Ω resistance as shown in the figure.

With 7.2 Ω and 4.8 Ω in series

$$R_s = 7.2 + 4.8 = 12\,\Omega$$

Total circuit resistance = 12 ohms

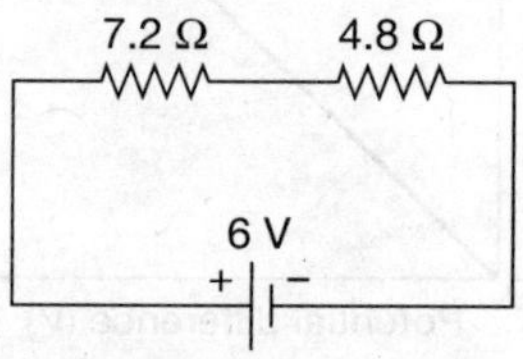

(ii) For total current

Total circuit resistance, $R = 12$ ohm

Potential difference applied, $V = 6$V

Circuit, current, $I = ?$

From Ohm's law $R = \frac{V}{I}$

We have, $I = \frac{V}{R}$

Putting values, we get $I = \frac{6}{12} = 0.5$

Total circuit current $= 0.5$ A

(iii) For potential difference across R_1

From relation, $R = \frac{V}{I}$

We have, $V = IR$

or $V_1 = IR_1$

Putting values, we get $V_1 = 0.5 \times 7.2 = 3.6$ V

Potential difference across, $V_1 = 3.6$ V

ILLUSTRATION 6.6

Three resistances are connected as shown in the diagram. A current of 1 ampere flows through a resistance of 5 ohms.

(i) What is the current through the other two resistors?

(ii) What is the potential difference (p.d.) across AB and across AC?

(iii) What is the total resistance?

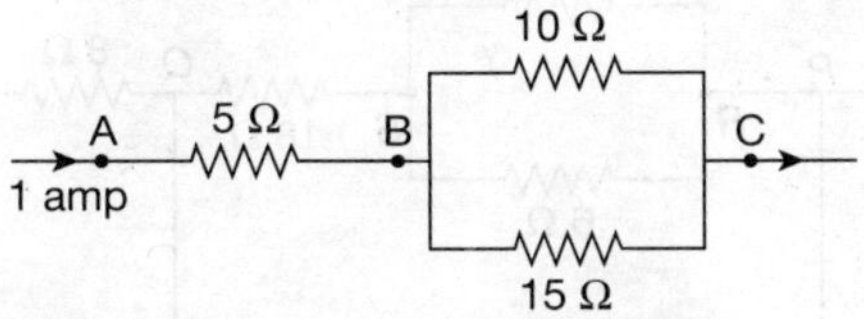

Solution

(i) For current in parallel resistors

For same potential difference across two parallel resistors,

$$V = I_1R_1 = I_2R_2$$

i.e., $$\frac{I_1}{I_2} = \frac{R_2}{R_1}$$

Current divides itself in inverse ratio of the resistances.

Also, total current, $I = I_1 + I_2$

From the above,

$$\frac{I_1}{I_2} = \frac{R_2}{R_1} = \frac{15}{10} = \frac{3}{2}$$

Also, $I_1 + I_2 = I = 1$ amp

Solving we get, $I_1 = 0.6$ A, $I_2 = 0.4$ A

and 0.4 A through 15 Ω

(ii) For p.d. across AB

From Ohm's law, $R = \frac{V}{I}$

We have, $V = IR$

Putting values, we get $V = 1 \times 5 = 5$V

P.D. across AB = 5V

For parallel combination of 10 Ω and 15 Ω P.D. across BC,

$$V = I_1R_1 = 0.6 \times 10 = 6V$$

P.D. across AC = PD across AB + P.D. across BC

$= 5 + 6 = 11$ V

(iii) For total circuit resistance

For 10Ω and 15Ω in parallel

$$R_p = \frac{10 \times 15}{10 + 15} = \frac{150}{25} = 6\Omega$$

Total resistance = 5 + 6 = 11Ω

Total circuit resistance = 11Ω

ILLUSTRATION 6.7

In the circuit diagram, find

(i) Total resistance

(ii) Current shown by the ammeter A.

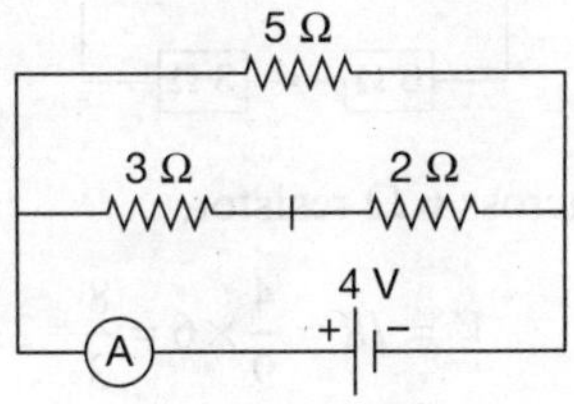

Solution

3Ω and 2Ω in series become 5Ω. The equivalent circuit is shown in the figure.

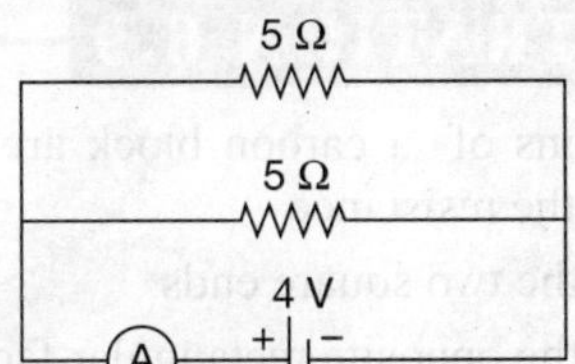

(i) For total resistance,

$R_1 = R_2 = 5\Omega$ are in parallel

$$R_p = \frac{5 \times 5}{5 + 5} = \frac{25}{10} = 2.5\Omega$$

Circuit resistance = 2.5 ohm.

(ii) For circuit current

Potential difference, $V = 4$ V

Circuit resistance $R_p = 2.5\Omega$

Circuit current, $I = ?$

From Ohm's law, $R = \frac{V}{I}$

We have, $I = \frac{V}{R_p}$

Putting values, we get, $I = \frac{4}{2.5} = 1.6$A

Circuit current = 1.6 A

ILLUSTRATION 6.8

For the circuit shown in the figure following diagram, what is the value of

(i) Current through 6 Ω resistor

(ii) Potential difference (p.d.) across 6 Ω resistor?

Solution

(i) For current through 6Ω

$$I = \frac{4}{6+3} = \frac{4}{9} = 0.44\text{A}$$

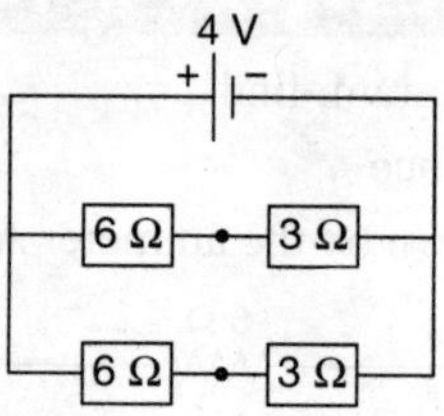

(ii) For p.d. across 6 Ω resistor,

$$V = IR = \frac{4}{9} \times 6 = \frac{8}{3} = 2.66\text{ V}$$

ILLUSTRATION 6.9

The dimensions of a carbon block are 1.0 × 1.0 × 50 cm³. Find the resistance –

(i) between the two square ends

(ii) between the opposite rectangular faces

(The resistivity of carbon is 3.5×10^{-5} Ω-m).

Solution

(i) The area of each square end is 1.0 cm × 1.0 cm = 1.0 cm² = 1.0×10^{-4} m²

The distance between the two ends = 50 cm = 0.50 m

$$R = \rho \frac{\ell}{A}$$

$$R = \frac{(3.5 \times 10^{-5}) \times (0.50)}{(1.0 \times 10^{-4})}$$

$$R = 0.175\ \Omega$$

(ii) The area of each rectangular face

$$= 1.0 \times 50\text{ cm}^2 = 50 \times 10^{-4}\text{ m}^2$$

The distance between two opposite faces = 1.0 cm = 1.0×10^{-2} m

$$R = \rho \frac{\ell}{A}$$

$$R = \frac{(3.5 \times 10^{-5}) \times (1.0 \times 10^{-2})}{(50 \times 10^{-4})}$$

$$= 7.0 \times 10^{-5}\ \Omega$$

ILLUSTRATION 6.10

Find the equivalent resistance and current in the circuit when a DC source of 12V is applied between A and B.

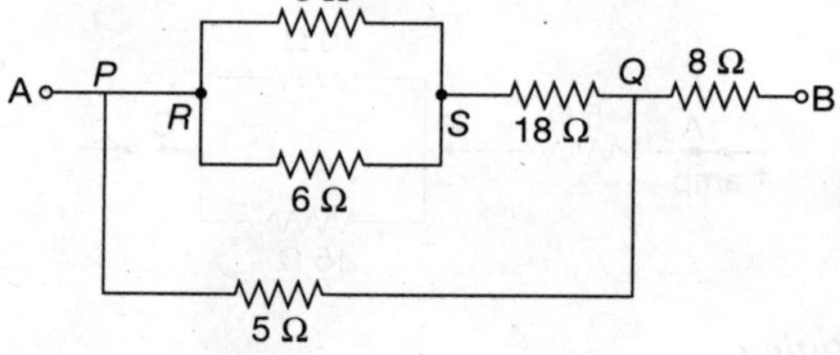

Solution

The equivalent resistance R_{AB} is calculated as –

Step I $\quad R_{RS} = \frac{3 \times 6}{3+6} = 2\ \Omega$

Step II $\quad R_{RQ} = 2 + 18 = 20\ \Omega$

Step III $\quad R_{PQ} = \frac{5 \times 20}{5+20} = 4\ \Omega$

Now $\quad \mathbf{R_{AB} = 4 + 8 = 12\ W}$

The current $i = \frac{V}{R_{AB}} = \frac{12}{12} = 1\text{ A}$

Internal Resistance

The resistance offered by an electrolyte to the flow of current in the cell is called internal resistance of cell. The internal resistance (r) of the cell depends upon

(a) distance between the electrodes.

(b) concentration of electrolyte.

(c) area of the electrodes.

(d) polarisation in the cell.

Terminal Potential Difference

When a cell is discharged, the potential difference V between its plates is

$$V = E - ir$$

or $\quad iR = E - ir \quad (\because V = iR)$

$$\Rightarrow \qquad i = \frac{E}{(R+r)}$$

When the cell is charged, $V = E + ir$

ILLUSTRATION 6.11

A 50 W bulb is connected to 200 V line. Determine the current flowing in it and its resistance. If 10% of total power is converted to light, then what will be the heat produced per second?

Solution

Current flowing in the bulb

$$i = \frac{P}{V} = \frac{50}{200} = 0.25\ \text{A}$$

Resistance of bulb

$$R = \frac{V^2}{P} = \frac{200 \times 200}{50} = 800\ \Omega$$

The heat is produced by 90% of the total power. Therefore,

$$P_{90\%} = 50 \times \frac{90}{100} = 45\ \text{W}$$

The heat produced per second is 45 J.

ILLUSTRATION 6.12

In the circuit shown in the figure, find the heat developed across each resistance in 2 s.

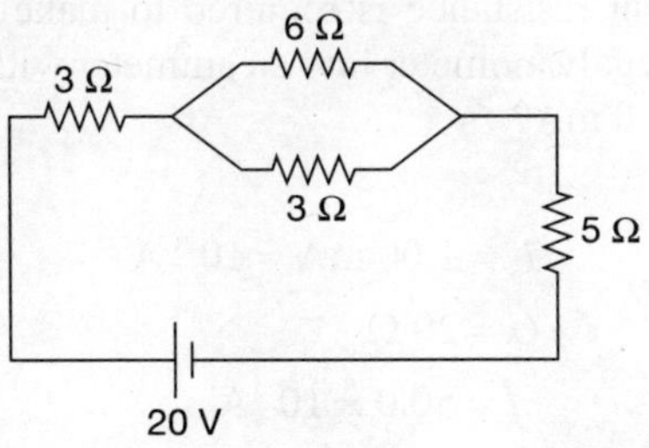

Solution

The 6 Ω and 3 Ω resistances are in parallel. So, their equivalent resistance is

$$R = \frac{6 \times 3}{3+6}, \qquad R = 2\ \Omega$$

The equivalent circuit can now be drawn as

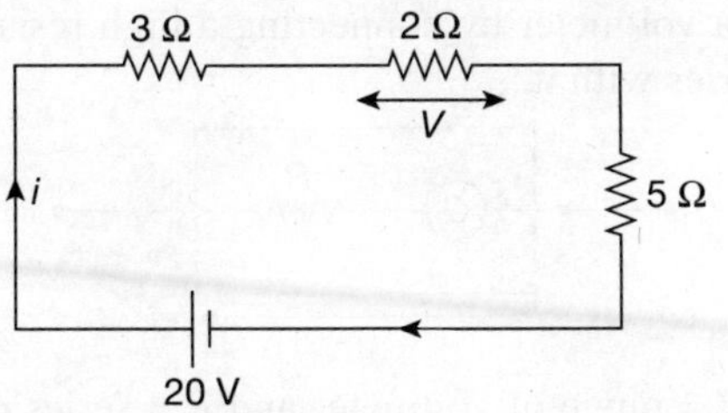

The current in the circuit

$$I = \frac{20}{3+2+5} = 2\ \text{A}$$

$$V = IR = 2 \times 2 = 4\ \text{V}$$

The potential difference across 6 Ω and 3 Ω resistances are 4 V.

Therefore, $H_{3\Omega} = I^2Rt = (2)^2\,(3)(2) = 24\ \text{J}$

$$H_{6\Omega} = \frac{V^2}{R}t = \frac{(4)^2}{6}(2) = \frac{16}{3}\ \text{J}$$

$$H_{3\Omega} = \frac{V^2}{R}t = \frac{(4)^2}{3}(2) = \frac{32}{3}\ \text{J}$$

and $H_{5\Omega} = I^2Rt = (2)^2\,(5)(2) = 40\ \text{J}$

Cell

The device which converts chemical energy into electrical energy is known as electric cell (Fig. 6.2). Cell is a source of constant emf but not constant current.

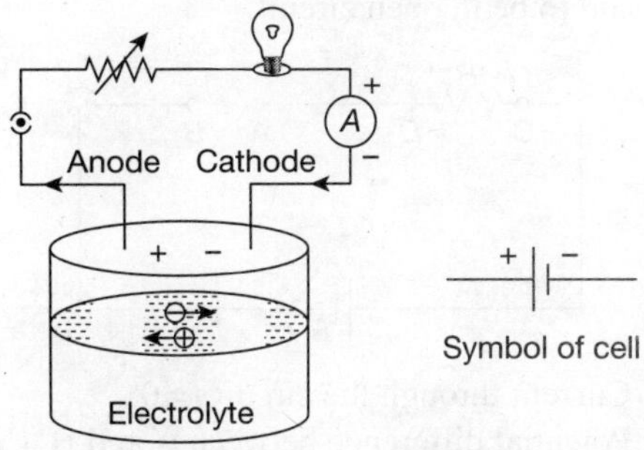

Figure 6.2 Basic structure of an electric cell

1. **Emf of cell (*E*)**: The potential difference across the terminals of a cell when it is not supplying any current is called its emf.
2. **Potential difference (*V*)**: The voltage across the terminals of a cell when it is supplying current to external resistance is called potential difference or terminal voltage. Potential difference is equal to the product of current and resistance of that given part, i.e., $V = iR$.
3. **Internal resistance (*r*)**: In case of a cell, the opposition of electrolyte to the flow of current through it is called internal resistance of the cell. The internal resistance of a cell depends on the distance between electrodes ($r \propto d$), area of electrodes [$r \propto (1/A)$] and nature, concentration ($r \propto C$) and temperature of electrolyte [$r \propto (1/\text{Temp.})$].

A cell is said to be ideal, if it has zero internal resistance.

Cell in Various Positions

1. **Closed circuit**: A cell supplies constant current in the circuit.

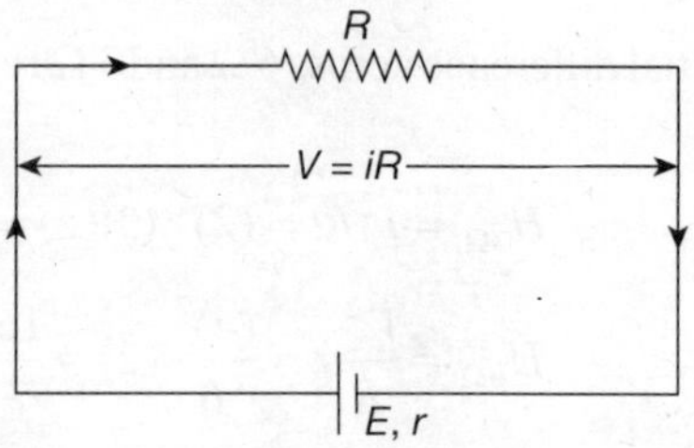

(i) Current given by the cell $i = E/(R+r)$
(ii) Potential difference across the resistance $V = iR$
(iii) Potential drop inside the cell = ir
(iv) Equation of cell $E = V + ir$ $(E > V)$
(v) Internal resistance of the cell $r = (E/V - 1).R$
(vi) Power dissipated in external resistance (load)

$$P = Vi = i^2R = V^2/R = (E/(R+r))^2.R$$

2. **Open circuit**: When no current is taken from the cell it is said to be in open circuit.

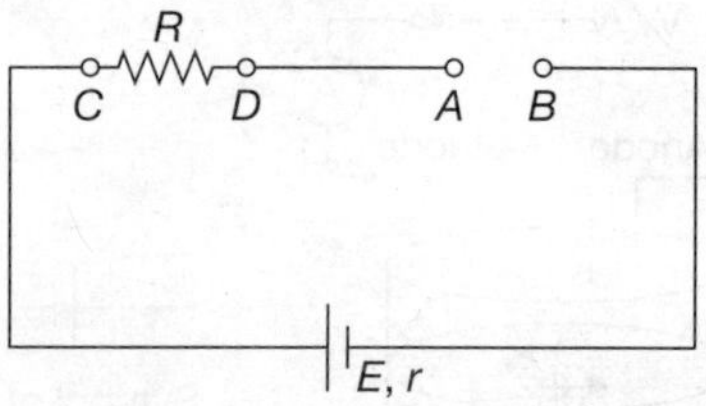

(i) Current through the circuit $i = 0$
(ii) Potential difference between A and B, $V_{AB} = E$
(iii) Potential difference between C and D, $V_{CD} = 0$

3. **Short circuit**: Two terminals of a cell are joined together by a thick conducting wire.

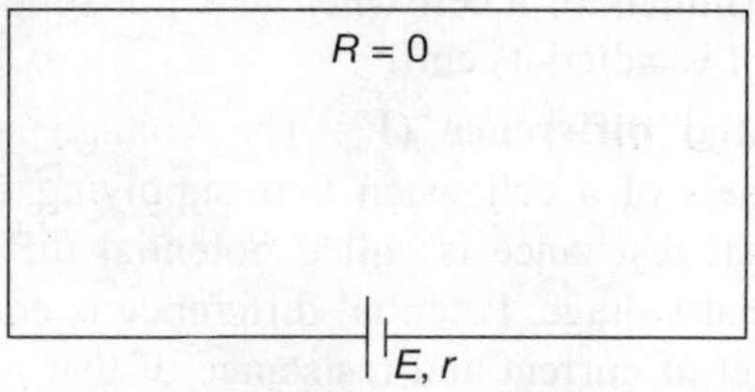

(i) Maximum current (called short-circuit current) flows momentarily, $i_{SC} = E/r$
(ii) Potential difference $V = 0$

Electrical Measuring Instruments

1. **Galvanometer:** An instrument used for the measurement of current and related quantities.

2. **Ammeter:** An instrument used for measurement of current is called as ammeter. A galvanometer can be converted into an ammeter by connecting a small resistance S (called shunt) in parallel.

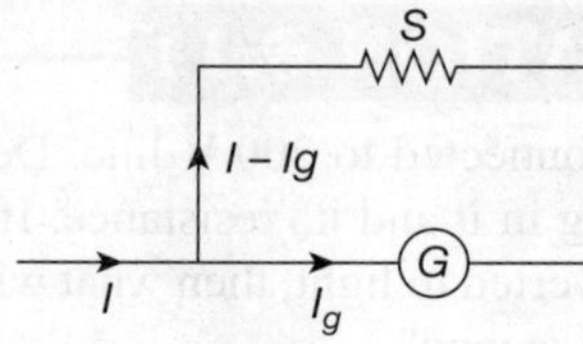

Let G = resistance of galvanometer
S = shunt resistance
I_g = value of current for which galvanometer gives full-scale deflection
I = range of ammeter

For full-scale deflection in galvanometer

$(I - I_g) \times S = I_g \times G$ $\quad$ ($\because$ S and G are in parallel)

$$S = \left(\frac{I_g}{I - I_g}\right)G$$

The reading of an ammeter is always lesser than actual current in the circuit.

The resistance (A) of ammeter is $A = \dfrac{GS}{G+S}$

ILLUSTRATION 6.13

What shunt resistance is required to make the 1.00 mA, 20 Ω galvanometer into an ammeter with a range of 0 to 50.0 mA?

Solution

$I_g = 1.00 \text{ mA} = 10^{-3}$ A

$G = 20\ \Omega$

$I = 50.0 \times 10^{-3}$A

Now $\quad S = \left(\dfrac{I_g}{I - I_g}\right)G = \dfrac{(10^{-3})(20)}{(50.0\times10^{-3}) - (10^{-3})}$

$= 0.408\Omega$

3. **Voltmeter:** An instrument used to measure voltage is called voltmeter. A galvanometer can be converted into a voltmeter by connecting a high resistance (R) in series with it.

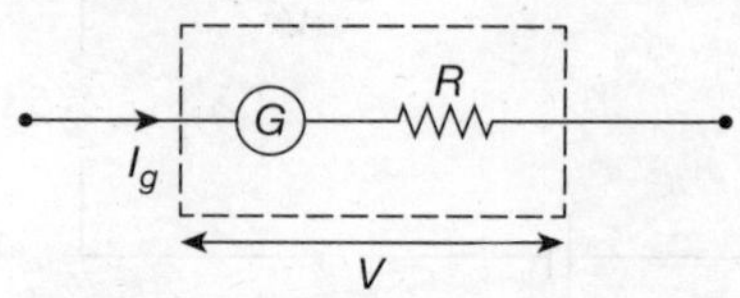

Let V = range of voltmeter and R = series resistance with galvanometer

Then, $V = I_g(R+G) \Rightarrow R = \frac{V}{I_g} - G$

The reading of the voltmeter is always lesser than the true value.

The resistance (R_v) of voltmeter is $R_v = R + G$

ILLUSTRATION 6.14

How can we make a galvanometer with $G = 20\ \Omega$ and $I_g = 1.0$ mA into a voltmeter with maximum range of 10 V?

Solution

Since, $$R = \frac{V}{I_g} - G$$

$$R = \frac{10}{10^{-3}} - 20 = 9980\Omega$$

Thus, a resistance of 9980 Ω is to be connected in series with the galvanometer to convert it into the voltmeter of desired range.

Magnetic Effect of Electric Current

There is long list of applications of magnetism. For example, large electromagnets are used to pick up heavy loads. Magnets are used in devices such as meters, motors and loud speakers. Magnetic tapes are used in sound and video recording, and for computer data storage. Intense magnetic fields generated by superconducting are currently being as a means of containing plasmas at temperatures of the order of 10^8 K used in controlled nuclear fusion research. As we study magnetism in this chapter, we find that subject cannot be separated from electricity. For example, magnetic fields affect moving electric charge, and moving charges produce magnetic fields.

Hans Christian Oersted played an important role in understanding **electromagnetism.** In 1820, he demonstrated that a compass needle got deflected when a metallic conductor carrying current is placed nearby it. According to Hans Christian Oersted, then needle kept near the current-carrying wire is deflected due to the magnetic field produced by the electric current. A change in the direction of current will show a variation in deflection. Thus, **electricity** and **magnetism** are phenomena that are related to each other.

Oersted Demonstration of Magnetic Effect of Current

Oersted set up a circuit with the help of a battery, plug key and compass and horizontally placed a small compass near to the XY wire. The circuit was switched on by inserting the key into the plug. As the current passed through the wire. The compass needle got deflected (Fig. 6.3).

He concluded that a current-carrying conductor behaves like magnet as a magnetic needle could be deflected only by a magnetic field. Thus, he established the magnetic effect of current.

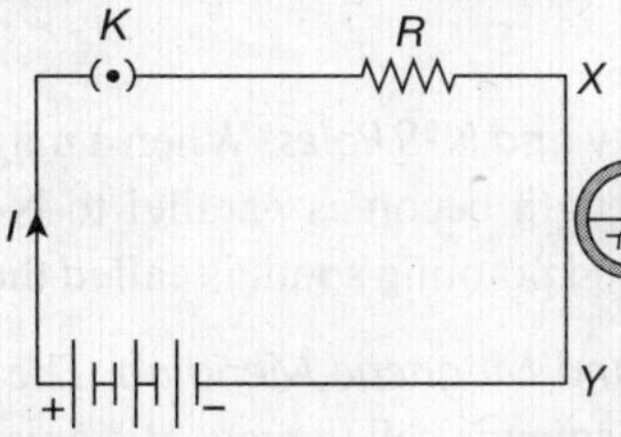

Figure 6.3 Oersted's experiment

Maxwell's Right-hand Thumb Rule

When one grasps the current element in the right hand such that the thumb points in the direction of the current, then curled fingers give the direction of the magnetic field due to current.

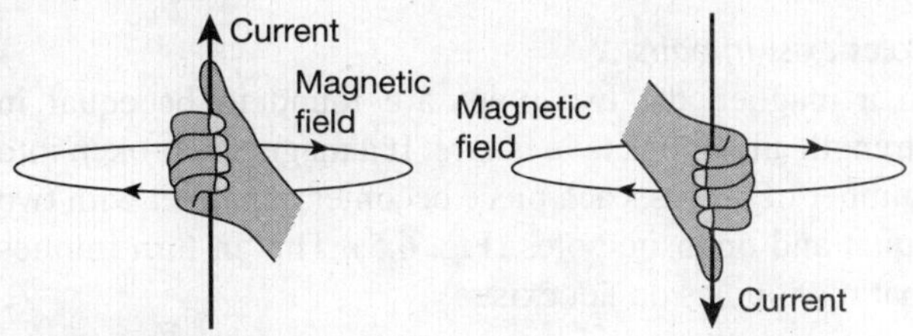

Magnetism

Much before the Christian era, it was discovered that a piece of the iron ore-magnetic (Fe_3O_4) found in Magnesia state has the property of attracting certain other substances and points in the North-South direction when freely suspended.

The molecular theory of magnetism

The molecular theory of magnetism was first given by Weber and Ewing. According the theory, every molecule of a magnetic substance (whether magnetized or not), is a complete magnet in itself. However, in magnetic substances the molecular magnets are randomly oriented to give net zero magnetic moment. On magnetizing, the molecular magnets are realigned in a specific direction leading to a net magnetic moment (Fig. 6.4).

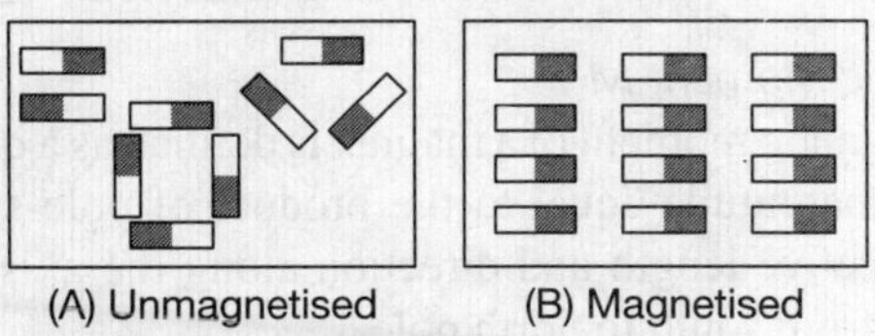

Figure 6.4 Molecular theory of magnetism

Properties of Magnets

Attractive Property and Poles

When a magnet is dipped into iron filings, it is found that the concentration of iron filings (attracting power of the magnet) is maximum at two points near the ends and minimum at the centre. The places in a magnet where its attracting power is maximum are called poles while the place of minimum attracting power is called the natural region.

Directive Property and N-S Poles When a magnet is freely suspended, its length becomes parallel to N-S direction. The pole at the end pointing south is called the south pole.

Magnetic Axis and Magnetic Meridian The line joining two poles of a magnet is called magnetic axis and the vertical plane passing through the axis of a freely suspended or pivoted magnet is called magnetic meridian.

Magnetic Length (2l) The distance between two poles along the axis of a magnet is called its effective or magnetic length. As poles are not exactly at the ends, the effective length is lesser than the actual length of the magnet.

Poles exist in pairs

In a magnet, the two poles are found to be equal in strength and opposite in nature. If a magnet is broken into number of pieces, each piece becomes a magnet with two equal and opposite poles (Fig. 6.5). This in turn implies that monopoles do not exist.

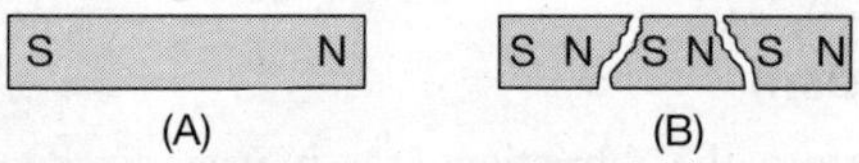

Figure 6.5 Pairs of magnetic poles

Repulsion is a Sure Test of Polarity

A pole of a magnet attracts the opposite pole, while repels similar poles. However, a sure test of polarity is repulsion and not attraction, as attraction can take place between opposite poles or a pole and a piece of unmagnetized magnetic material due to 'induction effect'.

Pole-strength m

The pole-strength of poles is conventionally represented as ***m*** and is a scalar quantity with dimension [AL] and S.I. unit A-m. The pole-strength of a magnet depends upon the nature of the material, area of cross-section of the magnet and also on the state of magnetization.

Magnetic Moment M

The magnetic moment of a magnet is defined as a quantity having magnitude equal to the product of pole-strength with effective length and direction along the axis of the magnet from south to north pole.

If a magnet is cut into two equal parts either along its length or perpendicular to its length, the magnetic moment of each part becomes half of the original value.

Magnetic Induction

A magnet attracts certain other substances through the phenomenon of magnetic induction, i.e., by inducing opposite pole in a magnetic material on the side facing it as shown in Fig. 6.6.

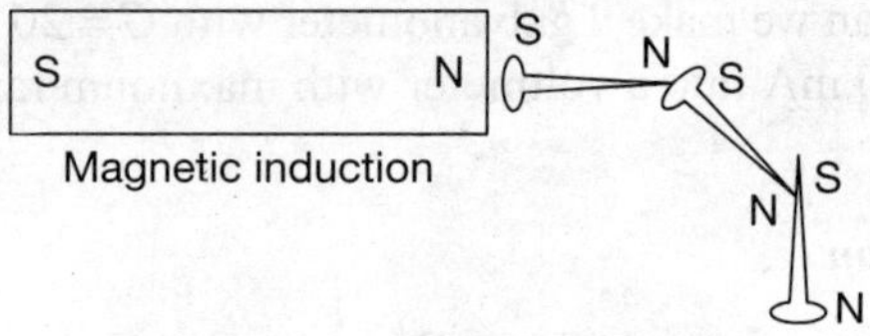

Figure 6.6 Magnetic induction

Magnetic Lines of Force and Their Characteristics

The space surrounding a magnet or magnetic configuration in which its effects are perceptible is called the magnetic field of the given magnet or magnetic configuration.

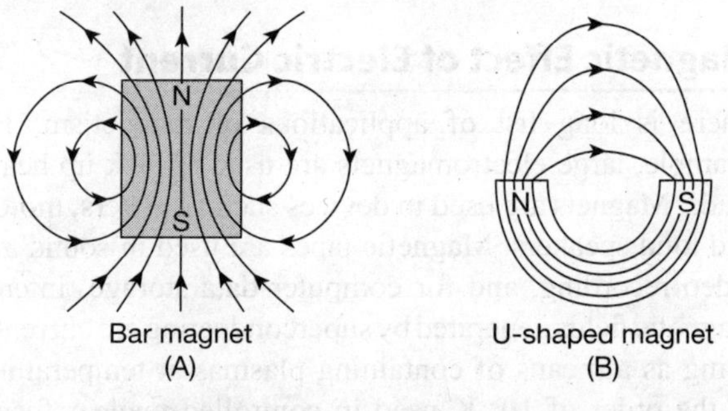

Figure 6.7 Types of magnets

In order to visualise a magnetic field graphically, Michael Faraday introduced the concept of lines of force. According to him, a line of force is an imaginary curve the tangent to which at a point gives the direction of the field at that point.

Regarding magnetic lines of force, it is worth noting that:

1. Outside a magnet, lines of force are from North to South Pole while inside from south to north, i.e., magnetic lines of force are closed curves, i.e., they appear to converge or diverge at the poles.
2. Magnetic lines of force can never intersect each other because if they intersect at a point, the intensity at that point will have two directions, which is absurd.
3. Magnetic lines of force have a tendency to contract longitudinally like a stretched elastic string

(producing attraction between opposite poles) and repel each other laterally (resulting in repulsion between similar poles).

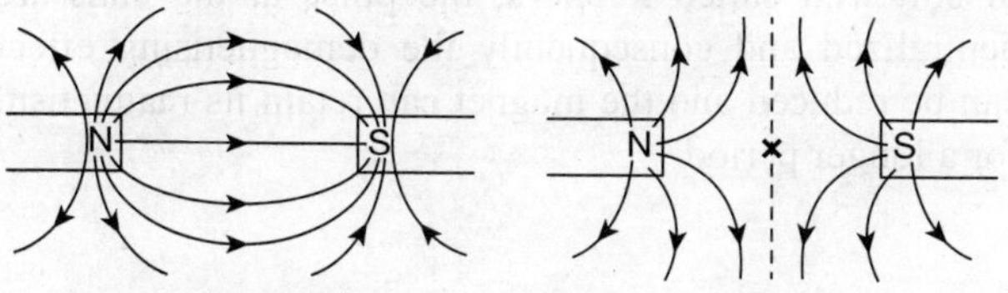

Figure 6.8 Attraction and repulsion in magnets

4. The number of lines of force per unit area, normal to the area at a point, represents the magnitude of the field at that point. So crowded lines represent a strong field while distant lines represent a weak field. Further, if the lines of force are equidistant and straight, the field is uniform; otherwise, it is not.

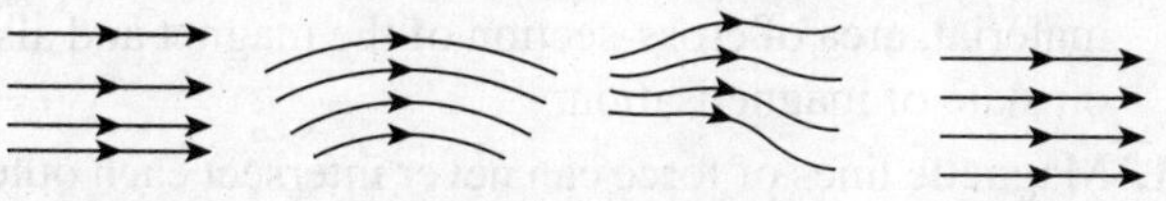

Figure 6.9 Uniform and non-uniform fields

5. In a region of space where there is no magnetic field, there will be no lines of force. This is why, at a neutral point (where resultant field is zero) there cannot be any line of force.
6. Magnetic lines of force emanate from or enter into the surface of a magnetic material at any angle.
7. Magnetic lines of force exist inside all magnetized materials.

Earth's Magnetism (Terrestrial Magnetism)

Dr William Gilbert in sixteenth century first gave the idea of the earth's magnetism.

The following are some basic terminologies regarding the earth's magnetic field

(i) Magnetic Meridian: It is a vertical plane passing through the axis of a freely suspended magnet at that place.

(ii) Geographical Meridian: The geographical meridian at a place is a plane containing the place and the earth's axis of rotation.

Elements of Earth's Magnetic Field

(i) Angle of declination: The angle between magnetic meridian and geographical meridian at a place is called angle of declination. It is generally denoted by angleθ.

(ii) Angle of inclination or angle of dip (S): It is the angle made by the resultant of earth's magnetic field with the horizontal in magnetic meridian at the place. At the poles $S = 90^\circ$, At equator $S = 0^\circ$.

(iii) Horizontal component of earth's magnetic field (B_H): It is defined as the component of earth's magnetic field in the horizontal direction on the magnetic meridian.

Electromagnet

An electromagnet is a solenoid coil that attains magnetism due to the flow of current. It works on the **principle of magnetic effect of current**. It consists of a long insulated cooper wire wound around a soft iron core. The magnetic effect remains only till the current is passing through the solenoid.

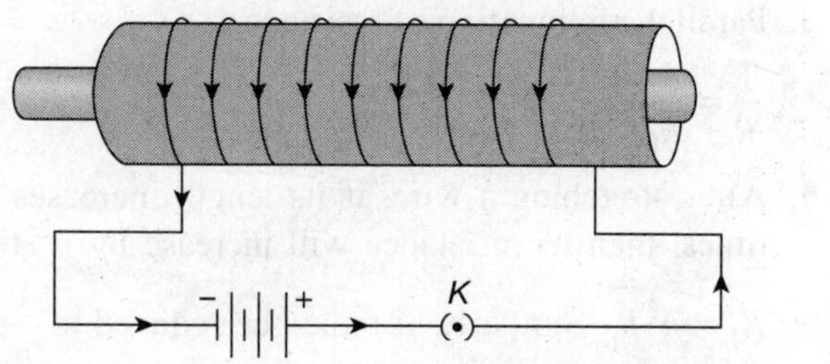

Soft iron is used for making electromagnet or temporary magnet. The properties of electromagnet are as below:

(i) The electromagnetism of temporary magnets is not a permanent property.

(ii) The electromagnet behaves as having poles at its ends.

(iii) They lose magnetism once the current is switched off.

The strength of an electromagnet increases with

(i) the increase in numbers of turns in the coil.

(ii) the increase in the amount of current flowing.

(iii) the decrease in the air gap between poles.

Uses of Electromagnet

An electromagnet is used in electric bells, electric motors, telephone diaphragms, loudspeakers and for sorting scrap metal and for extraction in hospitals. Giant electromagnets are used in cranes to lift machinery and other parts.

Permanent Magnets

If a magnet retains its attracting power for a long time it is said to be **permanent**, otherwise it is **temporary**. Permanent magnets are made of steel, Alnico, Alcomax or Ticonal, while temporary of soft iron, mumetal or stalloy.

A magnet gets demagnetised, i.e., loses its power of attraction if it is heated, hammered or alternating current is passed through a wire wound over it.

A magnet tends to become weaker with age owing to self-demagnetisation due to the poles at the ends, which tend to neutralize each other. However, by using pieces of soft iron called **keepers**, the poles at the ends are neutralized and consequently the demagnetising effect can be reduced and the magnet can retain its magnetism for a longer period.

POINTS TO REMEMBER

1. The direction of current is taken to the direction of flow of positive charge.
2. Ohm's law states that the electric current flowing through a conductor is directly proportional to the potential difference (voltage) V across its ends (temperature remaining constant). i.e., Potential difference $\propto$ Current.
3. Series combination of resistances, $R = R_1 + R_2 + R_3 \ldots\ldots$
4. Parallel combination of resistances, $\frac{1}{R} = \frac{1}{R_1} + \frac{1}{R_2} + \frac{1}{R_3}$
5. After stretching a wire, if its length increases by n times, then its resistance will increase by n^2 times; $R_2 = n^2 R_1$. Similarly, if radius be reduced to $\frac{1}{n}$ times then area of cross-section decreases $\frac{1}{n^2}$ times, so the resistance becomes n^4 times, $R_2 = n^4 R_1$.
6. If a wire of resistance R is cut in n equal parts and then these parts are collected to from a bundle, then equivalent resistances of combination will be $\frac{R}{n^2}$.
7. In series combination, if one resistance gets opened the current in the whole circuit becomes zero, but in case of parallel combination, if one resistance is open, all others will work.
8. When a magnet is freely suspended, its length becomes parallel to N–S direction. The pole at the end pointing south is called the south pole.
9. The line joining two poles of a magnet is called magnetic axis and the vertical plane passing through the axis of a freely suspended or pivoted magnet is called the magnetic meridian.
10. The pole strength of poles are conventionally represented by ***m*** and is a scalar quantity with dimension [AL] and the S.I. unit A-m. The pole strength of a magnet depends upon nature of the material, area of cross-section of the magnet and also on state of magnetisation.
11. Magnetic lines of force can never intersect each other because if they intersect at a point, the intensity at that point will have two directions, which is absurd.
12. In a region of space where there is no magnetic field, there will be no lines of force. This is why, at a neutral point (where the resultant field is zero) there cannot be any line of force.
13. Magnetic lines of force emanate from or enter into the surface of a magnetic material at any angle.
14. Magnetic lines of force exist inside every magnetised material.
15. The angle between magnetic meridian and geographical meridian at a place is called angle of declination. It is generally denoted by angle θ.
16. Angle of inclination or angle of dip (S) is the angle made by the resultant of earth's magnetic field with the horizontal in magnetic meridian at the place. At pole $S = 90°$, At equator $S = 0°$.
17. Horizontal component of earth's magnetic field (B_H) is defined as the component of earth's magnetic field in the horizontal direction on the magnetic meridian.

COMMON MISCONCEPTION

Ohm's law is not universal as it does not hold good in case of gases, crystal-rectifiers, thermionic valves, transistor, etc. The devices or substances which do not obey Ohm's law are called non-ohmic or non-linear conductors and for these, the I versus V curve characteristic is not a straight line.

PRACTICE QUESTIONS

Very-short Answer Type Questions

1. You have two metallic wires of resistance 6 Ω and 3 Ω. How will you connect these wires to get the effective resistance of 2 Ω?

2. The figure shows three resistors.

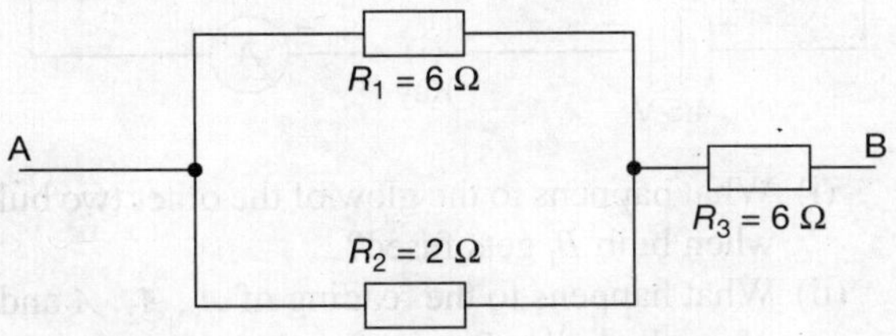

Find the equivalent resistance across AB.

3. What happens to the other bulbs in a series circuit, if one bulb blows off?

4. How can an ammeter be converted into a voltmeter?

5. Draw a circuit diagram using a cell of two batteries, two resistors of 3 Ω each connected in series, a plug key and a rheostat.

6. Find the minimum resistance that can be made using five resistors, each of 5 Ω.

7. What causes the potential difference between the two terminals of a cell?

8. The resistance of the incandescent filament of a bulb is comparatively much more than that when it is at room temperature. Why?

9. Give two examples for application of heating effect of electric current.

10. Name and define the smallest commercial unit of electricity.

11. In how much time will a bulb of 100 W consume energy of 2 kWh?

12. What do you understand by the term 'fuse' in an electric circuit?

13. Write one function of the instrument denoted by the given symbol.

14. Why is a closed path required for the flow of current?

15. When two ends of a metallic wire are connected across the terminals of a cell, some potential difference is set up between its ends. In which direction electrons are flowing through the conductors?

16. Suggest any two smaller units of current other than ampere.

17. What is the resistance of an air gap?

18. The resistance of a resistor is kept constant and the potential difference across its two ends is decreased to half of its former value. State the change that will occur in the current flowing through it.

19. Let the resistance of an electrical component remain constant while the potential difference across the two ends of the component decreases to half of its former value. What change will occur in the current through it?

20. Nichrome is used to make the element of electric heater. Why?

21. Alloys are used in electrical heating devices rather than pure metals. Why?

22. Out of the two wires P and Q shown below, which one has greater resistance? Justify.

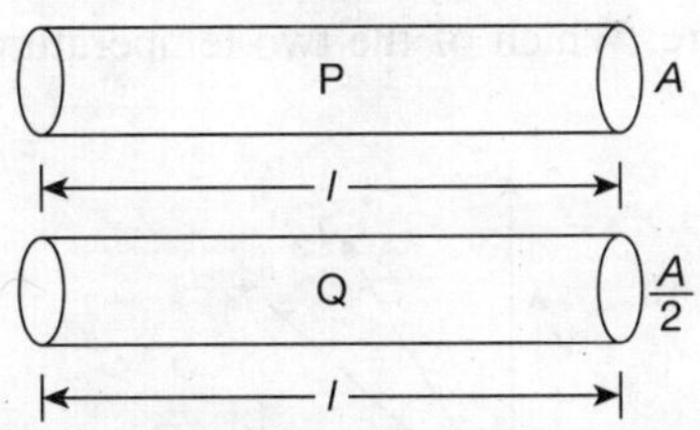

23. How does the resistance of an ohmic conductor depend on the applied voltage?

24. Draw the *V-I* graph for ohmic and non-ohmic conductor.

25. A student using the same two resistors, ammeter, voltmeter and battery makes two circuits connecting the two resistors, first in series and then in parallel. If the ammeter and voltmeter readings in both cases be I_1, I_2 and V_1, V_2, respectively. Write his observations.

26. How does the resistance of a wire vary with its area of cross-section?

27. On what factors does the pole-strength of a magnet depend?

28. If the shape of a magnet is changed, which of these will change: (a) pole-strength (b) effective length and (c) magnetic moment?

29. In what different ways can a specimen be (a) magnetised (b) demagnetised?

30. What are the sources of a magnetic field?

31. What is the nature of lines of force in a uniform field?

32. What is the angle of dip (a) in a plane perpendicular to the magnetic meridian (b) at the pole (c) at the equator?

Short-answer Type Questions

1. Will current flow more easily through a thick wire or a thin wire of the same material, when connected to the same source? Explain.

2. Why is the resistance more in series combination?

3. What are the advantages of connecting electrical appliances in parallel with the battery instead of connecting them in series?

4. Magnetic induction precedes attraction. Explain.

5. The voltage-current (*V-I*) graph of a metallic circuit at two different temperatures T_1 and T_2 is shown in the figure. Which of the two temperatures is higher and why?

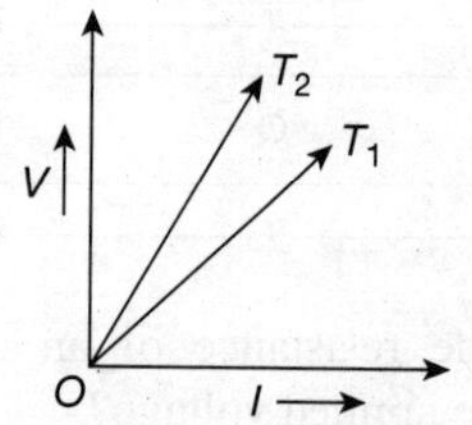

6. Repulsion is the sure test for magnetism. Explain.

7. What is horizontal component of the earth's magnetic field?

8. State the various methods of magnetisation.

9. Explain the important properties of a magnet.

Descriptive Type Questions

1. B_1, B_2 and B are three identical bulbs connected as shown in figure. Ammeters A_1, A_2 and A_3 are connected as shown. When all the bulbs glow, a current of 3 A is recorded by ammeter A.

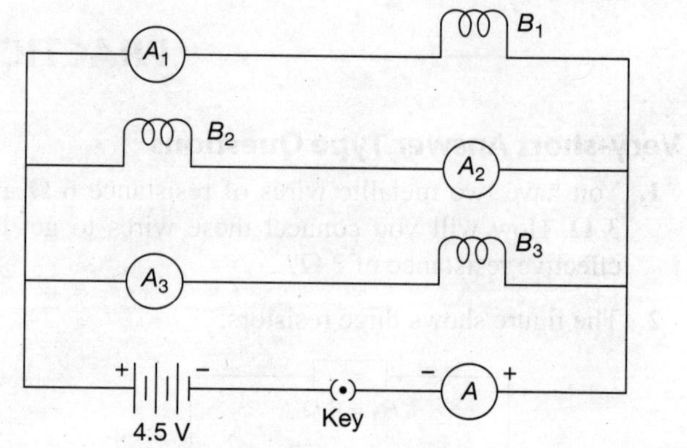

(i) What happens to the glow of the other two bulbs when bulb B_1 gets fused?

(ii) What happens to the reading of A_1, A_2, A and A when the bulbs B_2 gets fused?

(iii) How much power is dissipated in the circuit when all the three bulbs glow together?

2. The values of current I flowing in a given resistor for the corresponding values of potential difference V across the resistor are given as below:

I (amperes)	0.5	1.0
V (volts)	1.6	3.4

Plot a graph between V and I and also calculate the resistance of the resistor.

3. An ammeter and a resistor of 2 Ω are connected in series with a combination of two resistors (4Ω each) in parallel and a voltmeter across the parallel combination. Will the potential difference across the 2 Ω resistor be the same as that across the parallel combination of 4 Ω resistors? Give reasons.

4. What are magnetic series of force and describe its characteristics.

5. Demonstrate Oersted experiment to show magnetic effect of electric current.

Numericals

Level - I

1. A metallic wire of resistance R is cut into ten parts of equal length. Two pieces each are joined in series and then five such combinations are joined in parallel. What will be the effective resistance of the combination?

2. Draw a schematic diagram of a circuit consisting of a battery of three cells of 2 V each, a 5 Ω resistor, an 8 Ω resistor and a 12 Ω resistor and a plug key, all connected in series.

3. A piece of wire of resistance R is cut into five equal parts. These parts are connected in parallel. If equivalent resistance of this combination is R', then find the ratios R/R'.
4. How can three resistors of resistances 2 Ω, 3 Ω and 6 Ω be connected to give total resistance of (a) 4Ω (b) 1Ω?
5. What is (i) the highest and (ii) the lowest total resistance which can be secured by a combination of four coils of resistance 4 Ω, 8 Ω, 12 Ω and 24 Ω respectively?
6. A copper wire has diameter 0.5 mm and resistivity $\rho = 1.6 \times 10^{-8}$ Ω-m. What will be the length of its wire to make its resistance change if the diameter is doubled?
7. When a 12 V battery is connected across an unkno wn resistor, there is a current of 2.5 mA in the circuit. Find the value of resistance of the resistor.
8. A battery of 9 V is connected in series with resistors of 0.2 Ω, 0.3 Ω, 0.4 Ω, 0.5 Ω and 12 Ω, respectively. How much current would flow through the 12 Ω resistor?
9. How many 176 Ω resistors (in parallel) are required to carry 5 A on a 220 V line?

Level – II

1. A hot plate of an electric iron connected to a 220 V line has two resistance coils A and B, each of 24 Ω resistance, which may be used separately in series or in parallel. What are the currents in the three cases?
2. Judge the equivalent resistance when the following are connected in parallel:
(i) 1 Ω and 10^6 Ω (ii) 1 Ω,10^3 Ω and 10^6 Ω.
3. An ammeter reads up to 1 A. Its internal resistance is 0.81Ω.To increase the range to 10 A, what is the value of shunt required?
4. Two wires have lengths, diameters and specific resistances, all in the ratio 1 : 2. The resistance of the first wire is 10 ohm. What is the resistance of the second wire?

Fill in the Blanks

1. A device that is used to break an electric circuit is called _______.
2. Electric cell is a device which converts energy of chemicals into _______ energy.
3. The electric energy which is supplied in our homes comes from electric _______ houses.
4. An electric _______ is a continuous path along which the current flows.
5. The resistance varies inversely as _______ of radius of the wire.
6. A freely suspended magnet lies in the horizontal plane at _______ of earth.
7. The angle between resultant intensity of the earth's magnetic field and its horizontal component at a given place is called ______ at that place.
8. The direction of magnetic field due to a current carrying conductor can be determined by Ampere's ______ rule.
9. The effective length of a bar magnet is equal to _______.

True or False

1. If two resistors of resistance R each are connected in parallel, their equivalent resistance is equal to R only.
2. The direction of flow of electrons is from negative terminal to positive terminal in a circuit.
3. According to Ohm's law, current is inversely proportional to the potential applied.
4. A galvanometer is used to detect the presence of current in a circuit.
5. Resistivity is reduced to half, if area of cross-section of a wire is halved.

COMPETITION VISTA

Match the Columns

1. Match Column I with Column II.

Column I	Column II
(A) Current	(P) Ampere
(B) Potential	(Q) Ohm
(C) Resistance	(R) Coulomb
(D) Charge	(S) Volt

2. Match Column I with Column II.

Column I	Column II
(A) Easy flow of electric current	(P) Semiconductor
(B) Does not allow electricity to pass	(Q) Conductor
(C) Very small current	(R) Insulator
(D) Germanium	(S) Semiconductor

3. Match Column I with Column II.

Column I	Column II
(A) Current vs. Potential graph – Straight line	(P) Resistivity
(B) Specific resistance	(Q) Cell
(C) Chemical energy into electrical energy	(R) Ohm's law
(D) *VIt*	(S) Joule

4. Match Column I with Column II.

Column I	Column II
(A) EMF	(P) Opposition of electrolyte
(B) KWH	(Q) Potential difference across the terminals of a cell when there is no current through the cell
(C) Internal resistance	(R) Work done
(D) Potential × Charge	(S) One unit

Assertion–Reason Questions

Read the assertions and reason carefully to mark the correct option out of the options given below:

(a) If both assertion and reason are true and the reason is the correct explanation of the assertion.
(b) If both assertion and reason are true but reason is not the correct explanation of the assertion.
(c) If assertion is true but reason is false.
(d) If assertion is false but reason is true.

1. **Assertion:** A person touching a high-power line gets stuck with the line.
   Reason: The current-carrying wires attract the man towards it.
2. **Assertion:** Electric appliances with metallic body have three connections, whereas an electric bulb has a two-pin connection.
   Reason: Three-pin connections reduce heating of connecting wires.
3. **Assertion:** Bending a wire does not affect electrical resistance.
   Reason: Resistance of a wire is proportional to the resistivity of the material.

Comprehension Type Questions

1. Read the following information.
   i. Resistivity of copper is lower than that of aluminium, which in turn, is lower than that of constantan.
   ii. Six wires labelled as A, B, C, D, E and F have been designed as per the following parameters.

Wire	Length	Diameter	Material	Resistance
A	l	$2d$	Aluminium	R_1
B	$2l$	$d/2$	Constantan	R_2
C	$3l$	$d/2$	Constantan	R_3
D	$l/2$	$3d$	Copper	R_4
E	$2l$	$2d$	Aluminium	R_5
F	$l/2$	$4d$	Copper	R_6

Answer the following questions using the above data.

(a) Which of the wires has maximum resistance and why?
(b) Which of the wires has minimum resistance and why?
(c) Arrange R_1, R_3 and R_5 in ascending order of their values. Justify your answer.

2. Radhika is a student of class IX. Her mother was making tea in an old electric kettle having metal case. When she switched on the power supply to the electric kettle, she got a severe electric shock. Radhika put off the main switch quickly and found that the connecting cord was torn where her mother touched the metal case of the kettle. She also found that the red and black wires of connecting cord were firmly connected to the two lower terminals of the power plug but the green wire of cord was not connected to the upper terminal of the plug. Radhika replaced the torn connecting cord and also connected to the three wires of cord firmly to the power plug terminals.
On the basis of the above passage, answer the following questions.
   (a) Why did Radhika put off the main switch quickly?
   (b) Which wire red, black or green, touched the metal case of electric kettle when Radhika's mother got electric shock?
   (c) What values are displayed by Radhika in this incident?

3. Vishwajeet was doing an experiment by using an ammeter. Suddenly, it fell from his hand and broke. He was fearing that he might be scolded by his teacher. His classmate advised him not to tell the teacher but he refused and told his teacher. On listening to him patiently, the teacher did not scold him as it was just an incident and used the opportunity to show the whole class the internal structure of ammeter.
On the basis of above passage, answer the following questions.
   (a) What are the values displayed by Vishwajeet?
   (b) What is the use of ammeter? How is it connected in the circuit?
   (c) State the aim of any one experiment where Vishwajeet could have used the ammeter.

Objective Questions

One Option Correct

1. Four resistors of value 4 Ω are connected as shown in the figure. The equivalent resistance between the points A and B is ______.

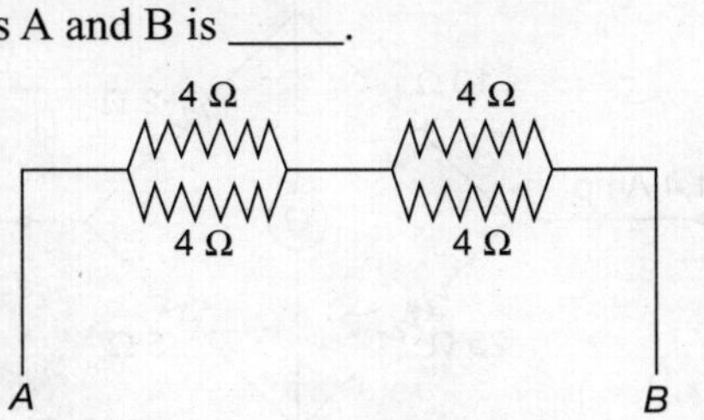

   (a) 2 Ω (b) 4 Ω
   (c) 8 Ω (d) 16 Ω

2. The diagram given below is the symbol of ______.

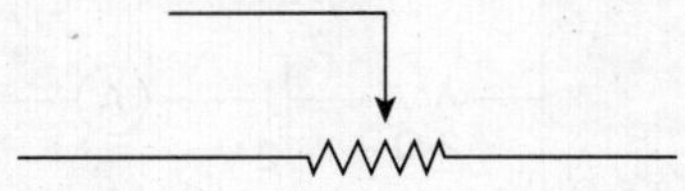

   (a) ammeter (b) voltmeter
   (c) rheostat (d) electric bulb

3. What is the effective resistance between the points P and Q in the resistor set given below?

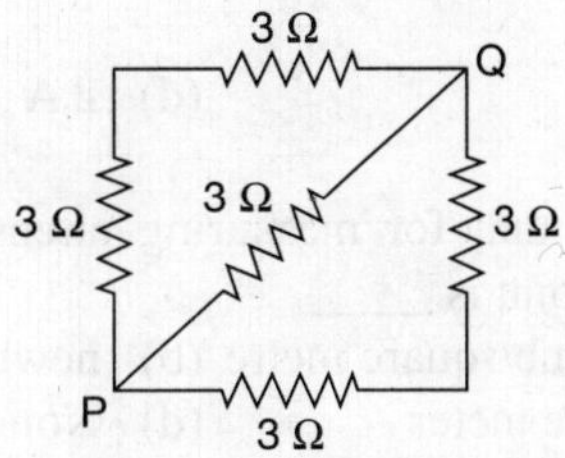

   (a) 0.15 Ω (b) 0.66 Ω
   (c) 1.5 Ω (d) 1.75 Ω

4. The current through the battery in the circuit shown in figure is ______.

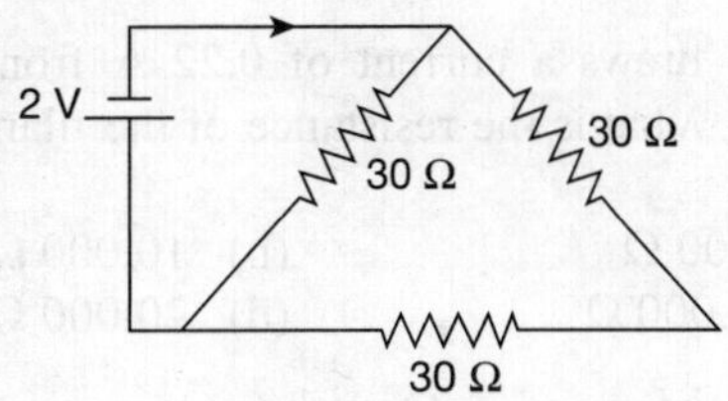

   (a) $\frac{1}{5}$ A (b) $\frac{1}{10}$ A
   (c) $\frac{1}{15}$ A (d) $\frac{1}{45}$ A

5. Which of the following is not a correct statement?
   (a) Resistance of a 40-watt bulb is less than the resistance of a 60-watt bulb.
   (b) When joined in series, a 40-watt bulb glows brighter than a 60-watt bulb
   (c) Resistivity of electrolytes decreases on increasing temperature.
   (d) Resistance of mercury falls on decreasing its temperature.

6. To make a voltmeter from a sensitive galvanometer,
   (a) a low resistance is connected in series.
   (b) a higher resistance is connected in series.
   (c) a low resistance is connected in parallel.
   (d) a high resistance is connected in parallel.

7. The reading of the ammeter in the circuit shown below is ______.

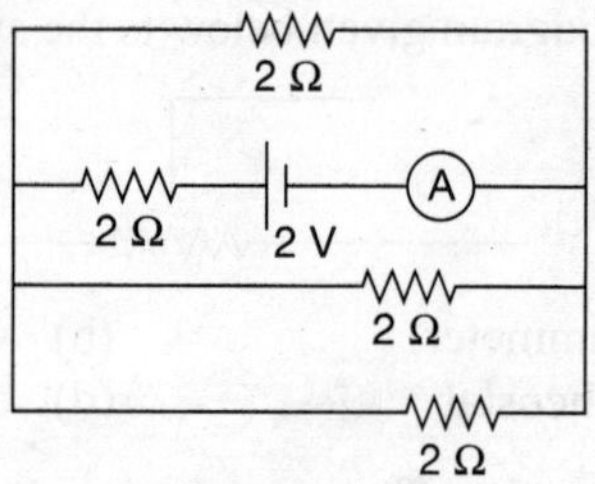

(a) $\frac{1}{2}$ A (b) $\frac{1}{8}$ A

(c) $\frac{3}{4}$ A (d) 2 A

8. A suitable unit for measuring intensity of electric field at a point is ______

(a) coulomb/square metre (b) newton/coulomb

(c) ampere meter (d) None of these.

9. If 600 C of electric charge flows through a circuit, what will be the current drawn by the filament of an electric bulb in 10 minutes?

(a) 0.5 A (b) 1 A

(c) 1.5 A (d) 2 A

10. A bulb draws a current of 0.22 A from a 220 V source. What is the resistance of the filament of the bulb?

(a) 1000 Ω (b) 10,000 Ω

(c) 15,000 Ω (d) 20,000 Ω

11. Two resistors of 4 Ω and 6 Ω are connected in parallel. The combination is connected across a 6V battery of negligible resistance. What is the current flowing through the battery?

(a) 1 A (b) 1.5 A

(c) 2 A (d) 2.5 A

12. Two resistors 2R and R are connected in series in an electric circuit. The ratio of heats dissipated in 2R to that in R is ______.

(a) 1 : 2 (b) 2 : 1

(c) 4 : 1 (d) 1 : 9

13. A rod has resistance 4Ω and it is turned a half-circle then the resistance along the diameter is ______.

(a) 1.56Ω (b) 2.44Ω

(c) 4Ω (d) 2Ω

14. In the figure given below, the current passing through the 6 Ω resistor is ______.

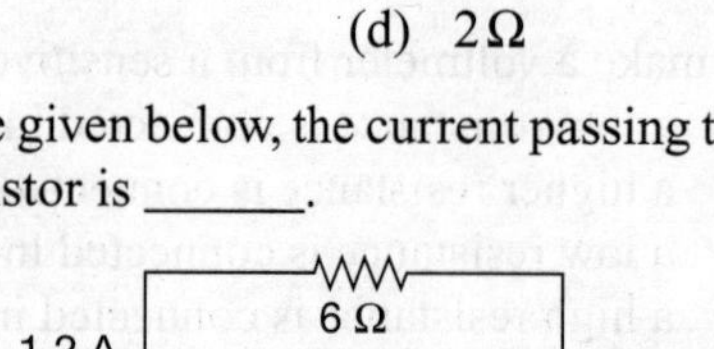

(a) 0.72 A (b) 0.80 A

(c) 0.48 A (d) None of these.

15. The effective resistance between the points A and B in the figure is ______.

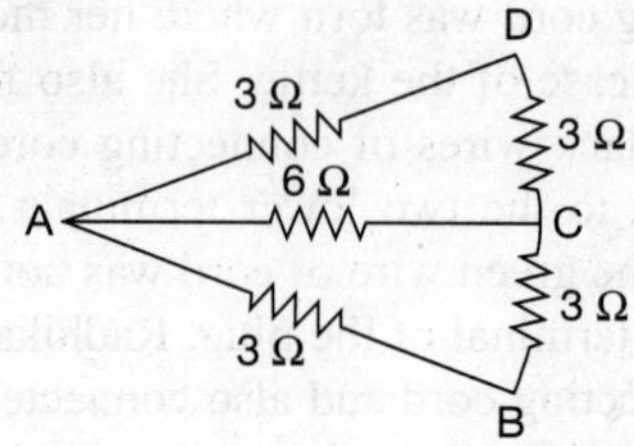

(a) 4 Ω (b) 3 Ω

(c) 2 Ω (d) 5 Ω

16. (1) The product of a volt and a coulomb is a joule.

(2) The product of a volt and an ampere is a joule/second.

(3) The product of volt and watt is horse power.

(4) Watt-hour can be measured in terms of electron volt.

State if

(a) All four are correct

(b) (1), (2) and (4) are correct

(c) (1) and (3) are correct

(d) (3) and (4) are correct

17. What will be the equivalent resistance between points P and Q in the network shown here?

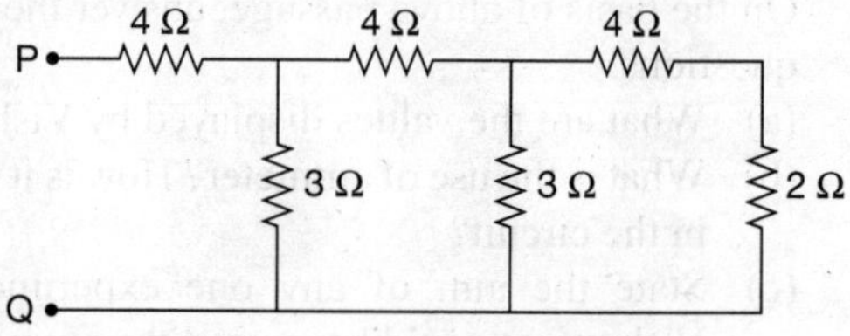

(a) 4 Ω (b) 6 Ω

(c) 24 Ω (d) 32 Ω

18. In the given circuit, the current in 2 Ω resistor is ______.

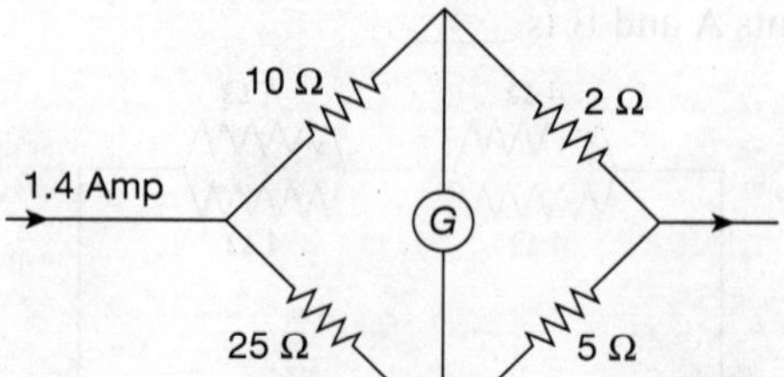

(a) 1.4 A (b) 1.2 A

(c) 0.4 A (d) 1.0 A

19. A parallel combination of two resistors of 1Ω each, is connected in series with a 1.5Ω resistor. The total combination is connected across a 10 V battery. The current flowing is circuit is

(a) 2A (b) 20A
(c) 0.2A (d) 5 A

20. Five resistances are connected as shown in the figure. The effective resistance between points A and B is ______.

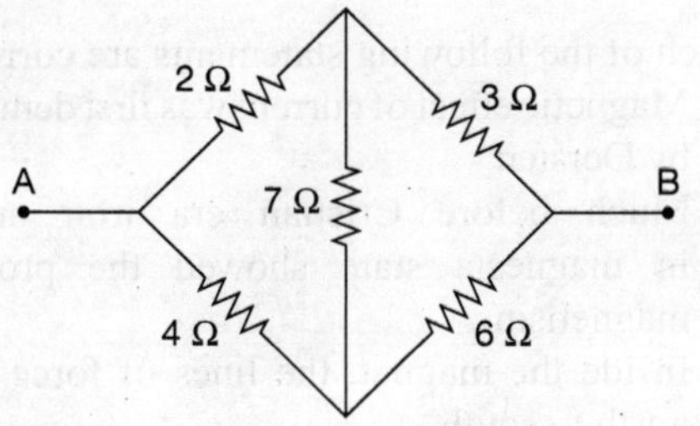

(a) $\frac{10}{3}\Omega$ (b) $\frac{20}{3}\Omega$
(c) 15 Ω (d) 6 Ω

21. The quantity in electricity analogous to temperature is______.

(a) charge (b) resistance
(c) potential (d) current

22. The quantity of electricity analogous to friction in linear motion is ______.

(a) resistance (b) potential
(c) charge (d) inductance

23. Which of the following materials would you use for making the coil of a resistance?

(a) Copper
(b) Constantan
(c) Brass
(d) Aluminium

24. For making an electrical room heater one should choose a wire of______.

(a) copper (b) nichrome
(c) platinum (d) tungsten

25. Which of the following graphs represent ohmic resistance?

(a) 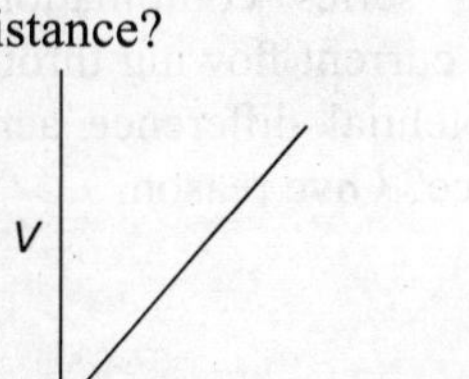

(b) 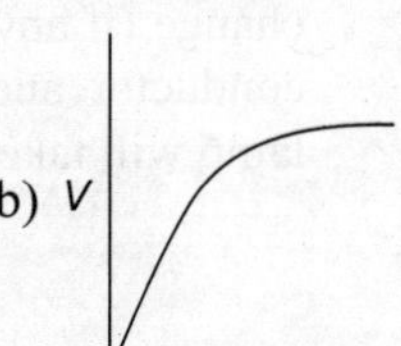

(c) 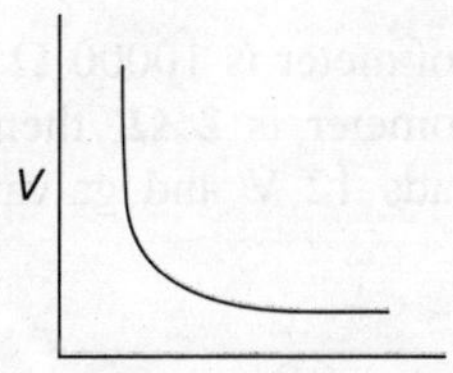

(d) 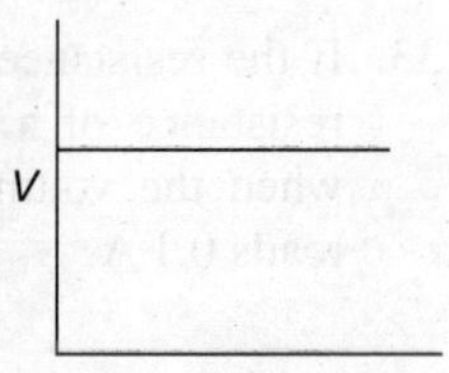

26. In Bohr's model of hydrogen atom, the electrons move around the nucleus in a circular orbit of radius 5×10^{-11}m with time period 1.5×10^{-6}s. The current associated with the electron motion is ______.

(a) Zero (b) 1.6×10^{-19} A
(c) 0.17 A (d) 1.07×10^{-3} A

27. The resistance of the human body is ______.

(a) 10Ω (b) 10 kΩ
(c) 10 MΩ (d) 100Ω

28. If the lines of force due to a current-carrying straight conductor are in anti-clockwise direction and lie on the plane of paper, the current is flowing

(a) parallel to the plane of paper.
(b) perpendicular to and directed into the paper.
(c) perpendicular to and directed out of the paper.
(d) parallel to the conductor.

29. The resistance of a rod is 1Ω. It is bent in the form of a square. What is the resistance across the adjacent corners?

(a) 1Ω (b) 3Ω
(c) $\frac{3}{16}\Omega$ (d) $\frac{3}{4}\Omega$

30. The effective length of a bar magnet is equal to

(a) the distance between its poles.
(b) the distance between any one pole and its centre.
(c) the distance between its geometric ends.
(d) the distance between any one geometric end and its centre.

31. Magnetisation can be done by ______

(a) heating (b) self-induction
(c) single-touch method (d) electrical method

32. A moving coil galvanometer has a resistance of 10 Ω and full-scale deflection of 0.01 A. It can be converted to 10 V full-scale by connecting a resistance of______.

(a) 9.9 Ω in series
(b) 10 Ω in series
(c) 990 Ω in series
(d) 0.10 Ω in series

33. If the resistance of a voltmeter is 10000 Ω and the resistance of a galvanometer is 2 Ω, then find *R* when the voltmeter reads 12 V and galvanometer reads 0.1 A.

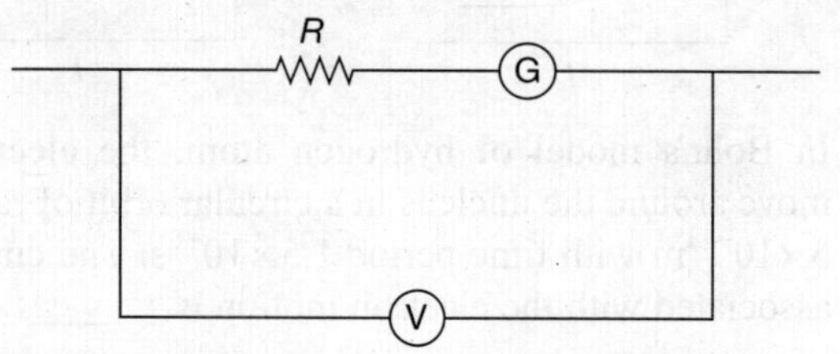

(a) 118 Ω (b) 120 Ω
(c) 124 Ω (d) 114 Ω

34. The magnetic field due to a bar magnet ______.
(a) has the same direction at any point
(b) is uniform.
(c) is non-uniform
(d) does not exist

35. Consider the following parameters A and B, and select the correct choice.
A: Repulsion is a sure test of magnetism.
B: Magnetic induction precedes attraction.
(a) Only A is true. (b) Only B is true.
(c) Both are true. (d) Both are false.

More Than One Option Correct

1. Which of the following statements are correct about charge?
(a) The SI unit of change is coulomb.
(b) Charge is conserved.
(c) Like charges attract each other.
(d) Unlike charges attract each other.

2. Which of the following statements are correct about potential?
(a) Potential is resistance per unit current.
(b) Potential is work done per unit charge.
(c) The unit of potential is volt.
(d) Voltmeter is used to measure the potential difference.

3. Which of the following statements are correct about Ohm's Law?
(a) Ohm's Law is a universal law.
(b) Ohm's Law is not a universal law.
(c) Mathematics formulation of Ohm's Law is Volt = Current × Resistance
(d) Mathematical formulation of Ohm's Law is Current = Volt × Resistance

4. Which of the following statements are correct?
(a) The unit of resistivity is Ohm m
(b) If two resistances R are connected in parallel, the equivalent resistance is R/2.
(c) Resistivity of a conductor is directly proportion to its length.
(d) Resistivity of a conductor is inversely proportion to its area of cross-section

5. Which of the following statements are correct?
(a) Magnetic effect of current was first demonstrated by Oersted.
(b) Much before Cristian era iron ore found in magnesia state showed the property of magnetism.
(c) Inside the magnet, the lines of force are from north to south
(d) The distance between the two poles along the axis of a magnet is called its effective length.

Scholar Zone

1. An electric lamp of 100 Ω, a toaster of 50 Ω and a water filter of resistance 500 Ω are connected in parallel to a 220 V source. What is the resistance of an electric iron connected to the same source that takes as much current as all three appliances and what is the current through it?

2. Draw a schematic diagram of a circuit consisting of a battery of three cells of 2 V each, a 5 Ω resistor, an 8 Ω resistor and a 12 Ω resistor, a plug key and ammeter. What would be the readings on the ammeter and voltmeter?

3. Show how would you connect three resistors, each of resistance 6 Ω, so that the combination has a resistance of (i) 9 Ω and (ii) 4 Ω?

4. (i) A current of 1 A flows in a series circuit having an electric lamp and a conductor of 5 Ω, when connected to a 10 V battery. Calculate the resistance of the electric lamp.

(ii) Now, if a resistance of 10 Ω is connected in parallel with this series combination, what change (if any) in current flowing through 5 Ω conductor and potential difference across the lamp will take place? Give reason.

Answer Keys

Objective Questions

One Option Correct

1. (a)	2. (c)	3. (c)	4. (b)	5. (a)	6. (b)	7. (c)	8. (b)	9. (b)	10. (a)
11. (d)	12. (b)	13. (c)	14. (c)	15. (c)	16. (b)	17. (b)	18. (d)	19. (a)	20. (a)
21. (c)	22. (a)	23. (b)	24. (b)	25. (a)	26. (d)	27. (b)	28. (c)	29. (c)	30. (a)
31. (c)	32. (c)	33. (a)	34. (c)	35. (c)					

More Than One Option Correct

1. (a), (b) and (d) 2. (b), (c) and (d) 3. (b) and (c) 4. (a), (b), (c) and (d) 5. (a), (b) and (d)

Hints and Solutions

Very-short Answer Type Questions

1. When the effective resistance is low, then these wires are connected in parallel.

$\therefore \quad 1/R_{eff} = 1/R_1 + 1/R_2$

$= 1/6 + 1/3 = (1+2)/6 = 3/6 = 1/2\ \Omega$

$\Rightarrow \quad R_{eff} = 2\ \Omega$

2. R_1 is parallel to R_2.

$\therefore \; 1/R = 1/R_1 + 1/R_2 = 1/6 + 1/2 = (1+3)/6 = 4/6 = 2/3\ \Omega$

$\Rightarrow \; R = 3/2\ \Omega$

Now, R and R_3 are in series.

$\therefore \quad R_C = R + R_3 = 3/2 + 6 = (3+12)/2$

$= 15/2 = 7.5\ \Omega$

3. In a series circuit, if one bulb blows off, then all other bulbs stop glowing.

4. An ammeter can be converted into a voltmeter by connecting a high resistance in series with it.

5. The required circuit diagram is given as below:

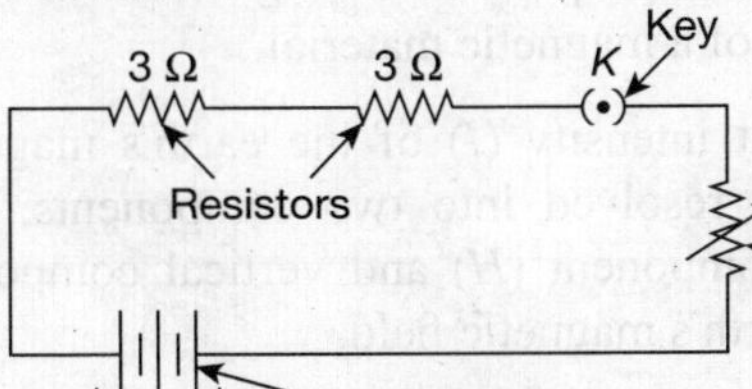

6. Minimum resistance can be obtained by connecting resistors in parallel.

$$R_{eq} = \frac{R}{n} = \frac{5}{5} = 1\Omega$$

7. The chemical reaction occurring inside the cell generates excess of electrons at the negative terminal and deficiency of electrons at the positive terminal.

8. This is because when the bulb is switched on, the temperature of the filament rises. As the temperature increases, the resistance of the conductor increases.

9. Electric iron and electric welding are two examples for application of heating effect of electric current.

10. The smallest commercial unit of electricity is watt hour. It is defined as the amount of electric energy that flows through a conductor in one hour, at power of 1 W.

11. Time, t = Energy / Power = $(2\text{ kWh})/(100\text{ W})$

$$= \frac{2000\text{ Wh}}{100\text{ W}} = 20\text{ h}$$

12. Fuse is the weakest link placed in series in an electric circuit, which melts when the circuit gets overloaded or short-circuited.

13. The given symbol is for a rheostat. It is used to change the magnitude of current in the circuit.

14. A closed path is required for the flow of current because it enables the movement of electrons in a particular direction.

15. Electrons are flowing through the conductors from its lower potential end to its higher potential end.

16. 1 mA and 1 μ A are two smaller units of current other than ampere.

17. The resistance of air gap is very large, almost infinite.

18. The current will reduce to half of its former value.

19. By Ohm's law $V \propto I$
If V becomes half, current flowing through it is also halved.

20. Nichrome is used to make the element of electric heater because it is an alloy with high resistivity and high melting point.

21. Alloys do not oxidise readily at high temperature, so they are most resistance to corrosion. This is the reason due to which they are used in electric heating devices.

22. Smaller the area of cross-section, greater will be resistance as
$R \propto 1/A$ [since, lengths are same]
So, wire Q has greater resistance.

23. The resistance of an ohmic conductor does not depend on the applied voltage, i.e., the temperature remains constant.

24. V-I graph for ohmic resistance is as given below:

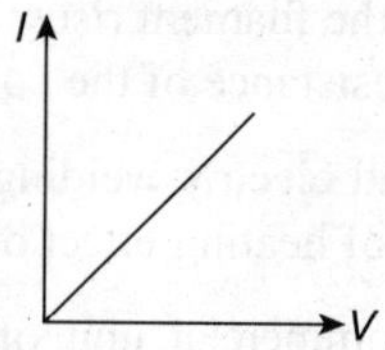

V-I graph for non-ohmic resistance is given as below:

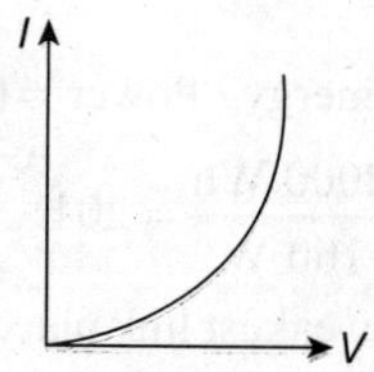

25. His observations are $I_1 < I_2$ because in series combination resistance is higher, so current is less. But, $V_1 = V_2$ because in both the cases battery is same.

26. Resistance is inversely proportional to the area of cross-section of the wire.
Thus, if the wire is thick, then resistance is less. If the wire is thin, then resistance is large.

27. Its cross-section, nature and state of magnetisation.

28. Effective length and magnetic moment will change while pole-strength will remain unchanged.

29. (a) By rubbing the specimen with a magnet or by burying it in earth, or by passing dc in a wire wound over it. (b) By heating or hammering, or by passing ac in a wire wound over it.

30. Magnet, magnetic effects of current, earth's magnetism, moving charge and changing electric field.

31. Equidistant, parallel straight lines

32. (a) 90° (b) 90° and (c) 0°.

Short-answer Type Questions

1. The current will flow more easily through the thick wire. As, resistance is inversely proportional to area of cross-section, then lesser the resistance, more is the current flowing through the wire.

2. We know that resistance is proportional to length. In series combination of resistors, the effective length of the cross-section of the conductor increases, so the resistance increases.

3. Following are the advantages of connecting electrical devices in parallel with the battery.
(i) The current is divided through the electrical devices. The total resistance in the circuit is decreased and devices work properly.
(ii) If one of the devices is in a parallel combination fails, then the other devices keep working without being affected.

4. When an iron bar is brought near a magnet it gets attracted by the magnet. Before attraction, poles are induced on the iron bar. The end of the iron bar facing the magnet, acquires opposite pole to that of the magnet and the far end of the iron bar acquires the same pole. Hence, the iron bar gets attracted by the magnet.

5. Resistance is equal to the slope of the V-I graph. Here, slope of graph for temperature T_2 is higher, so resistance at temperature T_2 is higher. Temperature T_2 is higher because resistance is proportional to temperature, i.e., $T_2 > T_1$ as $R \propto T$.

6. Repulsion is the property of a magnet; attraction is the property of a magnetic material.

7. The resultant intensity (I) of the earth's magnetic field can be resolved into two components, viz., horizontal component (H) and vertical component (V) of the earth's magnetic field.

8. Single-touch method, divided double touch method, double touch method, electrical method.

9. (i) Attractive property
(ii) Directive property
(iii) Pair property and
(iv) Law of magnetic poles

Descriptive Type Questions

1. Resistance of combination of three bulbs in parallel

$$R_{eq} = V/I = 4.5/3 = 1.5\ \Omega$$

If R is the resistance of each wire, then

$$1/R_{eq} = 1/R + 1/R + 1/R \text{ or } 1/R_{eq} = 3/R$$

or $\quad R = 3R_{eq} = 3 \times 1.5 = 4.5\Omega$

Current in each bulb,

$$I = \frac{V}{R} = \frac{4.5\ \text{V}}{4.5\ \Omega} = 1\ \text{A}$$

(i) When bulb B_1 gets fused, the currents in B_2 and B_3 remain same $I_2 = I_3 = 1\text{A}$, so their glow remains unaffected.

(ii) When bulb B_2 gets fused, the current in B_2 becomes zero and currents in B_1 and B_3 remain 1A.

Total current, $I = I_1 + I_2 + I_3 = 1 + 0 + 1 = 2\text{A}$

Current in ammeter $A_1 = I_1 = 1\text{A}$

Current in ammeter $A_2 = I_2 = 0$

Current in ammeter $A_3 = I_3 = 1\text{A}$

Current in ammeter $A = I = 2\text{A}$

(iii) When all the three bulbs are connected.

Power dissipated, $P = V^2/R_{eq} = (4.5)^2/1.5$

$= 13.5$ W

2.

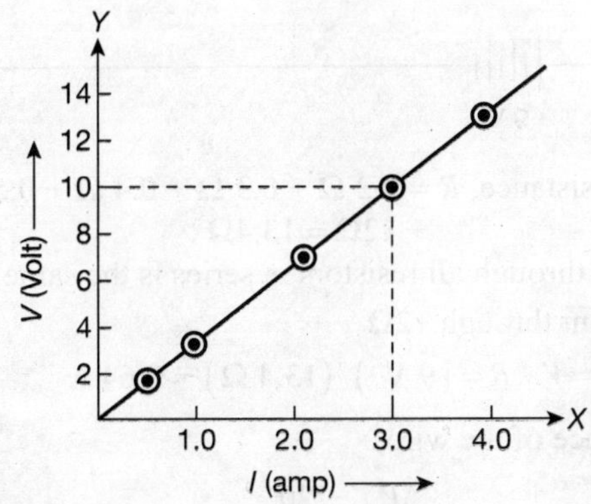

Resistance = Slope of graph

$$= \frac{\text{Y-intercept}}{\text{X-intercept}} = \frac{(10.2-0)\text{V}}{(3.0-0)\ \text{A}} = 3.4\ \Omega$$

3. The circuit is shown in figure:

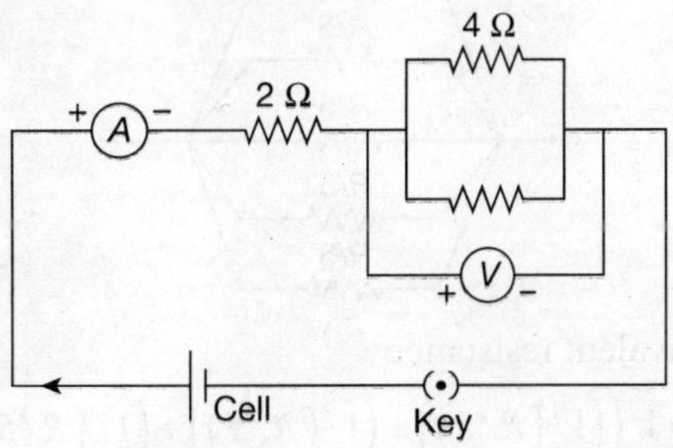

Effective resistance of combination of two resistors (4 Ω each) in parallel is

$$R_{eff} = (4 \times 4)/(4+4) = 2\ \Omega$$

Since the resistor of 2 Ω and parallel combination of two 4 Ω resistors are in series, same current will flow through these.

Hence, the potential difference across 2 Ω resistor is same as that across the combination of two resistors.

4. Refer theory.

5. Refer theory.

Numericals

Level - I

1. The resistance of a conductor is directly proportional to the length of the conductor. The resistance of each piece of wire, when it is cut into ten parts of equal length is

$$r = R/10$$

When two such pieces are joined in series, the equivalent resistance of these two parts

$$= r + r = 2r = 2R/10 = R/5$$

2. The schematic diagram is shown below:

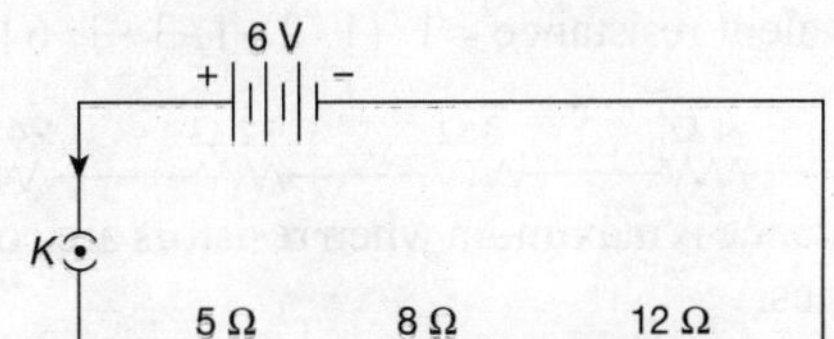

3. Resistance of complete wire is R.
If it is cut into 5 equal parts, then resistance of each part will be $R/5$. Five parts of resistance $R/5$ each are connected in parallel as shown in figure.

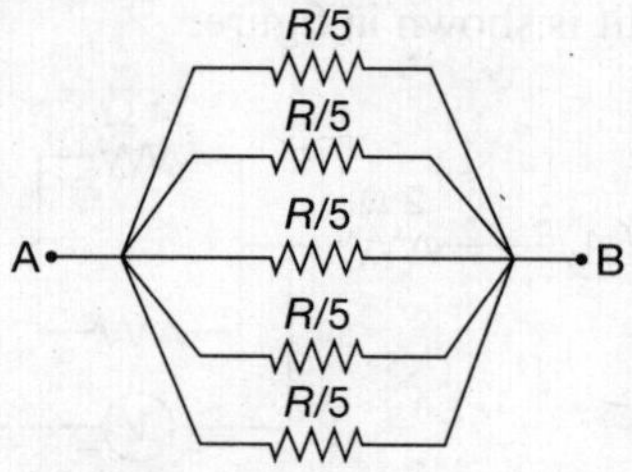

Equivalent resistance

$R' = 1/((1/(R/5))+(1/(R/5))+(1/(R/5))$
$+(1/(R/5))+(1/(R/5)))$
$= 1/(25/R) = R/25$

$\therefore$ Ratio, $R/R' = R/(R/25) = 25$

4. (a) If 3 Ω and 6 Ω are connected in parallel, thus equivalent resistance of parallel combination

$= 1/(1/3+1/6) = 2\,\Omega$

If this combination is connected in series with 2Ω resistance, then total equivalent resistance = 2Ω + 2Ω = 4Ω. The resistor connections are shown as below:

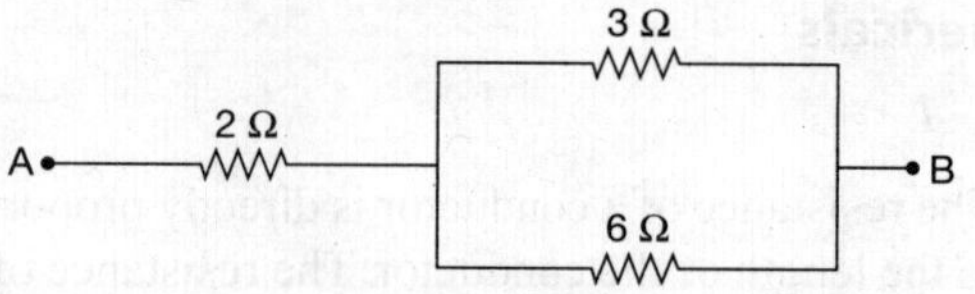

(b) Since equivalent resistance is less than the least value of resistance (i.e., 2 Ω), it means that all three resistors are connected in parallel.

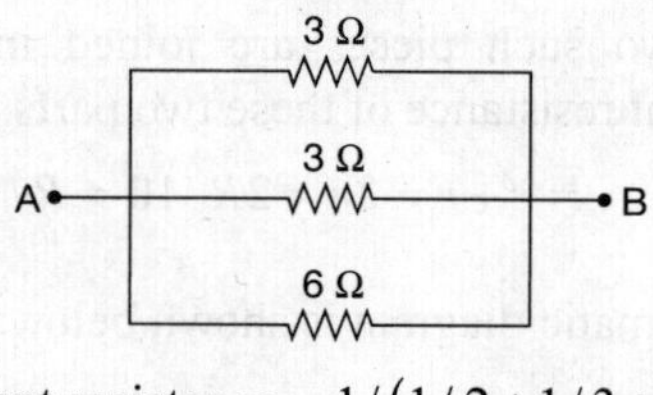

Equivalent resistance $= 1/(1/2+1/3+1/6) = 1\,\Omega$

5. (i) 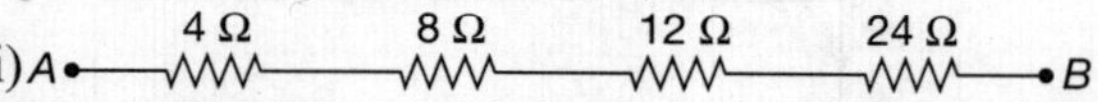

Resistance is maximum when resistors are connected in series.

$R_{max} = 4+8+12+24 = 48\Omega$

(ii)

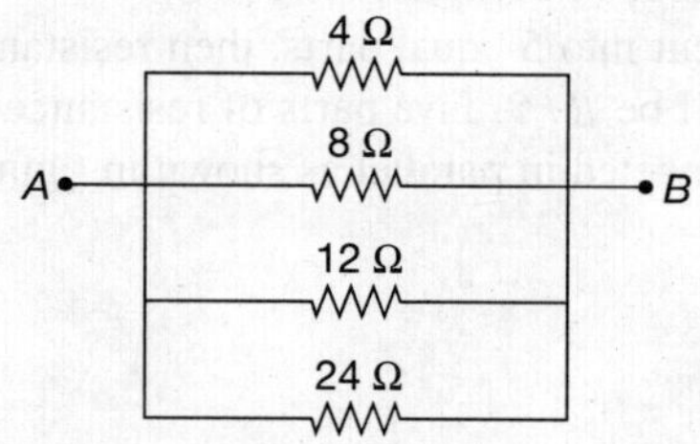

Resistance is minimum, when resistors are connected in parallel.

$R_{min} = 1/\begin{pmatrix} 1/R\_1+1/R\_2+1/R\_3 \\ +1/R\_4 = 1/(12/24) \end{pmatrix}\Omega$

6. Given, radius of wire = Diameter/2
$= 0.25$ mm $= 0.25 \times 10^{-3}$ m

$\rho = 1.6\times10^{-8}$ -m and $R = 10\ \Omega$

Resistance, $R = (\rho\iota)/A = \rho\iota/(\pi r^2)$

or $\iota = R\pi r^2/\rho$

$= \dfrac{(10\times3.14\times0.25\times0.25\times10^{-6})}{(1.6\times10^{-8})}$

$= 122.77$ m

If diameter is doubled, resistance becomes one-fourth of its original value.

7. Given, $V = 12$ V and $I = 2.5 \times 10^{-3}$ A

$R = V/I$ (By Ohm's law)

$= 12/(2.5\times10^{-3}) = 4.8\times10\,\Omega$

8. The circuit diagram for the given system of resistors is given as below:

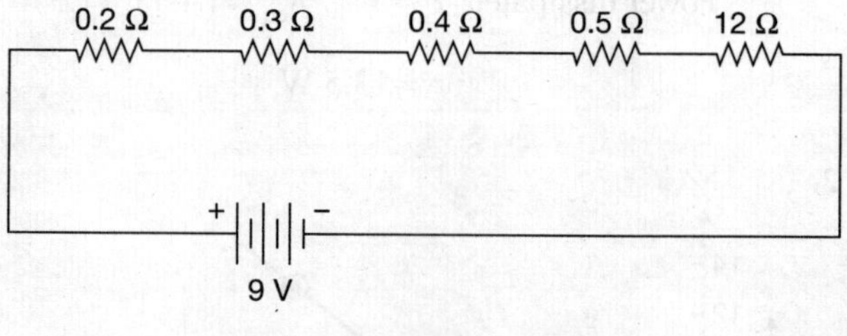

Total resistance, $R = 0.2\ \Omega + 0.3\ \Omega + 0.4\ \Omega + 0.5\ \Omega$
$+ 12\Omega = 13.4\Omega$

Current through all resistors in series is the same.

$\therefore$ Current through 12Ω

resistor $= V/R = (9\text{ V})/(13.4\ \Omega) = 0.64$ A

9. Resistance of the wire,

$R = \dfrac{V}{I} = \dfrac{220}{5} = 44\Omega$

The net resistance 44 Ω is less than the individual resistance 176 Ω, so individual resistances are to be connected in parallel order.

In parallel connection, equivalent resistance

$= 1/(1/R+1/R+1/R) = 1/(n/R) = R/n$

$\therefore$ 44 Ω = 176Ω/n or $n = 176/44 = 4$ resistors

Level - II

1. Given, $V = 220\,\text{V}$, $R_A = R_B = 24\,\Omega$
   (i) Current in plates when used separately
   $$I = V / R_A = V / R_B = 220/24 = 9.16\text{ A}$$
   (ii) Current in plates when connected in series
   Equivalent resistance in series
   $$R = R_A + R_B$$
   $$= 24 + 24 = 48\,\Omega$$
   Current flowing, $I = V / R = 220/48 = 4.58\text{ A}$
   (iii) Current in plates when connected in parallel.
   Equivalent resistance in parallel
   $R = (R_A R_B)/(R_A + R_B) = (24 \times 24)/48 = 12\,\Omega$
   Current flowing, $I = V / R = 220/12 = 18.32\text{ A}$

2. (i) $R_{eq} = (R_1 R_2)/(R_1 + R_2)$
   $= 10^6/(1 + 10^6)$, i.e., $R_{eq} < 1\,\Omega$
   (ii) $R_{eq} = (R_1 R_2 R_3)/(R_1 R_2 + R_2 R_3 + R_1 R_3)$
   $= 10^9/(10^3 + 10^9 + 10^6)$
   i.e., $R_{eq} < 1\,\Omega$
   In parallel combination, equivalent resistance is less than the least resistance.

3. $I_g G = I_s S$
   $$\frac{I_s}{I_g} = \frac{G}{S}$$

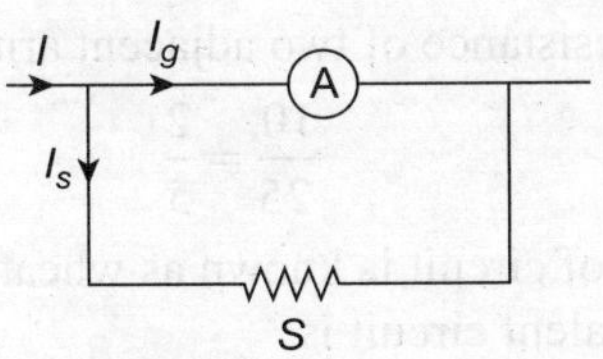

$$\Rightarrow \quad \frac{I_g + I_s}{I_g} = \frac{G + S}{S}$$
$$\frac{I}{I_g} = \frac{G + S}{S}$$
$$\frac{10}{1} = \frac{0.81 + S}{S}$$
$$9S = 0.81 \therefore S = 0.09\,\Omega$$

4. $R = \dfrac{4\rho l}{\pi d^2}$
   $$\frac{R_1}{R_2} = \frac{\rho_1 l_1}{\rho_2 l_2} \frac{d_2^2}{d_1^2}$$
   $$\frac{R_2}{10} = \frac{2}{1} \times \frac{2}{1}\left(\frac{1}{2}\right)^2$$
   $$\therefore R_2 = 10\Omega$$

Fill in the Blanks

1. switch
2. electrical
3. poles nearby
4. circuit
5. square
6. magnetic equator
7. angle of dip
8. right-hand grip
9. the distance between its two poles.

True or False

1. False
2. True
3. False
4. True
5. False

Match the Columns

1. (A) – (P), (B) – (S), (C) – (Q), (D) – (R)
2. (A) – (Q), (B) – (R), (C) – (P/S), (D) – (P/S)
3. (A) – (R), (B) – (P), (C) – (Q), (D) – (S)
4. (A) – (Q), (B) – (S), (C) – (P), (D) – (R)

Assertion-Reason Questions

1. (c) 2. (c) 3. (a)

Comprehension Type Questions

1. (a) Wire C has the maximum resistance because it has maximum length, least thickness and highest resistivity.
   (b) Wire F has the maximum resistance, since it has least length, maximum thickness and least resistivity [using $R = \rho l / A$]
   (c) $R_3 > R_5 > R_1$
   [using relation $R = \rho l / A$ and comparing]

2. (a) Radhika put off the main switch to cut-off the electricity supply to the faulty kettle to save her mother.

(b) Red wire which is at high potential of 220 V was touching the metal case of electric kettle.

(c) The values displayed by Radhika are
   (i) Presence of mind
   (ii) Concern for her mother
   (iii) Knowledge of household wiring and daily life activities.

3. (a) The values displayed by Vishwajeet are honesty, confidence and courage.

(b) An ammeter is used to measure electric current in the circuit. It is always connected in series in the circuit in which the current is to be measured.

(c) Vishwajeet can use the ammeter to verify Ohm's law.

Objective Questions

One Option Correct

1. (a) Series resistance $= 4+4+4 = 12\Omega$

$$\frac{1}{R} = \frac{1}{4} + \frac{1}{12} = \frac{1}{3}$$

$$\Rightarrow \quad R = 3\Omega$$

3. (c) $\frac{1}{R} = \frac{1}{6} + \frac{1}{3} + \frac{1}{6} = \frac{2}{3}$

$$R = \frac{3}{2} = 1.5\Omega$$

4. (b) In the circuit, two resistances in series are connected with another other resistance in parallel. Therefore,

$$\frac{1}{R_{eq}} = \frac{1}{30} + \frac{1}{60} = \frac{1}{20}$$

$$R_{eq} = 20\Omega$$

$$\text{Current } I = \frac{2}{20} = \frac{1}{10}\text{ A}$$

9. (b) $I = \frac{600}{10 \times 60} = 1\text{ C/s} = 1\text{A}$

10. (a) $R = \frac{V}{I} = \frac{220}{0.22} = 1000\Omega$

11. (d) $\frac{1}{R} = \frac{1}{4} + \frac{1}{6} = \frac{5}{12}$

$$\Rightarrow R = \frac{12}{5} = 2.4\Omega$$

$$I = \frac{V}{R} = \frac{6}{2.4}$$

$$= 2.5\text{ A}$$

12. (b) In series, heat dissipated is proportional to resistance; therefore, the ratio is $2R : R = 2:1$

13. (c) The total length remains same.

14. (c) The current divides in inverse ratio of resistance in a series circuit. Therefore, current in 6Ω resistance is $1.2 \times \frac{2}{5}$

15. (c) $\frac{1}{RAC} = \frac{1}{6} + \frac{1}{6} = \frac{1}{3}$

$$\Rightarrow RAC = 3\Omega$$

$$\frac{1}{RA} = \frac{1}{6} + \frac{1}{3} = \frac{1}{2}$$

$$RA = 2\Omega$$

16. (b) $W = qV$ Also, $P = i \times V = \frac{W}{t}$.

17. (b) In the last branch, 4Ω and 2Ω resistance are in parallel with 3Ω resistance, equivalent resistance R_1

$$\frac{1}{R_1} = \frac{1}{6} + \frac{1}{3} = \frac{1}{2}$$

$$R_1 = 2\Omega$$

Now, in second-from-last branch,

$$\frac{1}{R_2} = \frac{1}{6} + \frac{1}{3} = \frac{1}{2}$$

$$R_2 = 2\Omega$$

$$R_{PQ} = 4+2 = 6\Omega$$

18. (d) The resistance of two adjacent arms are equal

$$\frac{10}{25} = \frac{2}{5}$$

This type of circuit is known as wheat stone bridge
The equivalent circuit is

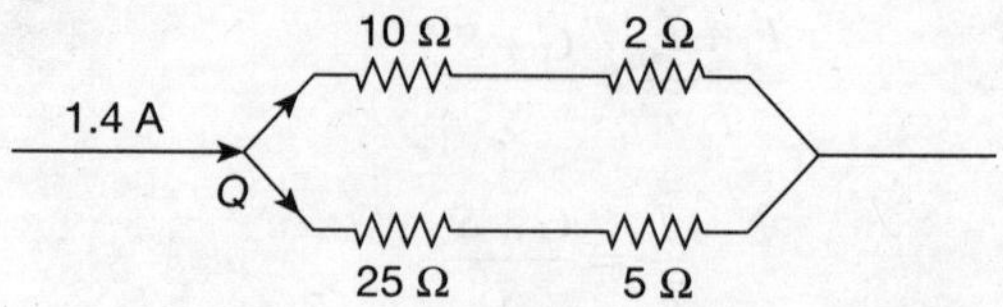

In parallel branch, the current divides in inverse ratio.

So current flowing 2Ω resistance is $1.4 \times \frac{30}{42} = 1\text{ A}$

19. (a) $i = \frac{10}{1.5 + (1\|1)} = \frac{10}{1.5 + 0.5} = 5\text{ A}$

20. (a) Similar to the solution of Question 10.

21. (c) As temperature variance leads to the flow of heat, variation of potential leads to the flow of electricity.

22. (a) As friction opposes the motion in mechanics, resistance opposes the flow of charge in electricity.

23. (b) Constant (because of high resistivity)

24. (b) Nichrome (because of high resistivity)

25. (a) $V \propto I$ (ohm's law)

26. (d) $I = \dfrac{q}{T} = \dfrac{1.6\times10^{-19}}{1.5\times10^{-6}}$

$= 1.070\times10^{-13}$ A

27. (b) 10 kΩ is human body resistance.

28. (c) If the lines of force due to a current-carrying straight conductor are in anti-clockwise direction, then according to right-hand grip rule the current flow is perpendicular to the plane of paper and directed out of the paper.

29. (c)

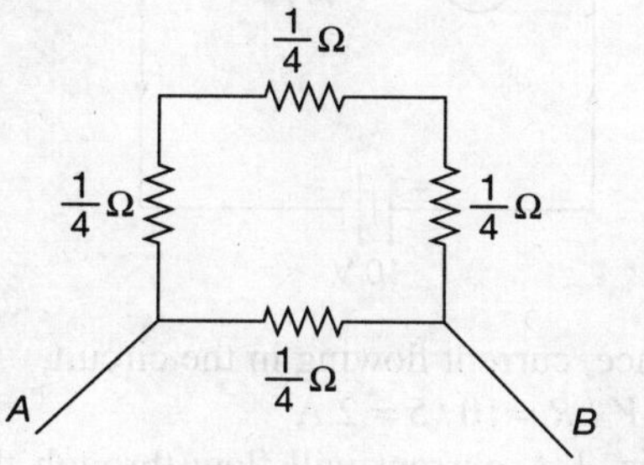

$$R = \frac{1}{4}+\frac{1}{4}+\frac{1}{4} = \frac{3}{4}$$

$\frac{3}{4}$ and $\frac{1}{4}$ are in parallel.

$$R_{eq} = \frac{\left(\frac{3}{4}\right)\left(\frac{1}{4}\right)}{\frac{1}{4}+\frac{3}{4}} = \frac{3}{16}\,\Omega$$

30. (a) The distance between the North Pole and the South Pole of a bar magnet is called effective length of the magnet.

31. (c) Magnetization can be done by single-touch method.

32. (c)

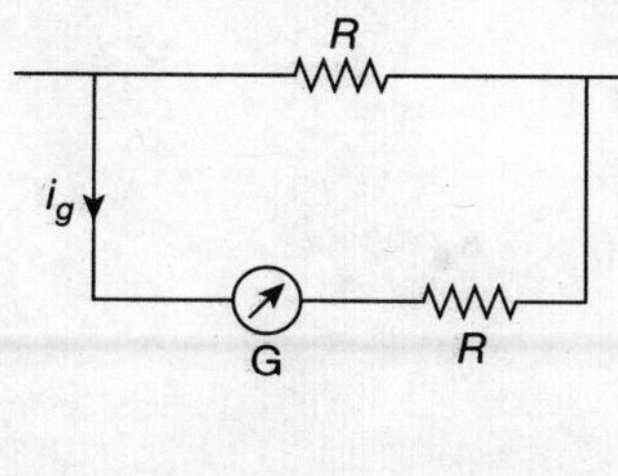

$$i_g = \frac{V}{G+R}$$

$G = 10\Omega$

$i_g = 0.01$

$V = 10$

$\therefore \quad R = 990\Omega$

33. (a) $I = \dfrac{V}{R}$

$$0.1 = \frac{12}{R+2}$$

$R = 118$

Scholar Zone

1. Let resistance of lamp be $R_1 = 100\ \Omega$

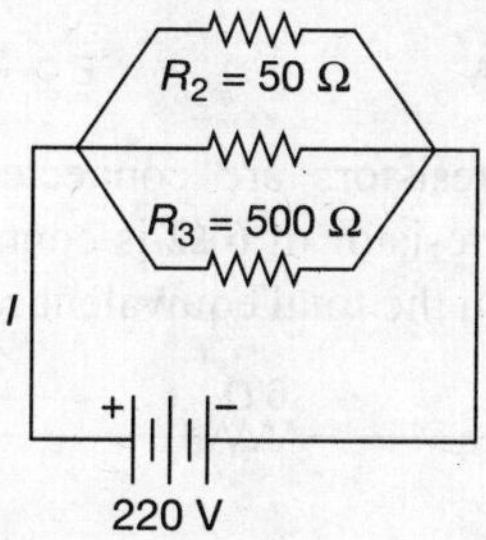

Resistance of toaster be $R_2 = 50\ \Omega$
Resistance of filter be $R_3 = 500\ \Omega$
Net resistance
$1/R = 1/R_1 + 1/R_2 + 1/R_3$ [parallel combination]
$1/R = 1/100 + 1/50 + 1/500 = 16/500$
Or $R = 500/16 = 31.25\ \Omega$
So, resistance of the iron to take the same current as drawn by all the appliances should be 31.25 Ω
Current through circuit $= V/R = 220/31.25 = 7.04$ A
Thus, current through iron is 7.04 A.

2.

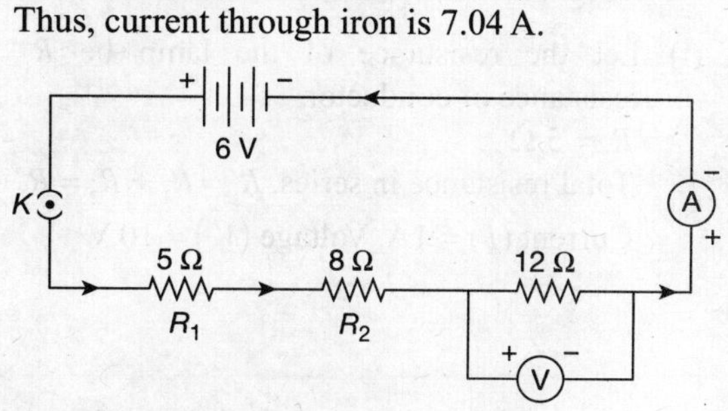

Equivalent resistance of the circuit
$R = R_1 + R_2 + R_3 = 5 + 8 + 12 = 25\ \Omega$
[hence R_1, R_2 and R_3 are connected in series]
In series combination, current flowing through all the resistance is same and equal to the total current flowing through the circuit.
$\therefore$ Current in 12 Ω resistor, $I = V/R = 6/25 = 0.24$ A

$\therefore$ Ammeter reading = 0.24 A
Potential across 12 Ω resistance,
$V = I\ R = 0.24 \times 12 = 2.88$ V
$\therefore$ Voltmeter reading is 2.88 V.

3. (i) If two 6Ω resistors are connected in parallel, then the equivalent resistance is $(6/2) = 3\ \Omega$. This combination is connected in series with a 6 Ω resistor to get overall equivalent resistance of $(6 + 3) = 9\ \Omega$.

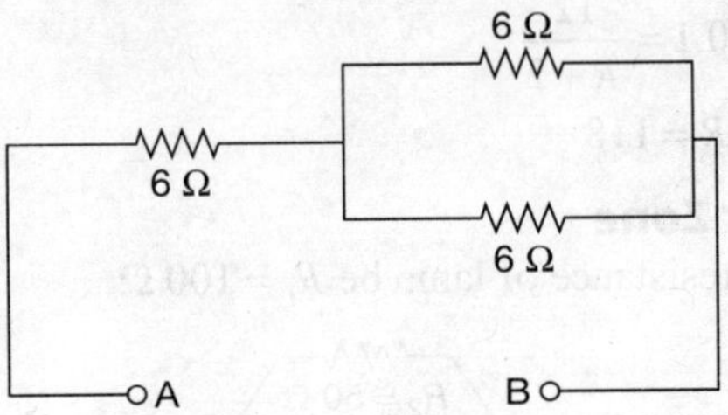

(ii) If two resistors are connected in series and another resistor of 6 Ω is connected in parallel to it, then the total equivalent resistance is 4Ω.

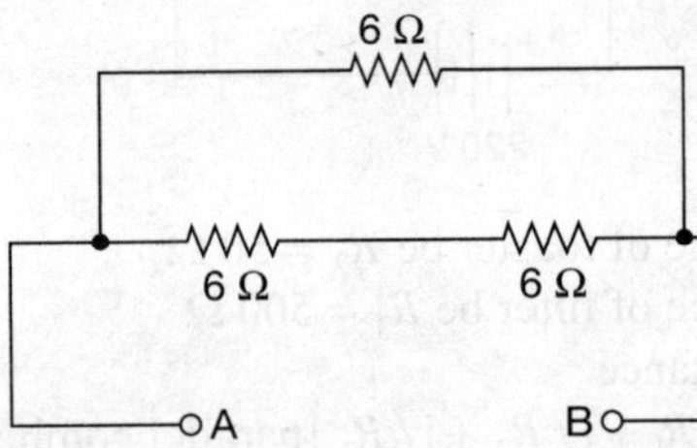

Equivalent resistance of 6Ω resistances connected in series, $R' = 6 + 6 = 12\ \Omega$. Now, 12Ω and 6Ω resistors are connected in parallel. Equivalent resistance,
$R_{eq} = (12 \times 6)/(12 + 6) = 72/18 = 4\ \Omega$

4. (i) Let the resistance of the lamp be R_1 and resistance of conductor,
$R_2 = 5\ \Omega$
Total resistance in series, $R_s = R_1 + R_2 = R_1 + 5$
Current $(I) = 1$A, Voltage $(V) = 10$ V

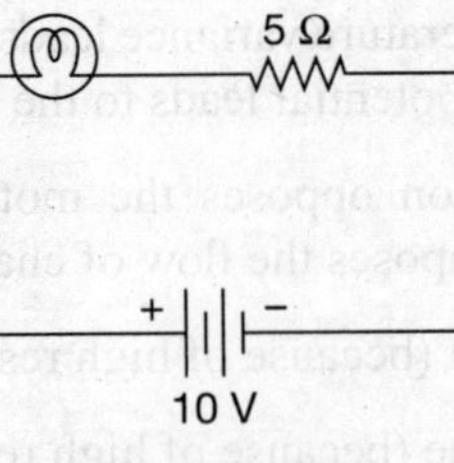

Using Ohm's law, $V = IR$
$$10 = 1(R_1 + 5)$$
$$\Rightarrow \quad R_1 = 5\ \Omega$$

(ii) Now, a resistance of 10 Ω is connected in parallel with the series combination. Therefore, the total resistance of the circuit is given by
$$1/R_p = 1/(R_1 + 5) + 1/10 \Rightarrow 1/R_p = 1/10 + 1/10$$
$$\therefore \quad R_p = 5\ \Omega$$

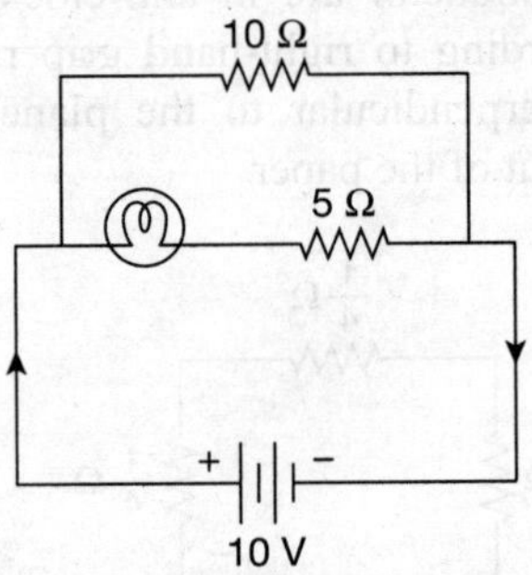

Hence, current flowing in the circuit,
$I = V/R = 10/5 = 2$ A
Thus, 1 A current will flow through the 10 Ω resistance and 1 A will flow through the lamp and conductor of 5 Ω resistance. Hence, there will be no change in current flowing through the 5 Ω conductor. Also, there will be no change in potential difference across the lamp.

CHAPTER

7 Fluids

LEARNING OBJECTIVES

At the end of this chapter, you will be able to:

- Define Fluid and Fluid Pressure
- Understand Atmospheric Pressure and variation of pressure in the atmosphere
- State Pascal's Law and understand its experimental proof
- Define Density and Relative Density
- Determine the relative density of a solid and liquid
- Describe buoyant force and Archimedes' Principle
- Learn about the laws of floatation and the factors affecting the stability of a floating body

Introduction

Matter is commonly divided into three phases: solid, liquid and gas. Solids and liquids (also called condensed matter) have a certain set of properties in common: they are relatively incompressible and their densities remain relatively constant with temperature variation (provided the pressure is kept constant) Gases are easily compressible and their density changes substantially with temperature provided the pressure is held constant.

We assign the word FLUID, collectively for liquids and gases from a different view: both can flow and assume the shape of any container that holds them. In this chapter we first study the properties of fluids at rest and the laws governing them and later we will discuss the different laws governing the fluids in motion.

Pressure

Fluids are unable to sustain shear when a force parallel to its surface is applied. Under static condition the only force component that need to be considered is one that acts in a direction perpendicular to the surface of the fluid.

The forces between the interior and exterior of the fluid are normal to the fluid boundary everywhere and is independent of the shape of the fluid. The magnitude of the *normal force per unit surface area is called pressure.* Consider Fig. 7.1. A small element of surface area can be represented by ΔA along the outward normal. The force ΔF exerted by the fluid lying within the bounding surface depends on the fluid pressure p according to

$$\Delta F = p\Delta A$$

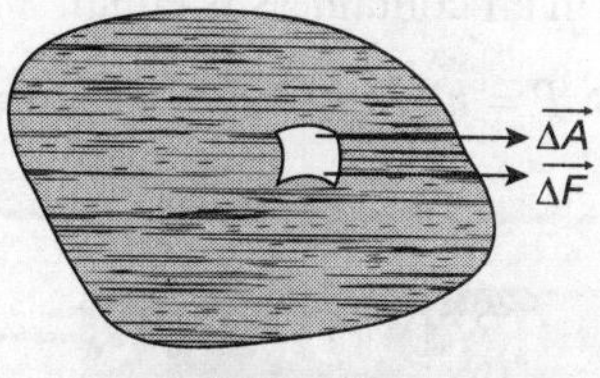

Figure 7.1

Since ΔF and ΔA are parallel, we can write the pressure in terms of scalar relationship

$$p = \frac{\Delta F}{\Delta A}$$

The unit of pressure is N/m^2 = 1 pascal

Pressure Exerted by a Liquid Column

Consider a liquid of density ρ contained in a cylinder of cross-sectional area A (Fig. 7.2). Let h be the height of the liquid column, The weight of the liquid column exerts a downward thrust and hence a downward pressure.

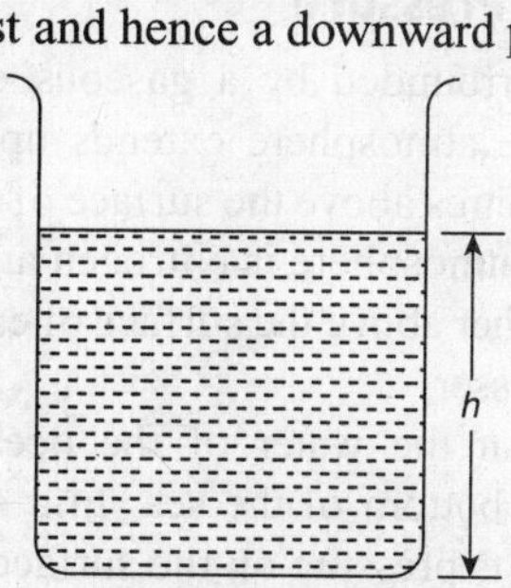

Figure 7.2

Weight of liquid column

= Volume of liquid column × Density of liquid × g

= $Ah\rho g$

$$\text{Pressure} = \frac{\text{Weight of liquid column}}{\text{Cross} - \text{sectional area}} = \frac{Ah\rho g}{A} = h\rho g$$

So, the pressure exerted by a liquid column at rest is proportional to (i) height of liquid column and (ii) density of liquid

ILLUSTRATION 7.1

Figure 7.3 shows four containers of olive oil. Rank them according to the pressure at depth h, greatest first.

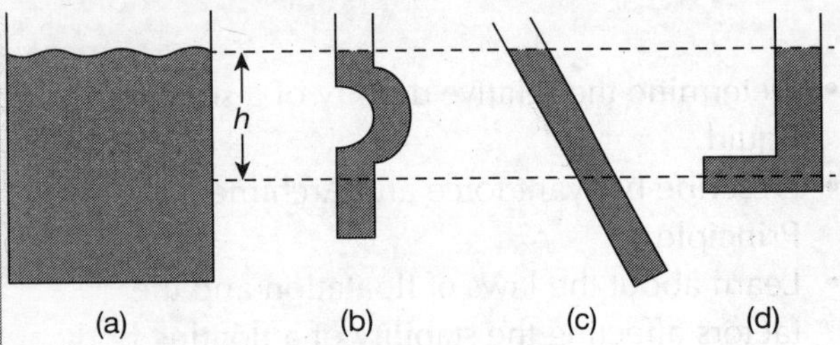

Figure 7.3

Solution

The pressure in all containers is equal.

Explanation: $P = h\rho g$; In each case, h, ρ and g are the same.

ILLUSTRATION 7.2

If the blood vessels in a human being acted as simple pipes (which they do not), what would be the difference in blood pressure between the blood in a 1.80 m-tall man's feet and in his head when he is standing? Assume the specific gravity of blood to be 1.06.

Solution

$$\Delta p = h\rho g$$

$$= 1.8 \times 1.06 \times 10^3 \times 9.8 \text{ Pa} = 18.7 \text{ kPa}$$

Atmospheric Pressure

The earth is surrounded by a gaseous envelope called atmosphere. The atmosphere extends up to a height of roughly 8 kilometres above the surface of earth. The main composition of atmosphere is Nitrogen and Oxygen.

As we go higher above the surface of earth, the density of air become lesser.

We know that the water of the ocean exerts great pressure on the bottom of the sea. In a similar way, the atmosphere exerts pressure on the surface of the surface of earth.

This pressure due to atmosphere is called atmospheric pressure. The SI unit of pressure is N/m². In pascal, this pressure is approximately 1.01×10^5 pascal at sea level.

Standard Atmospheric Pressure

If a standard atmosphere supports a column of height of exactly 76 cm of mercury, the standard atmospheric pressure can be expressed in pascals.

$$p_{atm} = h\rho g$$

Here, $h = 76 \text{ cm} = 0.76 \text{ m}$

$$\rho_{Hg} = 13.6 \times 10^3 \text{ kg/m}^3, \text{ and } g = 9.8 \text{ m/s}^2$$

$$\therefore \quad p_{atm} = 0.76 \times 13.6 \times 10^3 \times 9.8 \text{ N/m}^2$$

$$= 1.013 \times 10^5 \text{ Pa} \quad (\text{or, } 101.3 \text{ kPa})$$

Changes in atmospheric pressure are observed as changes in the height of mercury column. One standard atmosphere is a unit of air pressure expressed in height units as

$$1 \text{ atm} = 76 \text{ cm of Hg} = 760 \text{ mm of Hg}$$

In honour of Torricelli, a pressure supporting 1 mm of mercury is given the name **torr**:

$$1 \text{ mm Hg} = 1 \text{ torr and}$$

$$1 \text{ atm} = 760 \text{ torr.}$$

Variation of Pressure in the Atmosphere

The pressure of a gas in small vessel is taken as uniform since its density is very small and the difference in pressure at two nearby points is very small. However, this is not true if the height is large. As the vertical height from the sea level increases, the atmospheric pressure decreases.

The other factor on which the atmospheric pressure depends is temperature. As the temperature rises, the density of atmospheric air decreases. This reduces the atmospheric pressure at a height from the mean sea level. In a similar way if the temperature decreases, the density of atmospheric air increases so the atmospheric pressure also increases.

ILLUSTRATION 7.3

A U-tube open to the atmosphere at both ends contains some mercury. A quantity of water is carefully poured into the left arm of the U-tube until the vertical height of water column is 15.0 cm.

(a) What is the gauge-pressure at the water-mercury interface?

(b) Calculate the vertical height h from the free surface of mercury to the free surface of water column.

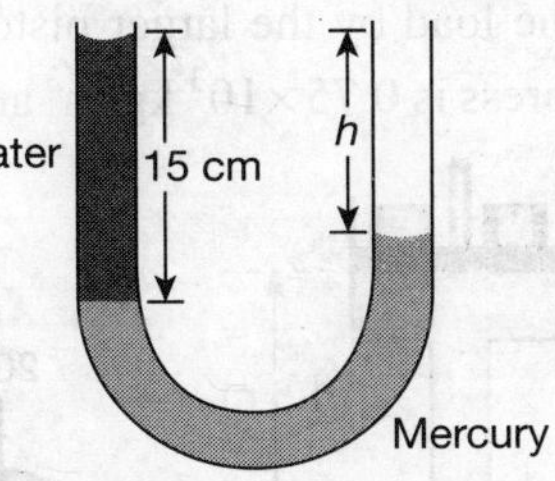

Solution

(a) Gauge pressure (which is the pressure above the atmospheric pressure) at water-mercury interface is

$$P = h_\omega \rho_\omega g = (15\times10^{-2})(1000)(10)$$
$$= 1500 \text{ Pa.}$$

(b) Equating the pressure at the same horizontal level of mercury in the two arms of the U-tube,

$$P_{\text{atm}} + h_\omega \rho_\omega g = P_{\text{atm}} + h_\text{m} \rho_\text{m} g$$

or, $\quad 15\times1000 = (15-h)\times13.6\times10^3$

or, $\quad h = 13.9$ cm.

Altimeter

The altimeter is an instrument which is used to measure altitudes in aircrafts. A fall in the atmospheric pressure by 1 cm of Hg in pressure corresponds approximately to an increase in altitude by 105 m (assuming the gravity is constant). An aneroid barometer with a calibrated scale is used to read the altitude in metres.

Weather Forecast

The atmospheric pressure at a given place depends on the weather conditions such as temperature and moisture, because these affect the density of air. A dip in atmospheric pressure is called depression. A fall in pressure due to rise in temperature when there is no increase in humidity portends a dust storm. On the other hand, a fall in pressure due to a rise in humidity indicates rainfall. A sudden fall in atmospheric pressure, forecast a storm or a cyclone.

Barometer is used to measure atmospheric pressure (Fig. 7.4).

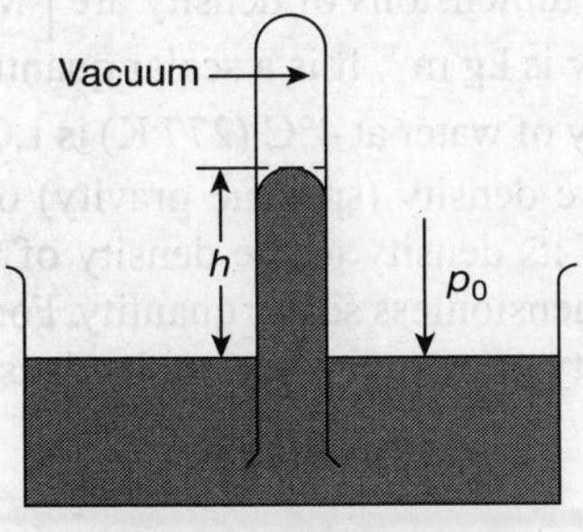

Figure 7.4 Barometer

$$p_0 = h\rho g$$

Manometer (Fig. 7.5) measures the pressure difference, i.e., gauge-pressure.

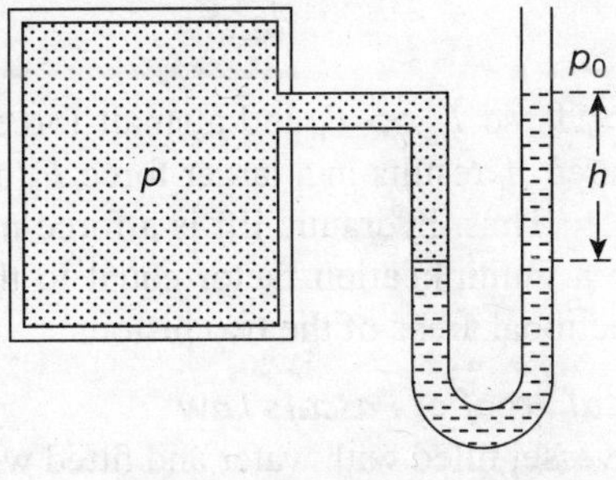

Figure 7.5 Manometer

$$p - p_0 = h\rho g$$

Pascal's Law

We know that fluid pressure depends on the depth h and is given by $p = p_0 + h\rho g$, where p_0 is the atmospheric pressure above the free surface of the liquid. A consequence of this fact used in hydraulic jack or lift is referred to as **Pascal's law** which states that:

"Pressure applied to a confined fluid (enclosed in a chamber) at rest is transmitted undiminished to every portion of the fluid and the walls of the containing vessel".

A well-known application of Pascal's law is the hydraulic *lift of jack* to support or lift heavy objects as shown in Fig. 7.6.

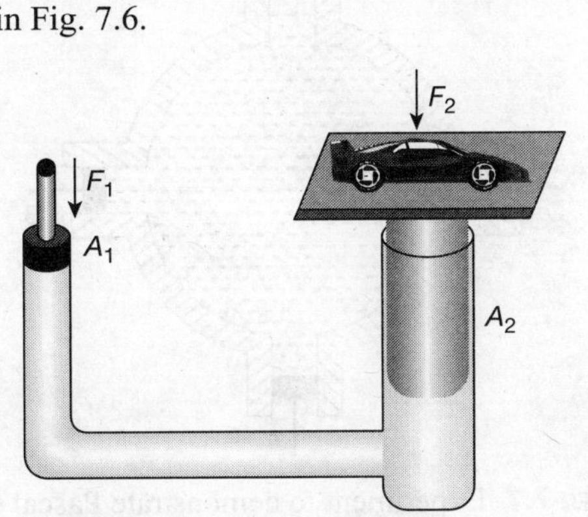

Figure 7.6 Application of Pascal's law

A piston with small cross-sectional area A_1 exerts a force F_1 on the surface of a liquid. The applied pressure $P = \dfrac{F_1}{A_1}$ is transmitted through the liquid in the connecting pipe to a larger piston of area A_2. The applied pressure is same in both the cylinders, hence

$$P = \frac{F_1}{A_1} = \frac{F_2}{A_2},$$

or,

$$F_2 = \frac{A_2}{A_1}F_1$$

Since $A_2 >> A_1$, so $F_2 >> F_1$ and a small force F_1 acting on a small area A_1 results in a larger force F_2 acting on a larger area A_2. Thus, hydraulic lift is a force multiplying device with a multiplication factor equal to the ratio of the cross-sectional areas of the two pistons.

Experimental Proof of Pascal's Law

Consider a vessel filled with water and fitted with pistons in different positions as shown in Fig. 7.7. The four pistons A, B, C and D have different cross-sectional areas A_1, A_2, A_3 and A_4 respectively. Suppose the piston A is pushed with a force F_1. The pressure on the piston is $\frac{F_1}{A_1}$.

It will be observed that the piston B, C and D can be held in their positions only if we apply forces F_2, F_3 and F_4 respectively such that

$$\frac{F_2}{A_2} = \frac{F_3}{A_3} = \frac{F_4}{A_4}$$

This experiment demonstrates that the pressure is transmitted undiminished in all directions, as required by Pascal's law.

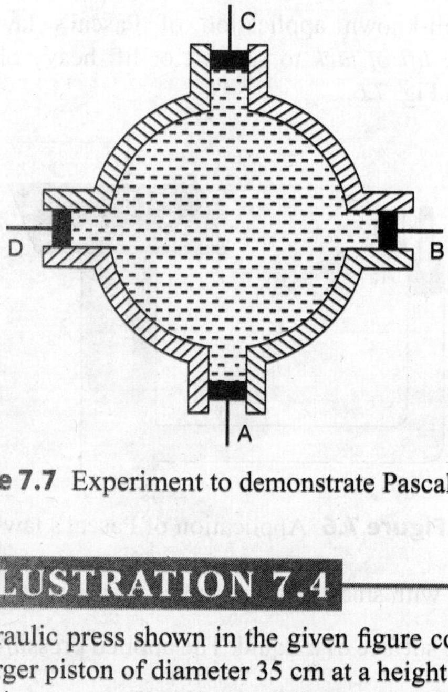

Figure 7.7 Experiment to demonstrate Pascal's law

ILLUSTRATION 7.4

A hydraulic press shown in the given figure consists of a larger piston of diameter 35 cm at a height of 1.5 m relative to a smaller piston of diameter 10 cm. The mass of the smaller piston is 20 kg. What is the force exerted on the load by the larger piston. The density of oil in the press is 0.75×10^3 kg/m^3 and $g = 9.8$ m/s^2

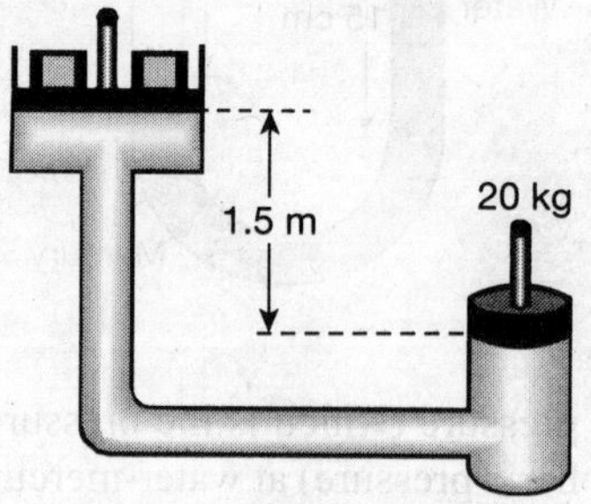

Solution

Fluid pressure on the smaller piston

$$P_1 = \frac{mg}{A} = \frac{20\times9.8}{\pi\times(5\times10^{-2})^2}\ \text{N/m}^2$$

Similarly, fluid pressure on the larger piston,

$$P_2 = \frac{F}{\pi(17.5\times10^{-2})^2}\ \text{N/m}^2.$$

The difference between the two pressure is balanced by $h\rho g$, where h = 1.5 m and density of fluid ρ = 0.75×10^3 kg/m^3

$$\therefore \quad \frac{20\times9.8}{\pi\times(5\times10^{-2})^2} - \frac{F}{\pi\times(17.5\times10^{-2})^2} = 1.5\times750\times9.8.$$

Simplifying we get,

$$F = 1.3\times10^3\ \text{N}.$$

Density and Relative Density

Density is defined as mass per unit volume.

For a fluid of mass m occupying volume V,

$$\rho = \frac{m}{V}$$

A liquid is largely incompressible and its density is therefore nearly constant at all pressures. Gases, on the other hand, exhibit a large variation in densities with pressure. The dimensions of density are $[ML^{-3}]$. The SI unit of density is kg m^{-3}. It is a scalar quantity.

The density of water at 4°C (277 K) is 1.0×10^3 kg m^{-3}.

The relative density (specific gravity) of a substance is the ratio of its density to the density of water at 4°C. It is also a dimensionless scalar quantity. For example, the specific gravity of mercury is 13.6. Its density is

$$13.6\times10^3\ \text{kg m}^{-3}.$$

Densities of some common fluids at STP

Fluid	ρ(kg m^{-3})
Water	1.00×10^3
Mercury	1.36×10^3
Whole Blood	1.06×10^3
Air	1.29
Oxygen	1.43
Hydrogen	9.0×10^{-2}

Mixing of two liquids

(i) Consider two liquids, each of same volume V but of different densities ρ_1 and ρ_2 mixed together.

Total mass $= V\rho_1 + V\rho_2$

Total volume $= 2V$

Density of mixture, $\rho = \dfrac{V\rho_1 + V\rho_2}{2V}$

or $\rho = \dfrac{\rho_1 + \rho_2}{2}$

(ii) Consider two liquids, each of same mass m but of different densities ρ_1 and ρ_2 mixed together.

Total mass $= 2m$

Total volume $= \dfrac{m}{\rho_1} + \dfrac{m}{\rho_2}$

Density of mixture, $\rho = \dfrac{2m}{\dfrac{m}{\rho_1} + \dfrac{m}{\rho_2}}$ or $\rho = \dfrac{2\rho_1\rho_2}{\rho_1 + \rho_2}$

(iii) Consider two liquids of different masses m_1 and m_2 mixed together. Let ρ_1 and ρ_2 be their respective densities.

Total mass $= m_1 + m_2$

Total volume $= \dfrac{m_1}{\rho_1} + \dfrac{m_2}{\rho_2}$

Density of mixture, $\rho = \dfrac{m_1 + m_2}{\dfrac{m_1}{\rho_1} + \dfrac{m_2}{\rho_2}}$

(iv) Consider two liquids of different volumes V_1 and V_2 mixed together. Let ρ_1 and ρ_2 be their respective densities.

Total mass $= V_1\rho_1 + V_2\rho_2$

Total volume $= V_1 + V_2$

Density of mixture $= \dfrac{V_1\rho_1 + V_2\rho_2}{V_1 + V_2}$

Relative density of a body

$$= \frac{\text{Weight of body in air}}{\text{Loss of weight of body in water}}$$

$$= \frac{\text{Weight of body in air}}{\text{Upthrust force}}$$

$$= \frac{\text{Weight of body in air}}{\text{Weight of equal volume of water}}$$

$$= \frac{\text{Density of body}}{\text{Density of water}}$$

ILLUSTRATION 7.5

What is the specific gravity of an iron bar weighing 192 g and having geometrical dimensions 12 cm × 2 cm × 1 cm?

Solution

$$\rho = \frac{192\text{ g}}{12 \times 2\text{ cm}^3} = 8\text{ g cm}^{-3}$$

In cgs system, specific gravity is numerically equal to density.

Determination of Relative Density of a Solid

Archimedes' principle states that when a body is immersed in a liquid, it experiences an apparent loss in its weight which is equal to the weight of the displaced liquid or weight of the liquid whose volume is equal to the volume of the solid. Based on this, relative density of a solid can be calculated as follows.

$$\text{Relative density (R.D.)} = \frac{d}{d_w},$$

where d and d_w are the densities of the solid and water, respectively.

$$\therefore \quad \text{R.D.} = \frac{\dfrac{m}{V}}{\dfrac{m_w}{V_w}};$$

Here m is the mass of solid, m_w is the mass of water, V is the volume of solid and V_w is the volume of water.

If $V = V_w$, then R.D. of the solid $= \dfrac{m}{m_w} = \dfrac{m}{m_w} \times \dfrac{g}{g} = \dfrac{W}{W_w}$

where W and W_w are the weight of the solid in air and the weight of water whose volume is equal to the volume of the solid, respectively.

i.e., W_w = weight of water displaced

$\therefore$ By Archimedes' principle w_w = apparent loss of weight of the solid in water.

$$\therefore \text{R.D.} = \frac{\text{Weight of the solid in air}}{\text{Apparent loss of weight of the solid in water}}$$

If W_1 and W_2 represent the weight of the solid in air and water, respectively, then the relative density of the solid

$$= \frac{W_1}{W_1 - W_2}$$

Determination of the Relative Density of a Liquid Using Archimedes' Principle

The relative density of a liquid can be determined by the following method:

1. Let the weight of solid in air be W_1.
2. Let the weight of the body in water be W_2.
3. Let the weight of the solid immersed in the given liquid be W_3.

The relative density of the given liquid is calculated as follows:

Relative density of a liquid

$$= \frac{\text{Mass of certain volume } (V) \text{ of liquid}}{\text{Mass of equal volume } (V) \text{ of water}}$$

$$= \frac{\text{Weight of the liquid displaced by the body}}{\text{Weight of the water displaced by the same body}}$$

$$= \frac{\text{Loss of weight of the body in the liquid}}{\text{Loss of weight of the same body in water}}$$

$$= \frac{W_1 - W_3}{W_1 - W_2}$$

ILLUSTRATION 7.6

A jeweller claims that he sells ornaments made of pure gold that has the relative density of 19.3. He sells a bangle weighing 25.250 gf to a person. The clever customer weighs the bangle when immersed in pure water and finds that it weighs 23.075 gf in water. Is the ornament made of pure gold?

Solution

Weight of ornament in air = 25.250 gf

Weight of ornament in water = 23.075 gf

Apparent loss of weight in water = $(25.250 - 23.075)$ gf = 2.175 gf

Relative density of the ornament

$$= \frac{\text{Weight in air}}{\text{Apparent loss of weight in water}}$$

$$= \frac{25.250}{2.175} = 11.61$$

Since the relative density of the ornament is much less than that of pure gold which is 19.3, the ornament is not made of pure gold.

ILLUSTRATION 7.7

A body weighs 20 N in water and its relative density is 3. Find the weight of the body in air.

Solution

Let W_1 be the weight of the body in air and W_2 be its weight in water

$\therefore$ $W_2 = 20$ N

R.D. of the body = 3

$$\therefore \quad \frac{W_1}{W_1 - W_2} = 3$$

$$\frac{W_1}{W_1 - 20} = 3 \Rightarrow W_1 = 30 \text{ N}$$

ILLUSTRATION 7.8

A body weighs 16 N in a liquid whose specific gravity is 6. Find the weight of the body in water if it weighs 40 N in air.

Solution

Let W_1, W_2 and W_3 be the weight of the body in air, water and liquid, respectively.

$W_3 = 16$ N; $W_2 = x$ (to be determined)

$W_1 = 40$ N

Specific gravity of the liquid = 6

$$\Rightarrow \quad \frac{W_1 - W_3}{W_1 - W_2} = 6 \Rightarrow \frac{40 - 16}{40 - x} = 6$$

$\Rightarrow x = 36$ N. Thus, weight of the body in water is 36 N.

Buoyancy and Archimedes' Principle

When an object is placed in a fluid, it will either sink or float. A similar effect is observed in gases as well. A falling object is sinking in the atmosphere, and objects such as hot-air balloons float.

Objects float because they are buoyant, or are buoyed up and thus require an upward net force. The upward force resulting from an object being wholly or partially immersed in a fluid is called the **buoyant force.**

Consider a buoyant object (piece of cork or wood) being held under the fluid surface as shown in Fig. 7.8. The fluid pressure on the upper and lower surfaces of the block are

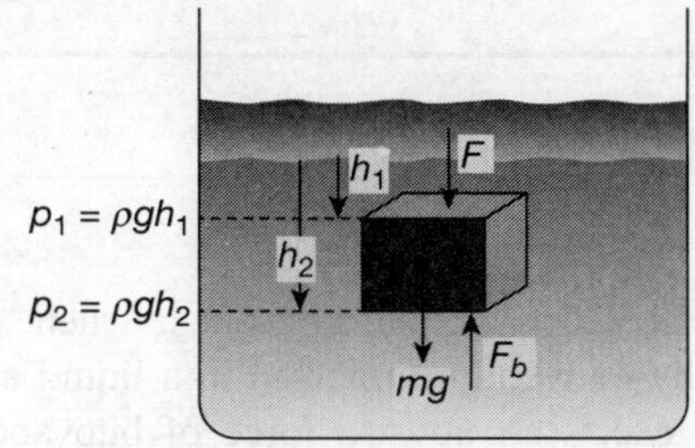

Figure 7.8

$$p_1 = \rho g h_1 \text{ and } p_2 = \rho g h_2,$$

where ρ is the density of the fluid. Thus, the pressure difference, $\Delta p = p_2 - p_1 = \rho g(h_2 - h_1)$ between the top and the bottom of the block gives rise to the buoyant force F_b. If the area of the top and bottom of the block area A, we have

$$F_b = (\Delta p)A = \rho g(h_2 - h_1)A$$
$$= \rho g V,$$

where V is the volume of the block, which is also the volume of the fluid displaced by the block. But ρV = mass of the fluid displaced by the block. Thus the buoyant force

$$F_b = m_{\text{fluid}} \times g = \text{weight of fluid displaced.}$$

This general result is known as **Archimedes' Principle** as stated below:

"A body immersed wholly or partially in a fluid is buoyed up by a force equal in magnitude to the weight of the volume it displaces"

Laws of Floatation

1. If the density of a floating body is more than the density of the fluid, the body will sink.
2. The body floats if its density is less than the density of fluid.
3. If the density of the body is equal to the fluid's density, the buoyant force is equal to weight of the body. The net weight of the body when fully immersed is zero.

ILLUSTRATION 7.9

What fraction of the total volume of an iceberg floating is sea-water is exposed? Given: density of ice = 917 kg m^{-3} and density of sea-water = 1024 kgm^{-3}.

Solution

The weight of the iceberg acting vertically downward is

$$W_{\text{ice}} = \rho_{\text{ice}} V_{\text{ice}} g,$$

where V_{ice} is the volume of the iceberg. If V_w be the volume of the submerged part of the iceberg, the buoyant force F_b is the weight of the displaced seawater.

$$\therefore \quad F_b = \rho_w V_w g$$

Since the iceberg is in equilibrium,

$$W_{\text{ice}} = F_b$$

or, $\quad \rho_{\text{ice}} \cdot V_{\text{ice}} g = \rho_w V_w \cdot g$

$$\therefore \quad \frac{V_w}{V_{\text{ice}}} = \frac{\rho_{\text{ice}}}{\rho_w} = \frac{917}{1024} = 0.896 = 89.6\%$$

$\therefore$ % of the iceberg exposed is 100 – 89.6 = 10.4%.

Stability of a Floating Body

The stability of a floating body depends on the positions of the effective point of application of the buoyant force and the centre of mass of the body. The buoyant force acts at the centre of mass of the displaced fluid, which is known as the centre of buoyancy. When the floating body rests in equilibrium, the centre of mass of the body G and the centre of buoyancy B lie along a vertical axis of the body as shown in Fig. 7.9 (a).

However, if the body tilts on one side, the line of action of the buoyant force F_B intersects the axis of the body at a point M, called the metacentre. If G is below M, the torque will restore the body to its stable equilibrium position (Fig. 7.9 (b)).

If G is above M, the torque will tend to rotate the body away from its equilibrium position and the body will be unstable

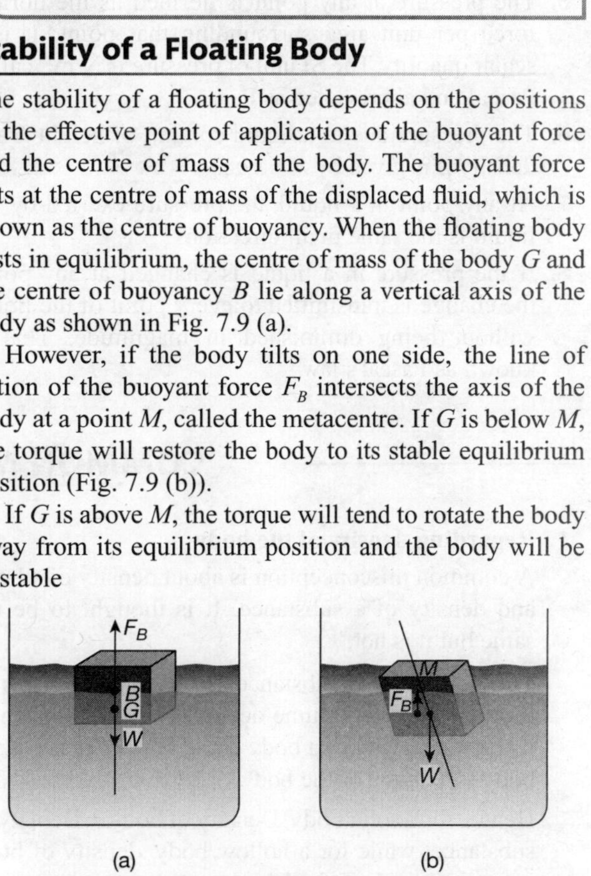

Figure 7.9

ILLUSTRATION 7.10

Explain why a small iron needle sinks in water while a large iron ship floats?

Solution

For floatation, the density of body must be lesser or equal to that of liquid. In case of iron needle, the density of needle, i.e., iron is more than that of water, so it will sink. However, the density of a ship due to its large volume is lesser than that of water, so it will float.

POINTS TO REMEMBER

1. A substance that can flow is called a fluid. It includes both liquids and gases.
2. The free surface of a liquid at rest is horizontal.
3. The density of a substance is defined as the mass per unit volume of that substance. Its unit is kg/m^3.
4. The relative density of a substance is defined as the ratio of the density of that substance to that of water at 4°C. As this is the ratio of two densities, it has no unit.
5. The pressure at any point is defined as the normal force per unit area surrounding that point. It is a scalar quantity. The SI unit of pressure is N/m^2, called Pascal and abbreviated as Pa.
6. The pressure due to a liquid column of height h and density ρ is given by $P = h\rho g$.
7. At any point in a liquid, the pressure exerted by the liquid is the same in all directions.
8. If the pressure in a liquid is changed at any point, the change is transmitted to every point of the liquid without being diminished in magnitude. This is known as Pascal's law.
9. Archimedes' principle states that when a body is partially or wholly immersed in a liquid at rest, the liquid exerts an upward force of buoyancy on the body, equal to the weight of the displaced liquid. Because of this upward force of buoyancy on the body, the apparent weight of a body in the liquid is reduced and the decrease in weight is equal to the weight of the liquid displaced.
10. A body floating in a fluid is in stable equilibrium if the centre of gravity of the displace fluid, called the centre of buoyancy, is vertically above the centre of gravity of the body.
11. When a solid body is dipped into a liquid, it experiences an upward force of buoyancy due to the liquid. If this force of buoyancy is equal to the weight of the solid, the solid will remain in equilibrium. This is known as floatation. When the density of the solid is less than that of the liquid, the solid floats with a part of it in the liquid. The fraction dipped is such that the weight of the displaced liquid equals the weight of the solid.

COMMON MISCONCEPTIONS

1. Regarding density of the body

A common misconception is about density of a body and density of a substance. It is thought to be the same but it is not.

The density of a substance is the ratio of mass of the substance to volume occupied by the substance, while the density of a body is the ratio of mass of the body to volume of the body.

Hence, for a solid body, Density of body = Density of substance, while for a hollow body, density of body is lesser than that of substance.

2. Regarding level of water in the pond when man drinks some water from the pond sitting on a boat

It is a general misconception that the level of water will go down, but it is the contrary.

The explanation is that when a man drinks from the pond, the weight of (Man + Boat) will increase by mg, so (Man + Boat) will displace more water for floating. So due to removal of water from the pond, the water level in the pond will fall but due to water displacement by the (Man + Boat) the water in the pond will rise. Therefore water removed from the pond is equal to the water displaced by the system. The level of water in the pond will remain same.

PRACTICE QUESTIONS

Very-short Answer Type Questions

1. State Archimedes' principle.
2. State Pascal's law of transmission of fluid pressure.
3. What is the condition required so that a solid body floats in a liquid with some part of the body above the surface of the liquid?
4. Mention two applications of Pascal's law of transmission of fluid pressure.
5. A container is filled with water to a height of 10 m. The pressure exerted by the water at the bottom of the container is ____.
6. State the laws of floatation.
7. If an object floats in water such that half of its volume is immersed in it, then the density of the object is ____ kg m^{-3}.
8. What is the SI unit of pressure.
9. We do not feel atmospheric pressure on our body. Explain the reason in brief.
10. State whether liquids and gases also exert pressure.

Short-answer Type Questions

1. Give the formula for the relative density of a body in terms of loss of weight.
2. Explain the cause of upthrust on a body immersed in a liquid.
3. Derive an expression for fluid pressure.
4. Explain the effect of force on the pressure when the area on which it is applied, decreases.
5. Double tyres are provided at the rear wheels of truck. Explain why.
6. Explain why the cutting instruments are sharpened.
7. Explain the variation of pressure exerted by a liquid with respect to the following factors:
   (a) Amount of liquid
   (b) Depth
   (c) Shape and size of container.
8. Mountaineers usually suffer from nose bleeding at high altitudes. Explain why.
9. An inflated gas balloon is placed in a jar which is connected to an evacuating pump. What happens if the air inside the jar is pumped out? Explain.
10. Suman stood on a weighing scale to measure her weight. She first stood on the scale with both feet. She then stood on scale with just one foot. Will the scale show any difference in Suman's weight in both the cases?
11. It is difficult to cut cloth using a pair of scissors with blunt blades. Explain.
12. A gas-filled balloon moves up. Is the upward force acting on it larger or smaller than the force of gravity?
13. The given figure shows a man in a parachute. Name the force which is responsible for his downward motion. Will he come down with the same speed without the parachute?

14. Two rods of the same weight and equal length have different thickness. They are held vertically on the surface of sand as shown in the figure. Which one of them will sink more? Why?

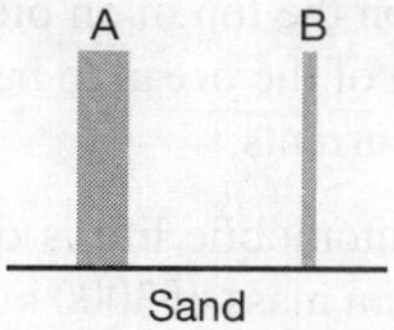

Descriptive Type Questions

1. Observe the vessels A, B, C and D as shown in figure carefully.

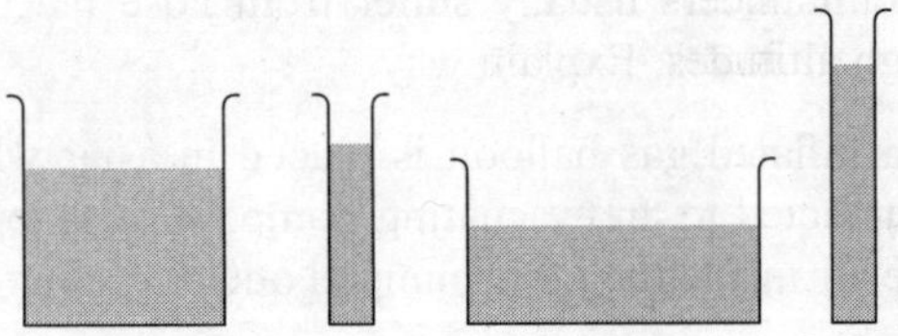

The volume of water taken in each vessel is as shown. Arrange them in the order of decreasing pressure at the base of each vessel. Explain.

2. A block of wood is floating on water at 0°C with a certain volume V outside the water-level. The temperature of water is slowly raised from 0°C to 20°C. How will the volume V change with rise in temperature?

3. Discuss the laws of floatation.

4. Discuss the conditions of equilibrium of a floating body.

5. State and prove Archimedes' principle.

6. Write a note on atmospheric pressure.

Numericals

Level - I

1. The reading of a spring balance when a block is suspended from it in air is 60 N. This reading is changed to 40 N when the block is submerged in water. Calculate the specific gravity of the block.

2. The density of the atmosphere at sea level is 1.29 kg m^{-3}. Assume that it does not change with altitude. Then how high would the atmosphere extend?

3. A vertical off-shore structure is built to withstand a maximum stress of 10^9 Pa. Is the structure suitable to be erected on the top of an oil-well in the ocean? Take the depth of the ocean to be roughly 3 km, and ignore ocean currents.

4. A hydraulic automobile life is designed to lift cars with a maximum mass of 3000 kg. The area of cross-section of the piston carrying the load is 425 cm^2. What is maximum pressure would the smaller piston have to bear?

5. In a hydraulic lift, the surface area of the input piston is 10 cm^2. The surface area of the output piston is 3000 cm^2. A 100 N force applied to the input piston raises the output piston by 1 m. Calculate the work done.

6. A body weighs 160 g in air, 130 g in water and 136 g in oil. What is the specific gravity of oil?

7. A fish weighs 348 g in air and 23 g in pure water. Calculate the relative density of fish.

8. A body weighs 20 N in water and 30 N in a liquid. If the actual weight of the body is 40 N, then find the specific gravity of the liquid.

9. A body weighs 20 N in water and 30 N in air. Find the density of the body. Take the density of water as 1.0 g cm^{-3}.

Level - II

1. A cube of wood supporting 200 g mass just floats in water. When the mass is removed, the cube rises by 2 cm. What is the size of the cube?

2. A U-tube contains water and spirit separated by mercury. The mercury columns in the two arms are in level with 10 cm of water in one arm and 12.5 cm of spirit in the other arm. What is the specific gravity of spirit?

3. Equal volume of a liquid is poured into containers A and B where area of cross-section of container A is double the area of cross-section of container B. If P_A and P_B are the pressures exerted at the bottom of the containers, then find $P_A : P_B$.

4. What is the pressure on a swimmer 10 m below the surface of a lake?

5. Water is filled in a flask up to a height of 20 cm. The bottom of the flask is circular and has an area of 1 m^2. If the atmospheric pressure is 1.01×10^5 Pa, then what force is exerted by water on the bottom? (Take g = 10 m s^{-2} and density of water = 1000 kg m^{-3})

6. An automobile is lifted by a hydraulic jack that consists of two pistons. The large piston is 1 m in diameter and the small piston is 10 cm in diameter. If W be weight of the car, how much smaller a force is needed on the small piston to lift the car?

7. In a car lift, compressed air exerts a force F_1 on a small piston having a radius of 5 cm. This pressure is transmitted to a second piston of radius 15 cm. If the mass of the car to be lifted is 1350 kg, what is F_1? What is the pressure necessary to accomplish this task?

8. A container is filled with two immiscible liquids A and B of densities 2 g cm^{-3} and 3 g cm^{-3}, respectively. A wooden cube of side 1 cm floats on the surface of liquid A such that one-fourth of its total length is immersed in this liquid A. Now, the wooden cube is completely immersed in liquid A by suspending a sinker of volume 10 cm^3 which is completely submerged in liquid B. Determine the weight of the sinker.

9. A concrete sphere of radius R has a cavity of radius r, which is packed with saw dust. The specific gravity of concrete and saw dust are respectively 2.4 and 0.3. For this sphere to float with its entire volume submerged under water, what will be the ratio of mass of concrete to mass of saw dust?

Fill in the Blanks

1. As we go high on the mountain, the ______ decreases.

2. A wooden plank immerses up to 50% in water. Then _____ % of it is immersed in a liquid of density 0.5 g cm^{-3}.

3. The difference between hydrostatic pressure and atmospheric pressure is known as ______.

4. In a liquid at the same level, the pressure will be _____ at all points.

5. When a body is dipped in a liquid, the loss in weight is equal to ______.

True or False

1. The force per unit volume is the pressure on that area.

2. The blood pressure in our body balances the atmospheric pressure acting on our body.

3. The nails penetrated into the planks easily because a large force acts on a small area.

4. The SI unit of relative density is kg m^{-3}.

5. When immiscible liquids of different densities are poured in a container, the liquid of higher density will be at the bottom while that of lower density will be at the top.

6. For a point at a depth h below the surface of a liquid of density ρ, hydrostatic pressure p is $h\rho g$.

COMPETITION VISTA

Match the Columns

1. Match Column I with Column II.

Column I	Column II
(A) Water supply in cities	(P) Pascal's law
(B) Hydraulic brake	(Q) Upthrust
(C) Sucking of a cool drink by using a straw	(R) Fluid pressure
(D) Fish weighs less in water than in air	(S) Atmospheric pressure

2. Match the Column I with Column II.

Column I	Column II
(A) Relative density of solids	(P) Law of floatation
(B) Hydrometer	(Q) Lactometer
(C) Density of milk	(R) Viscosity
(D) Mobility of liquids	(S) Archimedes' Principle

3. Match the Column I with Column II.

Column I	Column II
(A) Pressure	(P) Increase with depth
(B) Pressure of liquid column	(Q) 76 cm Hg
(C) Atmospheric Pressure	(R) Force
(D) Pressure × Area	(S) Pascal

Assertion–Reason Questions

Read the assertion and reason carefully to mark the correct option out of the options given below:

(a) If both assertion and reason are true and the reason is the correct explanation of the assertion.
(b) If both assertion and reason are true but reason is not the correct explanation of the assertion.
(c) If assertion is true but reason is false.
(d) If assertion is false but reason is true.

1. Assertion: Pascal's law is the working principle of a hydraulic lift.
Reason: Pressure is equal to thrust per unit area.

2. **Assertion:** The blood pressure in humans is greater at the feet than at the brain.
Reason: Pressure of liquid at any point is proportional to height, density of liquid and acceleration due to gravity.

3. **Assertion:** Hydrostatic pressure is a vector quantity.
Reason: Pressure is force divided by area, and force is a vector quantity.

4. **Assertion:** A man is sitting in a boat, which is floating on a pond. If the man drinks some water from the pond, the level of the water in the pond decreases.
Reason: According to Archimedes' principle the weight displaced by body is equal to the weight of the body.

5. **Assertion:** A piece of ice floats in water. The level of water remains unchanged when the ice' melts completely.
Reason: According to Archimedes' principle, the loss in weight of the 'body in the liquid is equal to the weight of the liquid displaced by the immersed part of the body.

Comprehension Type Questions

1. The manual of a car instructs the owner to inflate the tyres to a pressure of 200 kPa. (a) What is the recommended gauge pressure? (b) What is the recommended absolute pressure? (c) If, after the required inflation of the tyres, the car is driven to a mountain peak where the atmospheric pressure is 10% below that at sea level, what will the tyre gauge read?

2. A cubical block of iron, 5 cm on each side, is floating on mercury in a vessel. (a) What is the height of the block above mercury level? (b) Water is poured into the vessel so that it just covers the iron block. What is the height of water column? [RD of Hg = 13.6 and Fe = 7.2] (c) State the principle involved in the problem.

3. A block of wood floats in water with two-thirds of its volume submerged. In oil, the block floats with 0.90 of its volume submerged. Find the density of (a) wood and (b) oil, if density of water is 10^3 kg/m^3.

4. A manometer reads the pressure of a gas in an enclosure as shown in the figure (a). When some of the gas is removed by a pump, the manometer reads as in figure (b). The liquid used in the manometers is mercury and the atmospheric pressure is 76 cm of mercury.

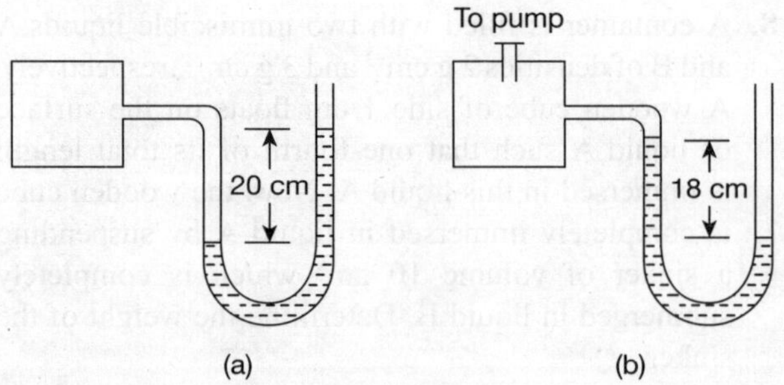

Give the absolute and gauge pressure of the gas in the enclosure for cases (a) and (b) in units of cm of mercury.

Objective Questions

One Option Correct

1. The pressure at a point inside a fluid is
(a) dependent on the height of the fluid column.
(b) dependent on the density of the fluid.
(c) equal in all directions.
(d) All the above are true.

2. The atmospheric pressure at a given place is dependent on
(a) the height of the air column.
(b) the temperature.
(c) humidity.
(d) All the above.

3. An object just floats in water. If common salt is added into the water,
(a) the volume of the object immersed in the liquid decreases.
(b) the object sinks.
(c) the object first sinks and then floats up.
(d) cannot be determined.

4. A substance floats in water, but sinks in coconut oil. The density of the substance
(a) is less than the density of water.
(b) is greater than the density of oil.
(c) Both (a) and (b).
(d) cannot be decided based on the given information.

5. When equal quantities of an oil, water and mercury are poured into a beaker, the order in which the liquids arrange themselves from bottom to top is
(a) mercury, water, oil.
(b) water, mercury, oil.
(c) water, oil, mercury.
(d) mercury, oil, water.

6. Sudden fall in atmospheric pressure by a large value indicates _____.
(a) arrival of storm
(b) arrival of dust-storm
(c) fair weather
(d) None of the above.

7. In a mercury barometer, if the tube containing mercury is tilted, then
(a) vertical height of the mercury column remains same.
(b) the length of mercury column in the tube increases.
(c) the vertical height of the mercury column decreases.
(d) Both (a) and (b).

8. At higher altitudes, the atmospheric pressure is lesser, as at higher altitudes
(a) the length of air column exerting the pressure is less.
(b) density of air decreases.
(c) acceleration due to gravity decreases.
(d) All the above.

9. The standard value of atmospheric pressure is _____.
(a) 78 cm of Hg (b) 76 cm of Hg
(c) 45 cm of Hg (d) 0.76 cm of Hg

10. Some mustard oil is kept in the beaker. It will exert pressure _____.
(a) upward (b) downward
(c) sideways (d) in all the directions

11. The pressure exerted by a liquid
(a) increases with depth.
(b) decreases with depth.
(c) is constant.
(d) first increases, then decreases.

12. The given figure shows the cross section of a dam and its reservoir. The widening of the wall, towards the bottom is because of

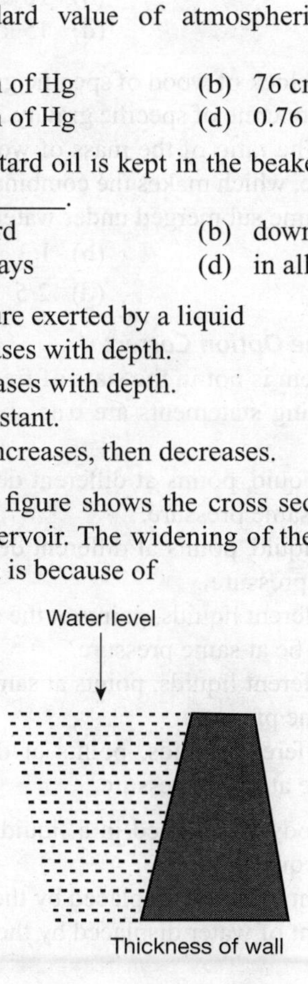

(a) increase in pressure with depth of water.
(b) decrease pressure with depth of water.
(c) change in density of water.
(d) increase in mass of the wall.

13. We use a straw or narrow pipe to drink juice from a glass. This is possible due to presence of:
(a) liquid pressure
(b) atmospheric pressure
(c) gravitational pull
(d) All of the above.

14. In cities, water supply from an overhead tank to the houses is done due to
(a) difference in pressure.
(b) gravitational force.
(c) decrease of friction in pipes.
(d) flow of liquid.

15. Which of the following is NOT true?
(a) A sharp knife can easily cut due to its small cutting surface.
(b) It is easier to walk in sand than on road.
(c) A battle tank can move easily on soft ground because its tracks have bigger surface.
(d) The pressure exerted by a needle is much more than that by the foot of an elephant.

16. If the weight of a body immersed in a fluid exceeds the buoyant force, the body will
(a) tend to move downward and finally it may sink
(b) float
(c) rise until its weight equals the buoyant force
(d) None of the above.

17. A force of 16 N is distributed uniformly on one surface of a cube of edge 8 cm. The pressure on this surface is _____.
(a) 3500 Pa (b) 2500 Pa
(c) 4500 Pa (d) 5500 Pa

18. Which of the following statement is NOT true?
(a) A freely falling body is acted upon by gravitational force.
(b) pressure on top of Mount Everest is much more that 1 atm.
(c) S.I. unit of pressure is pascal.
(d) The blood vessels in our body maintains the internal pressure equal to the atmospheric pressure.

19. Barometer is used for measuring _____.
(a) Liquid pressure (b) Thrust
(c) Atmospheric pressure (d) Air temperature

20. A steel ball is floating in a trough of mercury. If we fill the empty part of the trough with water, what will happen to the steel ball?
(a) It will continue in its position
(b) It will move up
(c) It will move down
(d) It will execute vertical oscillations

21. The unit of measuring pressure is ____.
(a) newton (b) newton/metre^2
(c) metre^2 (d) metre^2/newton

22. A liquid whose density is twice the density of mercury is used as a barometric liquid. Then one atmosphere pressure equals ____ cm of the liquid pressure.
(a) 76 (c) 152
(b) 38 (d) 380

23. A piece of ice having a stone frozen in it floats in a glass vessel filled with water. How will the level of water in the vessel change when the ice melts?
(a) The level will rise.
(b) The level will not change.
(c) The level will drop.
(d) The vessel will break

24. Which one of the following would a hydrogen balloon find easiest to lift?
(a) One kg of steel
(b) One kg of lightly packed feathers
(c) One kg of water
(d) All the same.

25. A body is just floating in a liquid (their densities are equal). If the body is slightly pressed down and released it will
(a) start oscillating.
(b) sink to the bottom.
(c) come back to the same position immediately.
(d) come back to the same position slowly.

26. A cork ball is floating on the surface of water in a beaker. The beaker is covered with a bell jar and the air is evacuated. What will happen to the ball?
(a) Sink a little (b) Rise a little
(c) Remain unchanged (d) Sink completely

27. A sample of metal weighs 210 gm in air, 180 gm in water and 120 gm in an unknown liquid. Then
(a) The density of the metal is 3 gm/cm^3.
(b) The density of the metal is 7 gm/cm^3.
(c) The density of metal is 4 times the density of unknown liquid.
(d) The metal will float on water.

28. A body of mass 2 kg is floating in water with half of its volume submerged. What would be the force required to wholly submerge it into water?
(a) 2 N (b) 9.8 N
(c) 19.6 N (d) 4.9 N

29. A metallic sphere floats in an immiscible mixture of water $\left(\rho_w = 10^3\,\text{kg/m}^3\right)$ and a liquid $\left(\rho_l = 13.5\times10^3\,\text{kg/m}^3\right)$ such that its $(4/5)$th portion is in water and $(1/5)$th portion in the liquid. The density of metal is ________.
(a) 4.5×10^3 kg/m^3
(b) 4.0×10^3 kg/m^3
(c) 3.5×10^3 kg/m^3
(d) 1.9×10^3 kg/m^3

30. The total weight of a piece of wood is 6 kg. In the floating state in water, its $(1/3)$ part remains inside the water. On this floating solid, what maximum weight is to be put such that the whole of the piece of wood is to be drowned?
(a) 12 kg (b) 10 kg
(c) 14 kg (d) 15 kg

31. A cubical block of wood of specific gravity 0.5 and a chunk of concrete of specific gravity 2.5 are fastened together. The ratio of the mass of wood to the mass of concrete, which makes the combination to float its entire volume submerged under water, is
(a) 1/5 (b) 1/3
(c) 3/5 (d) 2/5

More Than One Option Correct

1. If the system is not in the state of free fall, which of the following statements are true about hydrostatic pressure?
(a) In a liquid, points at different depths can never be at same pressure.
(b) In a liquid, points at different depths may be at same pressure.
(c) In different liquids, points at the same depth can never be at same pressure.
(d) In different liquids, points at same depth can be at same pressure.
(e) In different liquids, points at different depths can be at same pressure.

2. When a body is weighed in a liquid, the loss in its weight is equal to:
(a) Weight of liquid displaced by the body
(b) Weight of water displaced by the body

(c) The difference in weights of body in air and liquid
(d) The upthrust of liquid on the body

3. If a sample of metal weighs 210 g in air, 180 g in water and 120g in a liquid:
(a) RD of metal is 3
(b) RD of metal is 7
(c) RD of liquid is 3
(d) RD of liquid is (1/3)

4. When a body of density ρ and volume V is floating in a liquid of density σ:
(a) Its true weight is $V\rho g$.
(b) Loss in its weight is $V\sigma g$.
(c) Its apparent weight is zero.
(d) Its density ρ is lesser than of liquid σ.

5. A piece of wood is floating in water kept in a bottle. The bottle is connected to an air pump. When more air is pushed into the bottle from the pump:
(a) The wood piece will rise a little up.
(b) The thrust of air will increase.
(c) The thrust of water will increase.
(d) The total thrust will remain unchanged.

Scholar Zone

1. A piece of ice is floating in water. What will happen to the level of water when all ice melts? What will happen if the beaker is filled not with water but with liquid (a) denser than water (b) lighter than water?

2. A uniform rod of length of length 6 m is made of a material of specific gravity = 25/36. One end of the rod is tied to a string 5 m long. The other end of the string is tied to the floor of water-pool 10 m deep. Find the length of the rod lying above the water surface.

3. Explain why a soft plastic bag weighs the same when empty or when filled with air at atmospheric pressure? Would the weights be the same if measured in vacuum?

Answer Keys

Objective Questions

One Option Correct

1. (d)	**2.** (a)	**3.** (a)	**4.** (c)	**5.** (a)	**6.** (a)	**7.** (d)	**8.** (d)	**9.** (b)	**10.** (d)
11. (a)	**12.** (a)	**13.** (b)	**14.** (b)	**15.** (b)	**16.** (a)	**17.** (b)	**18.** (b)	**19.** (c)	**20.** (b)
21. (b)	**22.** (b)	**23.** (c)	**24.** (b)	**25.** (b)	**26.** (a)	**27.** (b)	**28.** (c)	**29.** (c)	**30.** (a)
31. (c)									

More Than One Option Correct

1. (a), (c) and (e) **2.** (a), (c) and (d) **3.** (b) and (c) **4.** (a) (c) and (d) **5.** All

Hints and Solutions

Very-short Answer Type Questions

1. When a body is partially or completely immersed in a fluid at rest, it experiences an upthrust which is equal to the weight of the fluid displaced by it.

2. An increase in pressure at any point inside a liquid at rest, is transmitted equally and without any change, in all directions to every other point in the liquid.

3. $D_{\text{liquid}} > D_{\text{solid}}$ body

4. (i) Bramah press or hydraulic press
(ii) Hydraulic brakes

5. 9.8×10^4 Pa ($P = h\rho g = 10\times10^3\times9.8 = 9.8\times10^4$)

6. (i) The weight of a floating body in a fluid is equal to the weight of fluid displaced by the body.
(ii) The centre of gravity of the floating body and the centre of buoyancy lie in the same vertical line.

7. 500

8. SI unit of pressure is N/m² or pascal.

9. We do not feel atmospheric pressure on our body. It is due to the fact that pressure inside our body is nearly equal to atmospheric pressure.

10. Yes, both liquids and gases exert pressure.

Short-answer Type Questions

1. R.D. of solid =

$$\frac{\text{Weight of solid in air}}{\text{Apparent loss of weight of the body in water}}$$

2. Pressure differences of top and bottom surfaces of a body immersed in a liquid.

3. $P = \frac{\text{Force}}{\text{Area}} = \frac{mg}{A}$

$$\text{Mass} = \text{Volume} \times \text{Density}$$

$$\therefore \quad P = \frac{Vdg}{A}$$

$$\text{Volume} = \text{Area} \times \text{Height}$$

$$\therefore \quad P = hdg$$

4. We know that pressure is defined as the force acting on a unit area of a surface. Then

$$\Rightarrow \quad \text{Pressure} = \frac{\text{Force}}{\text{Area}}$$

$$\therefore \quad p = \frac{F}{A}$$

$$\Rightarrow \quad p \propto \frac{1}{A}$$

So, according to the formula, pressure is inversely proportional to the area on which force is applied. Therefore, with decrease in area the pressure increases.

5. In order to carry the heavy loads, the rear wheels of heavy vehicles such as bus, truck, etc, are provided with double tyres. It is due to the fact that most of the weights of these vehicles is on their rear axle. So, the double tyre increases area of cross-section (for the tyre) and hence, decreases pressure on the road.

6. We know that pressure is inversely proportional to the area. So, in cutting instruments like knife, axe, blade, etc., the area of cross-section is made very small, which leads to the increase in pressure. Hence, they can easily penetrate the surface. That is why cutting instruments are usually made sharp.

7. The variation of pressure exerted by a liquid with respect to the given factors can be explained as
   (a) Pressure in a liquid increases with the amount of liquid.
   (b) Pressure in a liquid increases with depth.
   (c) Pressure of a liquid is independent of shape and size of container.

8. Since the atmospheric pressure becomes much less at higher altitudes than the blood pressure of our body, the blood is at higher pressure than outside pressure. Hence, some of the blood vessels in the body of mountaineers at high altitude may burst and nose-bleeding take place.

9. The balloon contains gas at a pressure which is more than the atmospheric pressure, i.e., outside pressure of the balloon is less. When the air in the jar is removed there is no opposition to the pressure exerted. So, when the air inside the jar is pumped out, the balloon will expand rapidly and burst.

10. Yes, the scale will show difference in weight in both the cases because in case of standing with both feet, the area will be more as compared to one foot. So, pressure will be different in both cases resulting in difference of weights.

11. It is difficult to cut cloth using a pair of scissors with blunt blades because the blunt blades have more area, due to which applied force produces very less pressure.
As we know that pressure is inversely proportional to area of cross-section, it is difficult to cut cloth with blunt blades.

$$\left(\because p = \frac{F}{A}\right)$$

12. The upward force is greater than the force of gravity. If it is less than gravity the gas balloon will fall down.

13. The name of the force responsible for the downward motion of the parachute is force of gravity.
No, because the air friction will become less.

14. We know that pressure exerted by the body is inversely proportional to the area where force is applied. So, the thinner rod, i.e., rod B will sink more because it has lesser area of cross-section in contact with the surface. Hence, it will exert more pressure on the sand with respect to the rod A.

Descriptive Type Questions

1. Pressure at a depth due to liquid column depends upon
   (i) gravitational acceleration (g)
   (ii) density of liquid
   (iii) height of liquid
And in this case, g and density are same, so only thing that can cause a change in pressure is height.
So, the order of pressure exerted by the fluid A, B, C and D at the base is given by

$$p_D > p_B > p_A > p_C$$

2. As the block of wood is floating partially submerged, $W = V_{in}\rho g$. Now as with rise in temperature, water contracts from 0°C to 4°C (anomalous expansion) and then expands, the density of water will increase from 0°C to 4°C and then will decrease. So, to provide proper upthrust (= mg) for floating, the volume V_{in} will decrease from 0°C to 4°C and then will increase, i.e., V (volume of body outside water) will increase from 0°C to 4°C and then will decrease with further increase in temperature.

3. Refer to theory.

4. Refer to theory.

5. Refer to theory.

6. Refer to theory.

Numericals

Level - I

1. $\text{Specific gravity} = \dfrac{\text{Weight of of block in air}}{\text{Loss of weight of block in water}}$

$$= \frac{60}{60-40} = 3$$

2. $\rho g h = 1.01\times10^5$ Pa

$$1.29\times9.81\times h = 1.01\times10^5 \text{ Pa}$$

$$h = 7981 \text{ m} \approx 8 \text{ km}$$

3. $\text{Pressure} = h\rho g = 3000 \text{ m}\times1000 \text{ kgm}^{-3}\times9.8 \text{ ms}^{-2}$

$$= 2.94\times10^7 \text{ Nm}^{-2}$$

This pressure is much lesser than the pressure which the structure can withstand. So, the structure is suitable.

4. $\text{Pressure on the piston due to car} = \dfrac{\text{Weight of car}}{\text{Area of piston}}$

$$P = \frac{3000\times9.8}{425\times10^{-4}} \text{Nm}^{-2} = 6.92\times10^5 \text{ Nm}^{-2}$$

This is also the maximum pressure that the smaller piston will have to bear.

5. Same pressure is transmitted to the output piston [Pascal's law]

$$\therefore \quad \frac{F}{3000} = \frac{100}{10} \quad \text{or} \quad F = 30{,}000 \text{ N}$$

Work done = 30 kN × 1 m = 30 kJ

6.

$$\text{Specific gravity of oil} = \frac{\text{Loss of weight of the body in oil}}{\text{Loss of weight of the body in water}}$$

$$= \frac{160-136}{160-130} = \frac{24}{30} = \frac{8}{10} = 0.8$$

7.

$$\text{Relative density} = \frac{\text{Weight of the fish in air}}{\text{Loss of weight of the fish in water}}$$

$$= \frac{348}{348-23} = \frac{348}{325} = 1.07$$

8. Weight of the body in air $W_1 = 40$ N

Weight of the body in liquid $W_3 = 30$ N

Weight of the body in water $W_2 = 20$ N

∴ Specific gravity of the liquid

$$= \frac{W_1 - W_3}{W_1 - W_2} = \frac{10}{20} = 0.5$$

9. Let W_1 and W_2 be the weights of the body in air and in water, respectively. Then $W_1 = 30$ N and $W_2 = 20$ N.

$$\text{Relative density of the body} = \frac{W_1}{W_1 - W_2} = 3$$

∴ Density of the body = 3 × Density of the water = 3 g cm^{-3}

Level - II

1. Let l be the length of the cube and cube rises 2 cm on removing the mass.

The weight of the body should be equal to the upthrust force provided by 2 cm height of the cube.

Therefore, $\quad mg = V\rho g$

$$200\times g = 2\times l^2\times1\times g$$

$$\Rightarrow \quad l = 10 \text{ cm}$$

Note: Here, all the values are in CGS system. The density is 1 gc^{-3}m and length l is in cm.

2. Since the mercury columns in the two arms are at same level, pressure due to water column = pressure due to spirit column.

$$h_w\rho_w g = h_s \rho s g$$

$$h_w\rho_w = h_s\rho_s$$

$$\rho_s = \frac{h_w\rho_w}{h_s}$$

$$= \frac{10 \times 1}{12.5}$$

$$= 0.8 \text{ g cm}^{-3}$$

Specific gravity of spirit is 0.8.

3. $V_A = V_B$

$$A_A \times h_A = A_B \times h_B$$

$$2A_B h_A = A_B \times h_B$$

$$h_B = 2h_A$$

$$P_B = h_B dg,\ P_A = h_A dg$$

$$P_B = 2P_A,\ P_A = \frac{P_B}{2}$$

4. Here $h = 10$ m and $\rho = 1000$ kg m^{-3}.

Now,

$$P = P_a + \rho gh$$

$$= 1.01 \times 10^5 + 1000 \times 9.8 \times 10$$

$$= 1.99 \times 10^5 \text{ Pa} \approx 2 \text{ atm}$$

This is a 100% increase in pressure from the surface level. At a depth of 1 km the increase in pressure is 100 atm. Submarines are designed to withstand such enormous pressure.

5. $P = P_0 + h\rho g$

$$= 1.01 \times 10^5 + 0.20 \times 1000 \times 10$$

$$= 1.01 \times 10^5 + 0.02 \times 10^5 = 1.03 \times 10^5 \text{ Pa}$$

$$\text{Force} = PA = 1.03 \times 10^5 \text{ N}$$

6. Both the pistons are walls of the same container. So, according to Pascal's principle, they should experience the same pressure. If P be the force required on the small piston, then

$$\frac{P}{\frac{\pi(0.1)^2}{4}} = \frac{W}{\frac{\pi(1)^2}{4}} \quad \left(A = \frac{\pi d^2}{4}\right)$$

$$\frac{4P}{\pi(0.1)^2} = \frac{4W}{\pi(1)^2} \text{ or } P = (0.1)^2 \text{ W or } P = 0.01 \text{ W}$$

Thus, the force on the pump piston should be only 1% of the weight of the car.

7. Since pressure is transmitted undiminished throughout the fluid,

$$F_1 = \frac{A_1}{A_2} F_2 = \frac{\pi(5 \times 5)}{\pi(15 \times 15)}(1350 \times 9.81)$$

$$\approx 1.5 \times 10^3 \text{ N}$$

The air pressure that will produce this force is

$$P = \frac{F_1}{A_1} = \frac{1.5 \times 10^3}{\pi\left(5 \times 10^{-2} \text{ m}\right)^2} = 1.9 \times 10^5 \text{ Pa}$$

This is almost double the atmospheric pressure. A similar principle is used in hydraulic brakes where the small forces applied by a person's foot is used to stop a car.

8. Apparent weight of the sinker = Extra upthrust

9. The weight of whole sphere will be equal to the upthrust. Hence,

$$\frac{4}{3}\pi\left(R^3 - r^3\right)\rho_1 + \frac{4}{3}\pi r^3 \rho_2 = \frac{4}{3}\pi R^3 \times 1$$

where ρ_1 and ρ_2 are the specific gravities of concrete and saw dust respectively.

$$\therefore \quad R^3\rho_1 - r^3\rho_1 + r^3\rho_2 = R^3$$

$$\text{or} \quad R^3(\rho_1 - 1) = r^3(\rho_1 - \rho_2)$$

$$\text{or} \quad \frac{R^3}{r^3} = \frac{(\rho_1 - \rho_2)}{\rho_1 - 1}$$

$$\text{or} \quad \frac{R^3 - r^3}{r^3} = \frac{\rho_1 - \rho_2 - \rho_1 + 1}{\rho_1 - 1}$$

$$\text{or} \quad \frac{\left(R^3 - r^3\right)\rho_1}{r^3\rho_2} = \left(\frac{1 - \rho_2}{\rho_1 - 1}\right) \times \frac{\rho_1}{\rho_2}$$

$$\text{or} \quad \frac{\text{mass of concrete}}{\text{mass of saw dust}} = \frac{1 - 0.3}{2.4 - 1} \times \frac{2.4}{0.3} = 4$$

Fill in the Blanks

1. pressure
2. 100
3. gauge pressure
4. same
5. weight of the liquid displaced by the body.

True or False

1. False, force per unit area is defined as pressure.
2. True
3. True
4. False, it has no unit as it is the ratio of two densities.
5. True
6. False, it is $p = p_0 + h\rho g$ where p_0 is atmospheric pressure.

Match the Columns

1. A → R, B → P, C → S, D → Q
2. A → S, B → P, C → Q, D → R
3. A → S, B → P, C → Q, D → R

Assertion-Reason Questions

1. (b) According to Pascal's law, if gravity effect is neglected, the pressure at every point of liquid in equilibrium at rest is same.

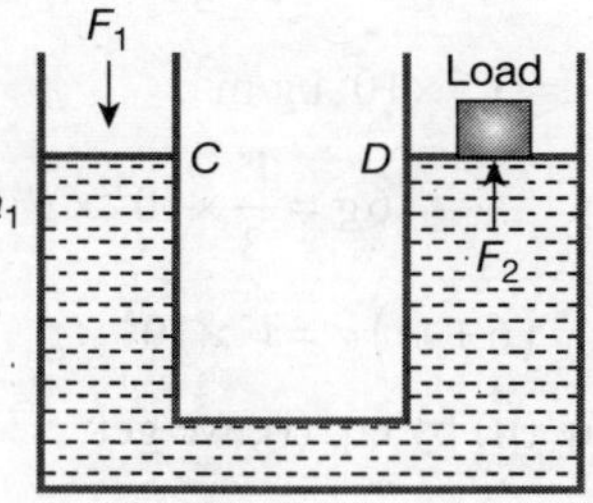

$$P_1 = P_2$$

i.e., $$\frac{F_1}{a_1} = \frac{F_2}{a_2}$$

or $$F_2 = \frac{a_2}{a_1} F_1$$

As $a_2 \gg a_1 \quad \therefore F_2 \gg F_1$

This shows that small force (F_1) applied on the smaller piston (of area a_1) will be appearing as a very large force on the larger piston.

2. (a) Height of the blood column in the human body is more at the feet than at the brain. As $P = h\rho g$. Therefore, blood exerts more pressure at the feet than at the brain.

3. (d) Since due to applied force on liquid, the pressure is transmitted equally in all directions inside the liquid. That is why there is no fixed direction for the pressure due to liquid. Hence hydrostatic pressure is a scalar quantity.

4. (d) The level of water does not change. The reason is that on drinking the water (say m gm), the weight of man increases by m gm and hence water displaced by man increases by m gm, tending to raise the water level. However, this much amount of water has already been consumed by the man. Therefore, the level of water in the pond remains same.

5. (a) This is due to the reason that ice displaces the same amount of water equal to its volume as it melts.

Comprehension Type Questions

1. (a) When the manual says pressure, it means gauge pressure P_g. Thus

$$P_g = 200 \text{ kPa}$$

(b) If P is the absolute pressure.

$$P - P_a = P_g$$

$$P = 101 \text{ kPa} + 200 \text{ kPa} = 301 \text{ kPa}$$

(c) At the mountain peak, the atmospheric pressure P_a' is 10% less. Thus

$$P_a' = 101 \times \frac{90}{100} = 90 \text{ kPa}$$

Assuming that the absolute pressure in the tyres has not changed during the drive,

$$P_g = P - P_a' = 301 - 90 = 211 \text{ kPa}$$

The tyre gauge reads gauge pressure and it will read 211 kPa.

2. (a) In case of floatation $W = Th$, i.e., $V\rho = V_{in} d$; so, if h is the height of the iron block above mercury and a is the side of iron cube,

$$(a \times a \times a)\rho = (a - h) \times a \times a \times d$$

i.e., $$(a - h) = a\frac{\rho}{d} \text{ or } h = a\left(1 - \frac{\rho}{d}\right)$$

So $$h = 5\left(1 - \frac{7.2}{13.6}\right) = 2.35 \text{ cm}$$

(b) Here upthrust is provided by both mercury and water and if h is the height of water-level,

$$V\rho g = Th_{Hg} + Th_w$$

i.e., $a \times a \times a \times \rho = a \times a(a - h)d_{Hg} + a \times a \times h d_w$

i.e., $a\rho = (a - h)d_{Hg} + h d_w$

or $$h(d_{Hg} - d_w) = a(d_{Hg} - \rho)$$

or $$h = a\frac{(d_{Hg} - \rho)}{(d_{Hg} - d_w)} = 5 \times \frac{(13.6 - 7.2)}{(13.6 - 1)} = 2.54 \text{ cm}$$

(c) Laws of floatation.

3. In case of floatation $w = Th$, i.e., $V\rho = V_{in} d$

So (a) $V\rho = (2/3)Vd_w$ [as $V_{in} = (2/3)V$]

or $$\rho = \frac{2}{3}d_w = \frac{2}{3} \times 10^3 = 667 \text{ kg/m}^3$$

(b) For oil, $V\rho = 0.9Vd_{\text{oil}}$

or $0.9Vd_{\text{oil}} = (2/3Vd_{\text{w}})$ [as $V\rho = (2/3)Vd_{\text{w}}$]

or $d_{\text{oil}} = \dfrac{2}{3\times 0.9}d_{\text{w}} = \dfrac{2}{2.7}\times 10^3 = 740 \text{ kg/m}^3$

4. (i) In Fig. (a) the absolute pressure P of the gas is greater than atmospheric pressure P_0.

$$P = P_0 + h_1\rho g$$

$$P = 76 \text{ cm of Hg} + 20 \text{ cm of Hg}$$

$$P = 96 \text{ cm of Hg}$$

Gauge pressure is the difference between the absolute pressure and atmospheric pressure.
Gauge pressure = 96 cm of Hg – 76 cm of Hg

= 20 cm of Hg

In Fig. (b) the absolute pressure P of the gas is less than atmospheric pressure.

$\therefore$ $P = P_0 - h_2\rho g$

= 76 cm of Hg – 18 cm of Hg

= 58 cm of Hg

Gauge pressure = 58 cm of Hg – 76 cm of Hg

= –18 cm of Hg

Objective Questions

One Option Correct

5. (a) $\rho_{\text{oil}} < \rho_{\text{water}} < \rho_{\text{Hg}}$.

Thus,
oil → top layer
water → middle layer
mercury → bottom layer

6. (a) A sudden fall in atmospheric pressure by a large amount indicates a storm.

7. (d) When a mercury barometer is tilted, then the vertical height remains constant but length of mercury column increases.

8. (d) As we move up, the air pressure decreases because the length of air column which exerts pressure, decreases.
Also, on moving up, the density of air and acceleration due to gravity decrease.
Thus, the decrease in pressure is due to decrease in length of air column, density of air and acceleration due to gravity.

26. (a) Here the upthrust due to air will be replaced by water vapour. We know that water vapour is lighter than air. Hence up thrust decreases. So, the ball sinks a little.

29. (c) Let ρ be the density of metal. Then

$$\rho = \left(\frac{4}{5}\times 10^3\right) + \left(\frac{1}{5}\times 13.5\times 10^3\right)$$

$$= 10^3\left[\frac{4}{5} + \frac{13.5}{5}\right] = 10^3\left[\frac{17.5}{5}\right]$$

$$= 3.5\times 10^3 \text{ kg/m}^3.$$

30. (a) Here, $6g = \dfrac{V}{3}\times 10^3 \times g$ (i)

And $(6+m)g = V\times 10^3 \times g$ (ii)

Dividing eq. (ii) by eq. (i), we get

$$\frac{6+m}{6} = 3 \text{ or } m = 18 - 6 = 12 \text{ kg}$$

Scholar Zone

1. If M g ice is floating in a liquid of density σ_{L}, then for its equilibrium
Weigh of ice = Thrust,

i.e., $Mg = V_{\text{D}}\sigma_{\text{L}}g$

So, the volume of liquid displaced by the floating ice,

$$V_{\text{D}} = (M/\sigma_{\text{L}}) \quad \text{...(i)}$$

If M g ice melts completely, the water formed will have mass M g (as mass is conserved). Now if σ_{W} is the density of water, the volume of water formed will be

$$V_{\text{F}} = (M/\sigma_{\text{W}}) \quad \text{...(ii)}$$

Here the liquid is water, i.e., $\sigma_{\text{L}} = \sigma_{\text{W}}$; so, water displaced by floating ice is equal to water formed by melting of whole ice and hence the level of water will remain unchanged. Furthermore:

(a) If $\sigma_{\text{L}} > \sigma_{\text{W}}, (M/\sigma_{\text{L}}) < (M/\sigma_{\text{W}})$, i.e., $V_{\text{D}} < V_{\text{F}}$, i.e., water displaced by floating ice will be lesser than water formed and so the level of liquid in the beaker will rise.

(b) If $\sigma_{\text{L}} \langle \sigma_{\text{W}}, (M/\sigma_{\text{L}}) \rangle (M/\sigma_{\text{W}})$, i.e., $V_{\text{D}} > V_{\text{F}}$ i.e., water displaced by floating ice will be more than water formed and so the level of liquid in the beaker will fall.

2. Let a length x of the rod remain exposed. The free-body diagram is shown in the figure. The forces acting are

(i) Weight of the rod $W = mg = AL\rho g$

(ii) Buoyant force $F_B = A(L-x)\rho_\omega g$ and

(iii) Tension in the string T.

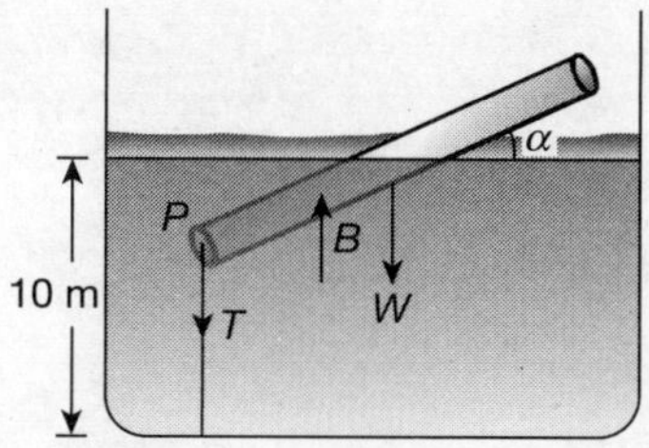

Since the system is in equilibrium, the moment of the forces about the point P must be zero. Hence,

$$W \cdot \frac{L}{2}\cos\alpha - F_B \frac{(L-x)}{2}\cos\alpha = 0$$

or, $$AL\rho g \frac{L}{2} = A(L-x)\rho_\omega g \frac{(L-x)}{2}$$

or, $$L^2\rho = (L-x)^2 \rho_\omega$$

Substituting $L = 6\text{ m}$, $\rho = \frac{25}{36}$, $\rho_\omega = 1$

$$6^2 \cdot \frac{25}{36} = (6-x)^2 \times 1$$

$\therefore$ $$x_1 = 1.0\text{ m.}$$

3. If the weight of the empty bag is W_0 and the volume of the bag is filled with air of density ρ at NTP, its weight will increase by $V\rho g$. Now, when the bag filled with air is weighed in air, the thrust of air $V\rho g$ will decrease its weight; so

$$W = W_0 + V\rho g - V\rho g = W_0$$

i.e., the weight of the bag remains unchanged when it is filled with air at NTP and weighed in air. However, if the bag is weighed in vacuum, its weight will be W_0 when empty and $(W_0 + V\rho g)$ when filled with air (as there is no upthrust), i.e., in vacuum an air-filled bag will weigh more than an empty bag.

APPENDIX

A Case-study Questions

Chapter 1: Motion

1. A car halts on a road. A chain-snatcher on a motorbike stops near the car and engages the lady driver in conversation. He snatches her gold chain and flees. At the same moment the car follows the motorbike in the same direction. The motorcycle accelerates uniformly at $1 m/s^2$ up to a speed of 36 kph and the car at $0.5 m/s^2$ up to a speed of 54 kph.
Based on above case, answer the following questions:

(i) Till what time does the motorcycle accelerate?
(a) 1.5 s (b) 10 s
(c) 5 s (d) 7.5 s

(ii) What is the distance covered by the motorcycle when it accelerates?
(a) 25 m (b) 12.5 m
(c) 50 m (d) 200 m

(iii) Till what time does the car accelerate?
(a) 20 s (b) 10 s
(c) 30 s (d) 15 s

(iv) The distance covered by the car during which it accelerates is _______.
(a) 150 m (b) 50 m
(c) 250 m (d) 225 m

(v) A what time will the car overtake the motorcycle?
(a) 35 s (b) 20 s
(c) 25 s (d) 50 s

Chapter 2: Force and Laws of Motion

1. A box sits on a sled which is being towed along by a donkey, as shown in the figure. The donkey starts from rest and begins to move forward slowly, accelerating steadily and pulling on the sled with a constant force.

The box has a mass of 12 kg, and the sled has a mass of 20 kg. Both the box and sled accelerate together at a rate of $2.0 m/s^2$. The coefficient of static friction between the box and the sled is 0.50.

(i) What is the friction force acting on the box?
(a) 20 N (b) 24 N
(c) 18 N (d) 10 N

(ii) What is the friction force between the road and sled?
(a) 64 N (b) 120 N
(c) 24 N (d) 18 N

(iii) What is the net force required to maintain an accelerating of $2m/s^2$?
(a) 36 N (b) 24 N
(c) 64 N (d) 48 N

(iv) How much force must the donkey apply to the sled to maintain this acceleration?
(a) 64 N (b) 128 N
(c) 150 N (d) 75 N

(v) What is the maximum acceleration of the donkey, hence the sled, that is possible before the box begins to slide off?
(a) $2.5 m/s^2$ (b) $10 m/s^2$
(c) $8 m/s^2$ (d) $5 m/s^2$

Chapter 3: Work, Power and Energy

1. A ball of mass 0.5 kg is rolled as shown in the three cases. In Cases I and II, the surface is smooth. Further, in Case I the initial velocity of the ball is 2m/s and in Case II, it is zero.

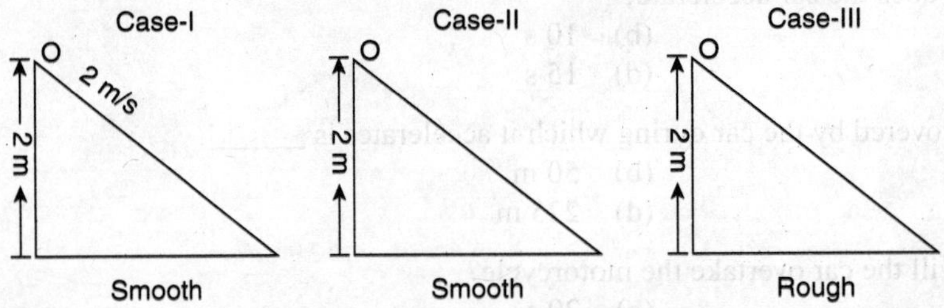

(i) The velocity of the ball when it reaches on the ground will be:
(a) More in Case I
(b) More in Case II
(c) Less in Case III as compared to Case II
(d) (a) and (c)

(ii) The kinetic energy of the ball on reaching the ground will be in the following order:
(a) Case I > Case II > Case II
(b) Case II > Case I > Case III
(c) Case I > Case II > Case III
(d) Case I = Case II = Case III

(iii) The velocity of the ball in Case I on reaching the ground is:
(a) 4.5 m/s (b) 6.63 m/s
(c) 9 m/s (d) 3.12 m/s

(iv) The velocity of the ball in Case II on reaching the ground is:
(a) $2\sqrt{5}$ m/s (b) $2\sqrt{10}$ m/s
(c) $2\sqrt{3}$ m/s (d) Data insufficient

(v) The velocity of the ball on reaching the ground in Case III:

(a) $2\sqrt{5}$ m/s (b) $2\sqrt{10}$ m/s

(c) $2\sqrt{3}$ m/s (d) Data insufficient

Chapter 4: Gravitation

1. An observer on earth measures the time period of two satellites, one moving from east to west and the other moving from west to east in a circular orbit of radius R near the equatorial plane of earth, with same velocity V_o.

(i) Which of the following option regarding velocity of the two satellites with respect to observer, is correct?

(a) The two velocities are equal.

(b) The velocity of the satellite moving from west to east is greater.

(c) The velocity of the satellite moving from east to west is greater.

(d) The satellites have zero velocities.

(ii) The velocity of the satellite moving west to east with respect to observer on earth is ______.

(a) $V_o + V_e$ (b) $V_o - V_e$

(c) $\dfrac{V_o + V_e}{2}$ (d) $\dfrac{V_o - V_e}{2}$

(iii) The velocity of the satellite moving from east to west is ______.

(a) $V_o + V_e$ (b) c

(c) $\dfrac{V_o + V_e}{2}$ (d) $\dfrac{V_o - V_e}{2}$

(iv) Which of the following is correct regarding time period of the satellite with respect to observer?

(a) Time period of the satellite moving from east to west is $\dfrac{2\pi R}{V_o - V_e}$.

(b) The time period of the satellite moving from east to west is $\dfrac{2\pi R}{V_o + V_e}$.

(c) Time period of the satellite moving from west to east is $\dfrac{2\pi R}{V_o + V_e}$.

(d) The two time periods are equal.

(v) Suppose one of the satellites is geostationary, then which of the following is incorrect about geostationary satellite?

(a) It should move from west to east.

(b) It is stationary with respect with an observer on earth.

(c) Its orbital period around earth 24 hours.

(d) It should move from east to west.

Chapter 5: Sound

1. Salman Khan has gone fishing and is sitting impatiently, waiting for a fish to bite, but nothing is happening. So, Salman turns on the radio to find some music to encourage the fish. The first piece of music Salman Khan finds is a bass guitar solo from station 'A', which has very low frequency. Salman Khan does not like the sound of that, so he finds another station 'B' which is playing a duet of much higher frequency.

(i) With increase in frequency, the wave length ______.
(a) increases (b) remains same
(c) decreases (d) Depends upon the quality of wave.

(ii) Which music will travel faster to fish?
(a) Music of station A.
(b) Music of station B.
(c) Both music will take same time to travel to fish.
(d) The data is insufficient to answer the question.

(iii) Which of the following is correct regarding sound wave in water?
(a) Velocity in water is greater and wavelength, smaller as compared to air medium.
(b) Velocity in water smaller and wavelength, greater as compared to air.
(c) Velocity as well as wavelength, greater in water as compared to air.
(d) Velocity as well as wavelength, smaller as compared to air.

(iv) Which physical phenomenon is involved here when the music in the form of sound wave travels from air to water?
(a) Reflection (b) Refraction
(c) Scattering (d) Acoustics

Chapter 6: Electricity and Magnetism

1. A student performing an experiment on lines of force by four different sources of magnetic field, found the following patterns of lines of force.

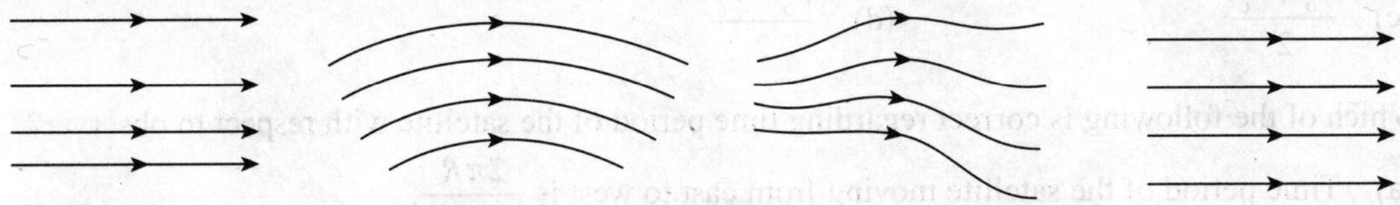

Case I Case II Case III Case IV

Based on above results answer the following questions.

(i) Which of the above cases represents the lines of force having same direction but different magnitude?
(a) Case I (b) Case II
(c) Case III (d) Case IV

(ii) In Case III,
(a) The lines of force are constant but in different directions.
(b) The lines of force have the same direction but different magnitudes.
(c) The lines of the force are neither constant in magnitude nor direction.
(d) The lines of the force have constant magnitude and direction.

(iii) Which of the above is a case of both constant magnitude as well as constant direction?
(a) Case I (b) Case II
(c) Case III (d) Case IV

(iv) In which of the above cases the direction of the lines of force is different but magnitude is constant?
(a) Case I (b) Case II
(c) Case III (d) Case IV

Chapter 7: Fluids

1. Raman, a boatman, was carrying a number of stones in a boat and approaching the shore of a river. At the destination, he unloaded the stones into water. The mass of the boat along with boatman is W kg and mass of stone is w kg. The density of stone is ρ. Based on the above case, answer the following questions:

(i) The volume of water displaced by the boat is _______.

(a) $\dfrac{W\text{m}^3}{1000^3}$ (b) $1000W\text{m}^3$

(c) $\left(\dfrac{W+w}{1000}\right)\text{m}^3$ (d) $(W-w)\text{m}^3$

(ii) The volume of water displaced by the boat when stones are unloaded is _______.

(a) $\dfrac{W\text{m}^3}{1000}$ (b) $9.8\times1000W\text{m}^3$

(c) $1000(W+w)\text{m}^3$ (d) $\dfrac{W\text{m}^3}{1000}$

(iii) Compare the volume of water displaced in the two cases.

(a) The volume of water displaced in the second case is less than the volume of water displaced in the case.

(b) The volume of water displaced in the second case is more than the volume of water displaced in the first case.

(c) The volume of water displaced in the first case is less than the volume of water displaced in the second case.

(d) The level of water in the two cases will remain same.

(iv) What happens to level of water?

(a) It will depend upon the density of the stone

(b) Increase

(c) Decrease

(d) No change.

Answer Keys

Chapter 1: Motion

1. (i) (b), (ii) (c), (iii) (c), (iv) (d), (v) (a)

Chapter 2: Force and Laws of Motion

1. (i) (b), (ii) (a), (iii) (c), (iv) (b), (v) (d)

Chapter 3: Work, Power and Energy

1. (i) (d), (ii) (c), (iii) (b), (iv) (b), (v) (d)

Chapter 4: Gravitation

1. (i) (c), (ii) (b), (iii) (a), (iv) (b), (v) (d)

Chapter 5: Sound

1. (i) (c), (ii) (c), (iii) (c), (iv) (b)

Chapter 6: Electricity and Magnetism

1. (i) (a), (ii) (c), (iii) (d), (iv) (b)

Chapter 7: Fluids

1. (i) (c), (ii) (a), (iii) (a), (iv) (b)

Hints and Solutions

Chapter 1: Motion

1. Time to which motorcycle accelerates

$$t_1 = \frac{v_1}{a_1} = \frac{10}{1} = 10 \text{ s}$$

and $s_1 = \frac{1}{2}(1)(10)^2 = 50$ m

Time to which the car accelerates

$$T_2 = \frac{v_2}{a_2} = \frac{15}{0.5} = 30 \text{ s}$$

and $s_2 = \frac{1}{2}(0.5)(30)^2 = 225$ m.

If overtaking occurs at time t, we use

$50 + 10(t - 10) = 225 + 15(t - 30)$

$10t - 50 = 15t - 225$

$5t = 175$

$t = \frac{175}{5} = 35$ s

At $t = 35$ s

Distance is

$s = 50 + 10(35 - 10)$

$= 300$ m

Chapter 2: Force and Laws of Motion

1. **(i)** (a) Friction force on the box $= 12 \times 2 = 24$ N

(ii) (a) Friction force between the sled and ground

$= \mu k(ms + mb)g$

$= 0.2(20 + 12) \times 10$

$= 64\ N$

(iii) (c) $F_{net} = (m_s + m_b)a$

$= (20 + 12) \times 2 = 64$ N

(iv) (b) $F_{net} = F_s + F_b$

or $F_s = F_{net} + F_b$

$= 64 + 64 = 128$

(v) (d) $a_{max} = \dfrac{F_{max}}{ms} = \dfrac{\mu sN}{mb}$

$= \dfrac{\mu s\, mb}{mb} g = \mu sg$

$= 0.2 \times 10 = 5N$

Chapter 3: Work, Power and Energy

1. (iii) (b) $\dfrac{1}{2}mv_2^2 = \dfrac{1}{2}mv_1 + mgh$

$v_2^2 = v_1^2 + 2 \times 9 \times h$

$= (2)^2 + 2 \times 10 \times 2$

$= 4 + 40 = 44$

$v_2 = 6.6 \text{ m/s}^2$

(iv) (b) $\dfrac{1}{2}mv^2 = mgh$

$v^2 = 2gh$

$v^2 = 2 \times 10 \times 2$

$v = 2\sqrt{10} \text{ m/s}$

(v) (d) Coefficient of friction as well inclination of the plane is not given.

Chapter 4: Gravitation

1. The earth is rotating on its axis from west to east. Therefore, the velocity of a satellite V_o moving towards east with respect to earth (Velocity V_e) would be $(V_o - V_e)$ and for the satellite moving towards west would be $(V_o + V_e)$.| So the period of the two satellite as observed by an observer on earth would be

$$T_1 = \left(\frac{2\pi R}{V_o - V_e}\right), T_2 = \left(\frac{2\pi R}{V_o + V_e}\right)$$

Hence, the time period of the two satellite will be different. The time period of the satellite moving from west to east will be greater than that of the satellite moving from east to west.

Chapter 5: Sound

1. (i) (c) With increase in frequency, the wavelength decreases as velocity remains same when the medium remains unchanged.

(ii) (c) Velocity does not depends upon frequency; hence both type of music will take same time to reach the fish.

(iii) (c) Velocity as well as wavelength is greater in water as compared to air because water is denser than air, i.e., the velocity of sound wave in water is greater than in air.

(iv) (b) Refraction.

Chapter 7: Fluids

1. (i) (c) Weight of water displaced by boat

$= mg = v \times 1000g$

$m = v\rho$

Or

$$v = \frac{m}{1000} = \left(\frac{W + w}{1000}\right) = \frac{W + w}{1000} \text{m}^3$$

(ii) (a) $\frac{W}{1000}\text{m}^3$

(iii) and (iv) If ρ be the density of the stone, the volume of water displaced by stone is $\frac{w}{\rho}$.

As $\rho > 1$ hence $\frac{w}{\rho} < w$

$$\left(W + \frac{w}{\rho}\right) < W + w$$

It means the volume of water displaced in the second case is less than the volume of water displaced in the first case

Hence, level of water will rise.